Fourier Analysis, Eigenfunction Expansions, and Partial Differential Equations

by

Satwindar Singh Sadhal
Professor of Aerospace & Mechanical Engineering
University of Southern California

Second Edition
Mathematics Education for Engineering
Yorba Linda, California

Dedicated to Aridaman Singh Thind
for his devotion to excellence in teaching,
and for inspiring me to learn calculus.

Fourier Analysis, Eigenfunction Expansions, and Partial Differential Equation
Second Edition

Mathematics Education for Engineering
4977 Avenida de las Estrellas
Yorba Linda, CA 92886-3196
USA
sadhal@att.net

The manuscript for this book was typeset with LaTeX as provided by Personal TeX, Inc., Mill Valley, California.

Second Edition, August 1, 2018

The author is grateful to the following persons for pointing out typographical errors in the first printing:
Zhixuan Cen, Christopher Cotner, Chen Cui, John Galiney, Giorgio Gaputan, Zhouyang Ge, Jianbo Guo, Stephanie Hoffer, John Hughet, Kevin Kiang, Dejuan Kong, Junlin Mu, Dan Mueller, Harsha Mylapilli, Alexander Roth, Daniel Scalese, Ramtin Sheikhhassani, Jialie Yan.

ISBN: 978-0-9913683-1-0

Contents

Preface

In this textbook, the basic principles of Fourier Series and Integrals are developed from an Engineering standpoint, and generalized to expansions of special functions. As such, the development is presented as eigenfunction expansions of second-order differential operators corresponding to various differential equations. For the sake of completeness, the solution procedure for various types of linear second-order differential equations is presented, leading on to the development of special functions.

We start out in Chapter 1 with a review of differential equations with constant coefficients, and the Cauchy-Euler equation which reduces to the constant-coefficients case. Nonhomogeneous equations are also treated with the method of undetermined coefficients as well the variation of parameters. Next, in Chapter 2, we go on to the solution of equations with non-constant coefficients by the Frobenius method (power-series solutions). We apply this technique for the development of the solutions for special functions such as Bessel functions, Legendre polynomials, Laguerre polynomials, and Hermite polynomials.

In Chapter 3, we develop the basic Fourier sine and cosine series as odd and even functions, respectively, and analyze some of their properties. This idea is extended to two variables double Fourier series and integral expansions. The extension to expansions of other special functions such as Legendre polynomials and Bessel functions is presented from the perspective of the Sturm-Liouville theory whereby these functions are orthogonal eigenfunctions of differential operators. In Chapter 4, some applications to partial differential equations in rectangular coordinates are presented. This is continued to cylindrical and spherical geometry in Chapters 5 and 6, respectively.

In this book, the aim has been to tie the basic Fourier sine and cosine series with the generalized eigenfunction expansions with a global perspective of the various classes of Fourier-type expansions, and to link the expansions to corresponding differential equations. It is hoped that this viewpoint will elucidate the broader basis of Fourier analysis and provide a clear understanding of the fundamental principles for application to engineering and physics problems.

I extend my thanks to several of my colleagues whose intellectual stimulation has inspired me towards furthering higher education. I am deeply grateful to my family, especially my wife Manjit, for the constant encouragement and unwavering support for my undertakings.

Satwindar Singh Sadhal
Yorba Linda, California
August 2018

Chapter 1

Review of Ordinary Differential Equations

1.1 Linear and Nonlinear Equations

A differential equation is a relationship between a function and its derivatives. For many physical systems such relationships usually describe the behavior of the corresponding systems.

If $f(x)$ is a function of x, then the following relationships can be considered to be differential equations:

$$\frac{d^2 f}{dx^2} + f\frac{df}{dx} + 6f = 0 \tag{1.1}$$

$$\frac{d^2 f}{dx^2} + x^2\frac{d^2 f}{dx^2} + 3f = 0 \tag{1.2}$$

$$\frac{d^3 f}{dx^3} + \left(\frac{df}{dx}\right)^2 \frac{d^2 f}{dx^2} + 4\frac{df}{dx} + f = 0 \tag{1.3}$$

$$\frac{d^2 f}{dx^2} + \sin x\frac{df}{dx} + \cos x f = 0 \tag{1.4}$$

Here, x is the independent variable and f is the dependent variable. Equations (1.2) and (1.4) are *linear* equations while (1.1) and (1.3) are considered to be *nonlinear*. The nonlinearity comes about through the presence of terms involving products of f and its derivatives. For example,

$$\left(\frac{df}{dx}\right)^2 \frac{d^2 f}{dx^2}, \quad f\frac{df}{dx}$$

are nonlinear terms. However, terms such as:

$$x^2\frac{df}{dx} \quad \text{and} \quad \sin x\frac{df}{dx}$$

do not make an equation nonlinear. Linear differential equations have an important property that when different solutions of the equation are added together, the sum also represents a solution. For example, the equation

$$\frac{d^2y}{dx^2} + 6\frac{dy}{dx} + 9y = 0 \tag{1.5}$$

has two solutions,

$$y_1(x) = \mathrm{e}^{-3x} \quad \text{and} \quad y_2(x) = x\mathrm{e}^{-3x}$$

At the same time, any linear combination of these two,

$$y = C_1 e^{-3x} + C_2 x e^{-3x}$$

is also a solution.

1.2 Linear Ordinary Differential Equations

An n-th order linear ordinary differential equation may be written in the following general form:

$$a_n(x)y^{(n)}(x) + a_{n-1}(x)y^{(n-1)}(x) + \ldots + a_1(x)y'(x) + a_0(x)y(x) = f(x), \tag{1.6}$$

where the superscript denotes derivatives as shown below:

$$y^{(k)}(x) = \left(\frac{d}{dx}\right)^k y(x).$$

The right-hand side of the equation, $f(x)$, is called the *non-homogeneous term* or the *forcing function*. The above differential equation becomes greatly simplified when the coefficients: $a_n, a_{n-1}, \ldots, a_2, a_1, a_0$ are constants. We shall first learn about these types of differential equations. In general we have the form

$$a_n y^{(n)}(x) + a_{n-1} y^{(n-1)}(x) + \ldots + a_1 y'(x) + a_0 y(x) = f(x)$$

The procedure for solving this differential equation is to first obtain a solution to the homogeneous equation (i.e., $f(x) = 0$) and then proceed with the non-homogeneous case. We shall begin with purely homogeneous differential equations first.

1.2.1 Homogeneous Equations with Constant Coefficients

Let us consider the following example:

Example 1.1

$$y''(x) + 7y'(x) + 12y(x) = 0. \tag{1.7}$$

This equation corresponds to $a_2 = 1, a_1 = 7$ and $a_0 = 12$. Since we have some kind of proportionality between $y(x)$ and its derivatives, it is reasonable to expect $y(x)$ to have exponential behavior. Therefore, we try:

$$y(x) = \mathrm{e}^{mx}$$

where m represents an unknown set of constants. Upon taking the derivatives of $y(x)$ with respect to x, we obtain:

$$y'(x) = m\mathrm{e}^{mx} \quad \text{and} \quad y''(x) = m^2\mathrm{e}^{mx}$$

The substitution of y, y' and y'' into equation (1.7) gives:

$$m^2\mathrm{e}^{mx} + 7m\mathrm{e}^{mx} + 12\mathrm{e}^{mx} = 0$$

or

$$(m^2 + 7m + 12)\mathrm{e}^{mx} = 0.$$

Since this should be valid for all values of x, we have the characteristic equation:

$$m^2 + 7m + 12 = 0$$

which, upon factorization, leads to:

$$(m + 3)(m + 4) = 0.$$

Clearly, the roots of the characteristic equation are $m = m_1 = -3$ and $m = m_2 = -4$. The exponential solutions for these values of m are therefore given by:

$$y_1(x) = \mathrm{e}^{-3x} \quad \text{and} \quad y_2(x) = \mathrm{e}^{-4x}$$

The general solution is therefore a linear combination of these two solutions:

$$y(x) = C_1\mathrm{e}^{-3x} + C_2\mathrm{e}^{-4x} \tag{1.8}$$

□

Here we had two distinct values of m giving rise to two linearly independent simple exponential solutions. However, if the roots of the characteristic equation for m are repeated, then one cannot find two simple exponentials that are linearly independent.
Let us consider an example with *repeated roots.*

Example 1.2 (repeated roots) Consider the equation

$$y'' + 4y' + 4y = 0 \tag{1.9}$$

By assuming exponential solutions of the type $y = \mathrm{e}^{mx}$ and substituting into equation (1.9) gives the following characteristic equation:

$$m^2 + 4m + 4 = 0,$$

which has repeated roots $m_1 = -2$ and $m_2 = -2$. Simple exponential solutions of the type:

$$y_1(x) = \mathrm{e}^{m_1x} \quad \text{and} \quad y_2(x) = \mathrm{e}^{m_2x}$$

in this case are just proportional to each other since $m_1 = m_2$. Since here one solution is just a multiple of the other, we really have only one solution. However, for a second order equation we must have two linearly independent solutions. To obtain the second solution, we use the reduction of order method. We let the second solution be:

$$y_2(x) = f(x)y_1(x) = f(x)e^{-2x}. \tag{1.10}$$

Differentiation with respect to x gives:

$$y_2' = (f' - 2f)e^{-2x}$$

and

$$y_2'' = (f'' - 4f' + 4f)e^{-2x}.$$

Substitution into equation (1.10) leads to:

$$[(f'' - 4f' + 4f) + 4(f' - 2f) + 4f]e^{-2x} = 0,$$

which upon simplification becomes:

$$f''e^{-2x} = 0$$

or

$$f'' = 0.$$

Upon integration we find:

$$f(x) = C_1x + C_2$$

Therefore, the second solution is:

$$y_2(x) = (C_1x + C_2)e^{-2x}$$

This is, in fact, the complete general solution and it contains the first solution. The two linearly independent solutions may be written as:

$$y_1(x) = e^{-2x} \quad \text{and} \quad y_2(x) = xe^{-2x}$$

□

Let us look at another example.

Example 1.3

$$y''' + 3y'' - 4y = 0. \tag{1.11}$$

By letting $y = e^{mx}$ and substituting into the differential equation, we obtain the following characteristic equation:

$$m^3 + 3m^2 - 4 = 0,$$

which, when factorized, becomes

$$(m - 1)(m + 2)^2 = 0.$$

The roots therefore are $m_1 = 1$, and $m_2 = m_3 = -2$. Here, we have:

$$y_2 = e^{-2x}$$

and we construct a general solution of the type:

$$y = f(x)e^{-2x}.$$

Upon taking derivatives we obtain:

$$y' = (f' - 2f)e^{-2x},$$
$$y'' = (f'' - 4f' + 4f)e^{-2x}$$

and

$$y''' = (f''' - 6f'' + 12f' - 8f)e^{-2x}.$$

The substitution of y and its derivatives into equation (1.11) leads to:

$$[(f''' - 6f'' + 12f' - 8f) + 3(f'' - 4f' + 4f) - 4f]e^{-2x} = 0$$

which, after some simplification, becomes:

$$f''' - 3f'' = 0$$

or

$$\left(\frac{df''}{dx} - 3f''\right) = 0.$$

The solution for f'' can easily be found to be

$$f'' = C_1e^{3x}.$$

Further integration gives

$$f' = \tfrac{1}{3}C_1e^{3x} + C_2$$

and finally

$$f = \tfrac{1}{9}e^{3x} + C_2x + C_3.$$

The general therefore is:

$$y = \left[\tfrac{1}{9}C_1e^{3x} + C_2x + C_3\right]e^{-2x}$$

or

$$y = C_1^*e^x + (C_2x + C_3)e^{-2x}.$$

□

We next consider a case of triply repeated roots.

Example 1.4 For the differential equation:

$$y''' + 3y'' + 3y' + y = 0,$$

the characteristic equation is:

$$m^3 + 3m^2 + 3m + 1 = 0.$$

Upon factorization,

$$(m + 1)^3 = 0,$$

we see that the roots are $m_1 = m_2 = m_3 = -1$. The reduction of order method will give the following general solution:

$$y = (C_1 + C_2x + C_3x^2)e^{-x}.$$

□

In some instances, the roots of the characteristic equation may be complex. Such cases are considered in the next example.

Example 1.5 (complex roots) For the equation

$$y'' + 4y = 0,$$

the characteristic equation is

$$m^2 + 4 = 0$$

or

$$m^2 = -4.$$

Therefore,

$$m = \pm 2i$$

or $m_1 = +2i$ and $m_2 = -2i$. With these characteristic values of m, the general solution is:

$$y = C_1 e^{2ix} + C_2 e^{-2ix},$$

or, we may choose to write it as

$$y = C e^{2ix} + C^* e^{-2ix},$$

where $C = A + iB$ and $C^* = A - iB$ is its complex conjugate. Using Euler's formula[1], we obtain:

$$\begin{aligned} y &= C(\cos 2x + i \sin 2x) + C^*(\cos 2x - i \sin 2x) \\ &= (A + iB)(\cos 2x + i \sin 2x) + (A - iB)(\cos 2x - i \sin 2x) \\ &= 2A \cos 2x - 2B \sin 2x \end{aligned}$$

This was the case of purely imaginary roots. In case the roots have both real and imaginary parts, we proceed as in the next example. Consider:

$$y'' + 8y' + 25y = 0.$$

Here the characteristic equation is:

$$m^2 + 8m + 25 = 0$$

The roots are:

$$\begin{aligned} m &= \frac{-8 \pm \sqrt{[64 - 100]}}{2} \\ &= -4 \pm 3i. \end{aligned}$$

The general solution therefore is:

$$\begin{aligned} y &= C e^{(-4+3i)x} + C^* e^{(-4-3i)x} \\ &= e^{-4x}\left(C e^{3ix} + C^* e^{-3ix}\right). \end{aligned}$$

[1]Euler's formula states the relationship: $e^{i\xi} = \cos\xi + i \sin\xi$

Again, using Euler's formula, we obtain:

$$y = e^{-4x}\left[(C + C^*)\cos 3x + (C - C^*)i \sin 3x\right].$$

If we chose $C = A + iB$ and $C^* = A - iB$, the above expression for $y(x)$ would become

$$y = 2e^{-4x}\left(A\cos 3x - B\sin 3x\right).$$

□

In the next example we treat the case of real as well as complex roots.

Example 1.6 For the differential equation

$$y''' - 3y'' + 12y' - 10y = 0,$$

the characteristic equation is

$$m^3 - 3m^2 + 12m - 10 = 0.$$

The roots are

$$m_1 = 1, \quad m_2, \, m_3 = -1 \pm 3i.$$

Therefore, the general solution can be written as

$$y = C_1 e^x + C_2 e^{(-1+3i)x} + C_2^* e^{(-1-3i)x},$$

or

$$y = C_1 e^x + 2e^{-x}\left[A\sin 3x - B\cos 3x\right]$$

□

Example 1.7 Mass-Spring System

In this example, we consider a mass-spring system as shown in Figure 1.1. Here we have a mass m held between two fixed points with a spring at one end and a damper at the other. The spring constant is k (N/m) and the damper coefficient is C (N-s/m). With a dynamic force balance with using Newton's laws, the differential equation for displacement $x(t)$ for this system is

$$m\ddot{x}(t) + C\dot{x}(t) + kx(t) = 0. \tag{1.12}$$

We shall consider here four cases with $m = 2$ kg for different values of k and C. The mass is initially displaced to the position $x(0) = 0.01$ m, and let go. The initial velocity therefore is zero, i.e., $\dot{x}(0) = 0$

CASE 1: $k = 12$ N/m and $C = 10$ N-s/m.
With these constant values, the differential equation (1.12) becomes

$$2\ddot{x} + 10\dot{x} + 12x = 0,$$

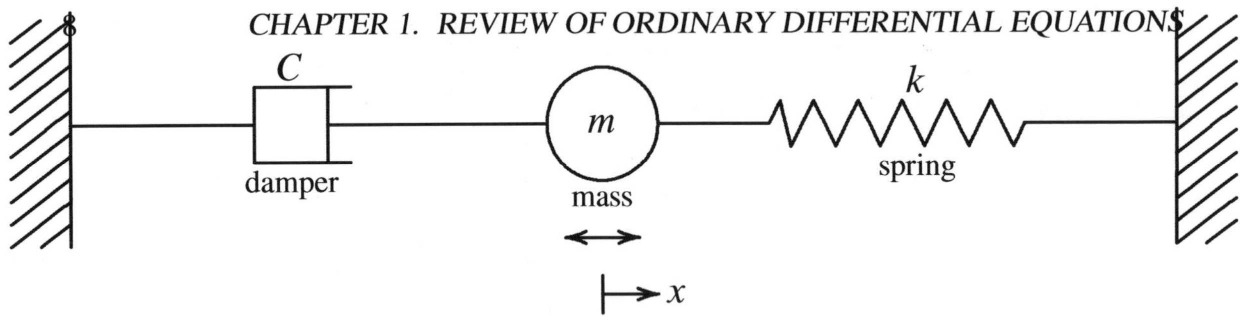

Figure 1.1: A schematic of damped mass-spring system

or equivalently,

$$\ddot{x} + 5\dot{x} + 6x = 0. \tag{1.13}$$

Assuming a solution of the form $x(t) = e^{\lambda t}$, and substituting into (1.13), we obtain the characteristic equation

$$\lambda^2 + 5\lambda + 6 = 0,$$

which factorizes in the form

$$(\lambda + 3)(\lambda + 2) = 0.$$

The roots therefore are

$$\lambda = -2, -3,$$

and the general solution may be written as

$$x(t) = Ae^{-2t} + Be^{-3t}.$$

Applying the initial conditions, $x(0) = 0.01$ and $\dot{x}(0) = 0$ leads to

$$\begin{aligned} A + B &= 0.01 \\ -2A - 3B &= 0 \end{aligned}$$

The solution to this set of algebraic equations is

$$A = 0.03 \qquad \text{and} \qquad B = -0.02.$$

The final solution may be written as

$$x(t) = 0.02\left(\tfrac{3}{2}e^{-2t} - e^{-3t}\right).$$

The mass-spring system for this case is considered 'overdamped.'

Case 2: $k = 20$ N/m and $C = 4$ N-s/m.
With these values, the differential equation (1.12) becomes

$$\ddot{x} + 2\dot{x} + 10x = 0. \tag{1.14}$$

Again, assuming a solution of the form $x(t) = e^{\lambda t}$, and substituting into (1.14), we obtain the characteristic equation

$$\lambda^2 + 2\lambda + 10 = 0,$$

This has roots

$$\begin{aligned}\lambda &= \frac{-2 \pm \sqrt{4-40}}{2} \\ &= \frac{-2 \pm \sqrt{-36}}{2} \\ &= \frac{-2 \pm 6i}{2} \\ &= -1 \pm 3i\end{aligned}$$

The general solution for this case is

$$x(t) = e^{-t}\left(A \cos 3t + B \sin 3t\right).$$

With the same initial conditions, $x(0) = 0.01$ and $\dot{x}(0) = 0$ we have

$$\begin{aligned}x(0) &= A = 0.01 \\ \dot{x}(0) &= e^{-t}\left[-\left(A \cos 3t + B \sin 3t\right) + \left(-3A \sin 3t + 3B \cos 3t\right)\right]_{t=0} \\ &= -A + 3B = 0 \\ B &= \tfrac{1}{3}A\end{aligned}$$

The final solution is

$$x(t) = 0.01e^{-t}\left(\cos 3t + \tfrac{1}{3}\sin 3t\right).$$

This solution has oscillatory behavior with attenuation in time. Such a system is considered to be 'underdamped.'

CASE 3: $k = 8$ N/m and $C = 8$ N-s/m.
The differential equation (1.12) now becomes

$$\ddot{x} + 4\dot{x} + 4x = 0. \tag{1.15}$$

In this case the characteristic equation is

$$\lambda^2 + 4\lambda + 4 = 0,$$

or, upon factorization

$$(\lambda + 2)^2 = 0.$$

This is a case of repeated roots,

$$\lambda = -2, -2,$$

leading to the general solution,

$$x(t) = e^{-2t}\left(A + Bt\right).$$

With the initial conditions, $x(0) = 0.01$ and $\dot{x}(0) = 0$ we have

$$\begin{aligned}x(0) &= A = 0.01 \\ \dot{x}(0) &= e^{-2t}\left[-2\left(A + Bt\right) + B\right]_{t=0} \\ &= -2A + B = 0 \\ B &= 2A = 0.02.\end{aligned}$$

The final form of the solution is

$$x(t) = 0.01e^{-2t}(1 + 2t).$$

This is a case of a 'critically damped' system. A slight increase in the spring constant, and/or a slight decrease in the damping constant will lead to oscillatory behavior. The plots for Cases 1, 2 and 3 are given in Figure 1.2

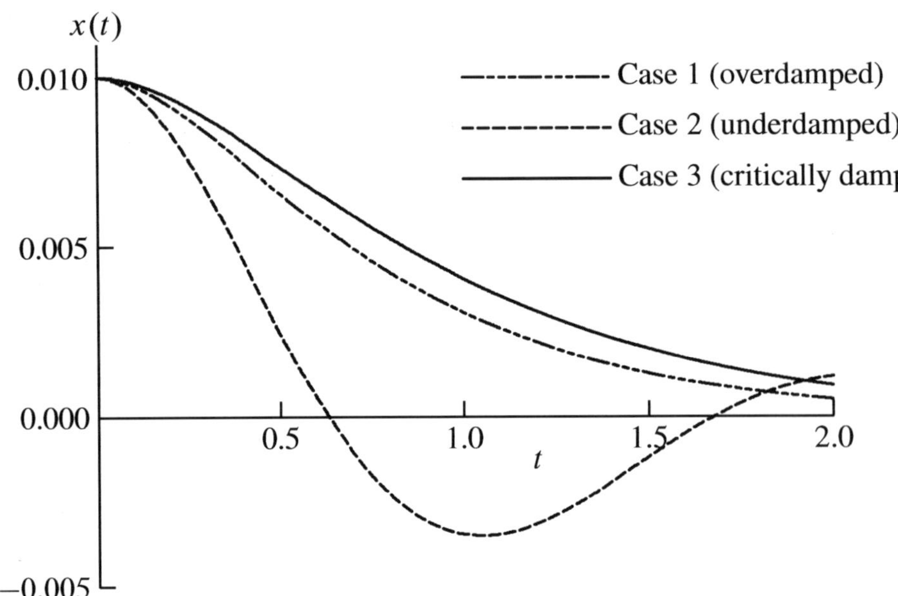

Figure 1.2: Plots of the displacement $x(t)$ as a function of time for Cases 1, 2 and 3

CASE 4: $k = 8$ N/m and $C = 0$.
Without the damping term, the differential equation for $x(t)$ takes on a much simplified form,

$$\ddot{x} + 4x = 0. \tag{1.16}$$

The characteristic equation for this case is

$$\lambda^2 + 4 = 0,$$

with roots,

$$\lambda = \pm 2i.$$

The general solution now is

$$x(t) = A\cos 2t + B\sin 2t,$$

and with the application of the initial conditions we end up with

$$x(t) = 0.01\cos 2t,$$

which is purely oscillatory, and undamped , as shown in Figure 1.3

□

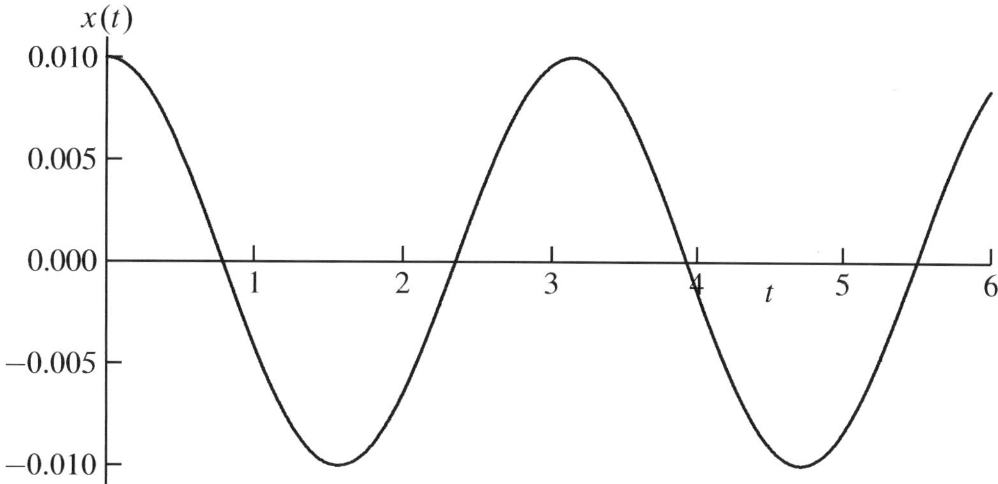

Figure 1.3: Displacement $x(t)$ as a function of time for Case 4: undamped oscillations

These examples cover all types of possibilities for second order homogeneous equations with constant coefficients. Higher order equations can be treated in a similar fashion. We shall now go on to non-homogeneous equations.

EXERCISES 1.1 *Obtain the general solutions to the following homogeneous differential equations:*

1. $y'' + 13y' + 40y = 0.$	2. $y'' + y' - 12y = 0.$
3. $y'' - 4y' - 21y = 0.$	4. $y'' + 6y' + 9y = 0.$
5. $y'' + 15y' + 44y = 0.$	6. $y'' + 2y' - 15y = 0.$
7. $y'' + 4y' + 13y = 0.$	8. $y'' + 16y = 0.$
9. $y'' + 2y' + 5y = 0.$	10. $y'' - 4y' + 5y = 0.$
11. $y'' - 6y' + 10y = 0.$	12. $y''' + 5y'' - 8y' - 12y = 0.$
13. $y''' - 7y' + 6y = 0.$	14. $y''' - 3y'' + 4y' - 12y = 0.$
15. $y'' - 9y' - 52y = 0.$	16. $y''' - 5y'' + 9y' - 5y = 0.$

1.2.2 Non-Homogeneous Differential Equations

For the general case as given by equation (1.6) the solution can be written as a sum of two parts:

$$y(x) = y_h(x) + y_p(x).$$

Here, y_h is the solution to the homogeneous equation and y_p represents the solution arising from the forcing function, $f(x)$. We have already studied the techniques for obtaining the homogeneous solution in the case of constant coefficients. For non-homogeneous cases of such equations if the forcing function is a simple exponential, the solution can be obtained in a straightforward manner by a technique called the *Method of Undetermined Coefficients*. Also polynomial expressions and combinations of polynomials and exponentials can be treated by this method.

The Method of Undetermined Coefficients

We shall first consider exponential forcing functions.

Exponential Forcing Functions

In the example,

Example 1.8

$$y'' - 4y' + 3y = \mathrm{e}^{-x}, \tag{1.17}$$

the forcing function, $f(x) = \mathrm{e}^{-x}$, and its derivatives are all proportional to each other. It is therefore quite reasonable to try:

$$y_p = C\mathrm{e}^{-x}$$

because, as we shall see, the substitution of this form of y_p into equation (1.17) will yield terms proportional to e^{-x} on both sides of the equation. Taking the derivatives, we obtain:

$$y_p' = -C\mathrm{e}^{-x}$$

and

$$y_p'' = C\mathrm{e}^{-x}.$$

Upon the substitution of y_p and its derivatives into equation (1.17) gives:

$$C\mathrm{e}^{-x} - (-4C\mathrm{e}^{-x}) + 3C\mathrm{e}^{-x} = \mathrm{e}^{-x},$$

which simplifies to:

$$8C\mathrm{e}^{-x} = \mathrm{e}^{-x}.$$

Therefore,

$$C = \tfrac{1}{8},$$

and

$$y_p = \tfrac{1}{8}\mathrm{e}^{-x}.$$

For the homogeneous equation:

$$y_h'' - 4y_h' + 3y_h = 0,$$

the characteristic equation is:

$$m^2 - 4m + 3 = 0,$$

with solutions $m_1 = 1$ and $m_2 = 3$. We may now write the solution to the homogeneous equation as:

$$y_h = C_1\mathrm{e}^{x} + C_2\mathrm{e}^{3x}.$$

The complete general solution is:

$$y = y_h + y_p = C_1\mathrm{e}^{x} + C_2\mathrm{e}^{3x} + \tfrac{1}{8}\mathrm{e}^{-x}.$$

□

In situations involving combinations of exponentials, simple superposition of particular solutions will work. Let us look at another example.

Example 1.9

$$y'' - 4y' + 3y = 3e^{\frac{3}{2}x} - 2e^{2x}$$

we should try a particular solution of the type:

$$y_p = C_1 e^{\frac{3}{2}x} + C_2 e^{2x}.$$

Upon taking derivatives, we obtain:

$$y_p' = \tfrac{3}{2} C_1 e^{\frac{3}{2}x} + 2C_2 e^{2x}$$

and

$$y_p'' = \tfrac{9}{4} C_1 e^{\frac{3}{2}x} + 4C_2 e^{2x}.$$

Substitution into the differential equation gives:

$$\tfrac{9}{4} C_1 e^{\frac{3}{2}x} + 4C2e^{2x} - 4\left(\tfrac{3}{2} C_1 e^{\frac{3}{2}x} + 2C_2 e^{2x}\right) + 3(C_1 e^{\frac{3}{2}x} + C_2 e^{2x}) = 3e^{\frac{3}{2}x} - 2e^{2x}.$$

After some simplification of the left side we obtain:

$$-\tfrac{3}{4} C_1 e^{\frac{3}{2}x} - C_2 e^{2x} = 3e^{\frac{3}{2}x} - 2e^{2x}.$$

Clearly, we can see that $C_1 = -4$ and $C_2 = 2$. The particular solution is therefore given by:

$$y_p = -4e^{\frac{3}{2}x} + 2e^{2x}.$$

□

If the forcing function is a sin or a cos function, it can be treated as a combination of exponentials. This is done in the next example.

Example 1.10

$$\begin{aligned} y'' + 2y' + y &= \cos 2x && (1.18) \\ &= \tfrac{1}{2}\left(e^{2ix} + e^{-2ix}\right), && (1.19) \end{aligned}$$

we assume a particular solution of the type:

$$\begin{aligned} y_p &= Ce^{2ix} + C^* e^{-2ix} \\ &= C(\cos 2x + i \sin 2x) + C^*(\cos 2x - i \sin 2x) \\ &= C_1 \cos 2x + C_2 \sin 2x. \end{aligned}$$

As we can see, the forcing function $f(x) = \cos 2x$ requires us to initially assume both $\cos 2x$ and $\sin 2x$ in the particular solution. Taking the derivatives of y_p, we obtain:

$$y_p' = 2\left[-C_1 \sin 2x + C_2 \cos 2x\right],$$

and

$$y_p'' = -4[C_1 \cos 2x + C_2 \sin 2x].$$

The substitution of y_p and its derivatives into equation (1.19) leads to:

$$-4[C_1 \cos 2x + C_2 \sin 2x] + 2\,[2(-C_1 \sin 2x + C_2 \cos 2x)] + C_1 \cos 2x + C_2 \sin 2x = \cos 2x.$$

Some simplification of the left side reduces the above equation to:

$$[-3C_1 + 4C_2] \cos 2x + [-4C_1 - 3C_2] \sin 2x = \cos 2x.$$

By comparing both sides, it is easy to see that

$$\begin{aligned} -3C_1 + 4C_2 &= 1, \\ -4C_1 - 3C_2 &= 0, \end{aligned}$$

which yields

$$C_1 = -\tfrac{3}{25} \quad \text{and} \quad C_2 = \tfrac{4}{25}.$$

The particular solution may now be written as:

$$y_p = \tfrac{1}{25}\left[-3 \cos 2x + 4 \sin 2x\right].$$

The homogeneous part of the differential equation has the characteristic equation:

$$m^2 + 2m + 1 = 0$$

which has repeated roots: $m_1 = m_2 = -1$. Thus the homogeneous solution is:

$$y_h = (A_1 + A_2 x)\mathrm{e}^{-x}.$$

The complete general solution is:

$$\begin{aligned} y &= y_h + y_p \\ &= (A_1 + A_2 x)\mathrm{e}^{-x} + \tfrac{1}{25}\left[-3 \cos 2x + 4 \sin 2x\right]. \end{aligned}$$

□

In situations involving linear combinations of several sin and cos functions in the forcing function, we assume y_p to be a sum of sin − cos pairs for each different argument. For example, in

$$y'' + 2y' + y = \sin 3x + 2 \cos 3x + \sin x,$$

we let:

$$y_p = C_1 \sin 3x + C_2 \cos 3x + C_3 \sin x + C_4 \cos x,$$

and then solve for the constants, $C_1,\ C_2,\ C_3,$ and C_4.

The idea of assuming y_p to be made up of terms simply proportional to each exponential term in the forcing function will work, provided the terms are not proportional to any of the homogeneous solutions. In case $f(x)$ contains a homogeneous solution, we refer to such a term as a *secular term*. In such situations, we have to assume a somewhat modified solution for y_p. To illustrate this point we consider another example.

Example 1.11

$$y'' + 4y' + 3y = e^{-x}, \tag{1.20}$$

has the homogeneous solution, $y_h = C_1 e^{-3x} + C_2 e^{-x}$. If we now try $y_p = C e^{-x}$, we obtain:

$$C e^{-x} - 4C e^{-x} + 3C e^{-x} = e^{-x},$$

which simplifies to the ridiculous result:

$$0 = e^{-x}.$$

This is to be expected because $y_p = C e^{-x}$ is one of the solutions to the homogeneous equation:

$$y'' + 4y' + 3y = 0,$$

and when substituted into equation (1.20) it will certainly give zero on the left hand side.

In this case we shall first carry out rigorous derivation of the particular solution. We assume a very general form for y_p, i.e.,

$$y_p(x) = g(x)e^{-x}.$$

The derivatives are:

$$y_p' = (g' - g)e^{-x}$$

and

$$y_p'' = (g'' - 2g' + g)e^{-x}.$$

Substitution into equation (1.20) leads to:

$$\left[(g'' - 2g' + g) + 4(g' - g) + 3g\right] e^{-x} = e^{-x},$$

which simplifies to:

$$g'' + 2g' = 1 \quad \text{or} \quad \frac{dg'}{dx} + 2g' = 1. \tag{1.21}$$

Solving for g', the homogenous solution can easily be found to be

$$g_h' = C_1 e^{-2x}.$$

For the particular solution we can, for the moment, look at the forcing function as $e^0 = 1$ and let g_p' be a function proportional to it. Therefore, by letting:

$$g_p' = C_2,$$

and substituting into equation (1.21) yields $C_2 = \frac{1}{2}$. The complete solution for g' can now be written as:

$$g' = C_1 e^{-2x} + \tfrac{1}{2},$$

which, after further integration, becomes:

$$g = -\tfrac{1}{2} C_1 e^{-2x} + \tfrac{1}{2} x + C_3.$$

We may therefore write $y(x)$ as:

$$\begin{aligned} y &= g(x)e^{-x} \\ &= \left[-\tfrac{1}{2} C_1 e^{-2x} + \tfrac{1}{2} x + C_3\right] e^{-x} \\ &= C_1 e^{-3x} + C_2 e^{-x} + \tfrac{1}{2} x e^{-x}. \end{aligned}$$

□

In this case we could have simply started with:

$$y_p = Cxe^{-x}$$

and substituted into equation (1.21) to obtain the particular solution. Let us look at another example.

Example 1.12

$$y'' - 5y' + 6y = e^{3x}. \tag{1.22}$$

We first check the solution to the homogeneous equation:

$$y_h'' - 5y_h' + 6y_h = 0,$$

which has the characteristic equation

$$m^2 - 5m + 6 = 0 \quad \text{or} \quad (m-3)(m-2) = 0,$$

with roots $m_1 = 3$ and $m_2 = 2$. The homogeneous solution is:

$$y_h = C_1e^{3x} + C_2e^{2x}.$$

We can see that the forcing function is proportional to one of the solutions to the homogeneous equation. In this case we let:

$$y_p = Cxe^{3x}.$$

The derivatives are:

$$y_p' = C(1+3x)e^{3x}$$

and

$$y_p'' = C(6+9x)e^{3x}.$$

Substitution into equation (1.22) gives:

$$C(6+9x)e^{3x} - 5C(1+3x)e^{3x} + 6Cxe^{3x} = e^{3x},$$

which simplifies to:

$$Ce^{3x} = e^{3x}$$

and yields $C = 1$. The particular solution is thus given by:

$$y_p = xe^{3x},$$

and the complete general solution is:

$$y = C_1e^{3x} + C_2^{2x} + xe^{3x}.$$

□

Let us look at another example.

Example 1.13

$$y'' + 2y' + y = \mathrm{e}^{-x}, \tag{1.23}$$

the two linearly independent solutions of the homogeneous equation are:

$$y_{h1} = \mathrm{e}^{-x} \quad \text{and} \quad y_{h2} = x\mathrm{e}^{-x}.$$

Therefore, the particular solution cannot be assumed to be a function proportional to any one of these solutions. In this case

$$y_p = Cx^2\mathrm{e}^{-x}$$

will work. The derivatives are:

$$y_p' = C(2x - x^2)\mathrm{e}^{-x}$$

and

$$y_p'' = C(2 - 4x + x^2)\mathrm{e}^{-x}$$

Upon the substitution y_p and its derivatives into equation (1.23), we obtain:

$$C(2 - 4x + x^2)\mathrm{e}^{-x} + 2C(2x - x^2)\mathrm{e}^{-x} + Cx^2\mathrm{e}^{-x} = \mathrm{e}^{-x}.$$

Simplification of the left hand side leads to:

$$2C\mathrm{e}^{-x} = \mathrm{e}^{-x} \quad \text{or} \quad C = \tfrac{1}{2}.$$

The general solution for equation (1.23) may be written as:

$$y = C_1\mathrm{e}^{-x} + C_2x\mathrm{e}^{-x} + \tfrac{1}{2}x^2\mathrm{e}^{-x}.$$

□

Next, we shall consider polynomial type forcing functions.

Polynomial Forcing Functions

We begin with the following example.

Example 1.14

$$y'' + 2y' + y = 3x^3 \tag{1.24}$$

Here, if we assume a particular solution proportional to x^3 the derivatives on the left side will generate terms like x^2 and x. To neutralize these terms we assume a more general particular solution including all integer powers lower than and including x^3, i.e.,

$$y_p = C_3x^3 + C_2x^2 + C_1x + C_0.$$

Taking derivatives with respect to x, we obtain:

$$y_p' = 3C_3x^2 + 2C_2x + C_1$$

and

$$y_p'' = 6C_3x + 2C_2$$

After substituting into equation (1.24), we find:

$$6C_3x + 2C_2 + 2\left[3C_3x^2 + 2C_2x + C_1\right] + C_3x^3 + C_2x^2 + C_1x + C_0 = 3x^3.$$

After some rearrangement in powers of x, we obtain:

$$C_3x^3 + (C_2 + 6C_3)x^2 + (C_1 + 4C_2 + 6C_3)x + (C_0 + 2C_1 + 2C_2) = 3x^3.$$

By matching the coefficients of like powers in x, we end up with the following set of algebraic equations:

$$\begin{aligned} C_3 &= 3 \\ C_2 + 6C_3 &= 0 \\ C_1 + 4C_2 + 6C_3 &= 0 \\ C_0 + 2C_1 + 2C_2 &= 0. \end{aligned}$$

The solution of this set yields:

$$C_3 = 3, \quad C_2 = -18, \quad C_1 = 54 \quad \text{and} \quad C_0 = -72,$$

and we can write the particular solution as:

$$y_p = 3x^3 - 18x^2 + 54x - 72.$$

□

For cases in which the lowest derivative in the equation is not the zeroth derivative, we have to be a little careful. This is illustrated in the next example.

Example 1.15

$$y'' + 2y' = 3x^2 + 2x \tag{1.25}$$

is really a differential equation for y'. Therefore, we assume a particular solution:

$$y_p' = C_2x^2 + C_1x + C_0.$$

and substitute into equation (1.25). As a result we obtain

$$2C_2x + C_1 + 2\left[C_2x^2 + C_1x + C_0\right] = 3x^2 + 2x,$$

or

$$2C_2x^2 + [2C_1 + 2C_2] + [2C_0 + C_1] = 3x^2 + 2x,$$

to yield

$$\begin{aligned} C_2 &= \tfrac{3}{2}, \\ C_1 &= -\tfrac{1}{2}, \\ C_0 &= \tfrac{1}{4}. \end{aligned}$$

Thus the particular solution in this case is

$$y_p' = \tfrac{3}{2}x^2 - \tfrac{1}{2}x + \tfrac{1}{4},$$

which may be integrated to give

$$y_p = \tfrac{1}{2}x^3 - \tfrac{1}{4}x^2 + \tfrac{1}{4}x.$$

An integration constant is not needed because it is one of the homogeneous solutions. □

We now go on to cases of products of polynomials and exponentials in the forcing function.

Products of Polynomials and Exponentials

Here, we begin with the example,

Example 1.16

$$y'' + 4y' + 3y = x^2 e^x. \tag{1.26}$$

It is better to obtain the solution for the homogeneous equation first, so that we may check for the presence of secular terms in the forcing function. The characteristic equation for this equation is:

$$m^2 + 4m + 3 = 0,$$

which has the solution $m_1 = -3$ and $m_2 = -1$. The two linearly independent solutions of the homogeneous equation are:

$$y_{h1} = e^{-3x} \quad \text{and} \quad y_{h2} = e^{-x}.$$

Since the forcing function in equation (1.26) does not contain either one of these solutions, we can say that there is no secular term. Therefore, we assume a particular solution:

$$y_p = (C_2 x^2 + C_1 x + C_0)e^x.$$

The derivatives of y_p are:

$$y_p' = \left[C_2 x^2 + (2C_2 + C_1)x + (C_1 + C_0)\right] e^x,$$

and

$$y_p'' = \left[C_2 x^2 + (4C_2 + C_1)x + (2C_2 + 2C_1 + C_0)\right] e^x.$$

The substitution of y_p and its derivatives into equation (1.26) leads to

$$\begin{aligned}\{C_2 x^2 &+ (4C_2 + C_1)x + (2C_2 + 2C_1 + C_0) \\ &+ 4\left[C_2 x^2 + (2C_2 + C_1)x + (C_1 + C_0)\right] \\ &+ 3(C_2 x^2 + C_1 x + C_0)\}e^x \\ &= x^2 e^x,\end{aligned}$$

which simplifies to

$$8C_2 x^2 + (12C_2 + 8C_1)x + (2C_2 + 6C_1 + 8C_0) = x^2,$$

and yields

$$\begin{aligned}C_2 &= \tfrac{1}{8} \\ C_1 &= -\tfrac{3}{16} \\ C_0 &= \tfrac{7}{64}.\end{aligned}$$

The particular can now be written as:

$$y_p = \left(\tfrac{1}{8}x^2 - \tfrac{3}{16}x + \tfrac{7}{64}\right) e^x.$$

□

In case the exponential part of the forcing function is proportional to one of the solutions to the homogeneous part, the polynomial part of the particular solution needs to higher order than the forcing function itself. The example below will show this.

Example 1.17

$$y'' + 2y' - 3y = x^2 e^x \tag{1.27}$$

the characteristic equation

$$m^2 + 2m - 3 = 0$$

gives the two linearly-independent solutions of the homogeneous part as

$$y_1 = e^x \quad \text{and} \quad y_2 = e^{-3x}.$$

Here, if we try a particular solution of the type

$$y_p = (C_2 x^2 + C_1 x + C_0) e^x$$

and substitute into equation (1.27) we obtain

$$(8C_2 x + 2C_2 + 4C_1) e^x = x^2 e^x,$$

which clearly shows that the type of particular solution assumed does not work. In this case we have to assume a particular solution of the form:

$$y_p = \left[C_3 x^3 + C_2 x^2 + C_1 x\right] e^x.$$

The derivatives are:

$$y_p' = \left[C_3 x^3 + (3C_3 + C_2)x^2 + (2C_2 + C_1)x + C_1\right] e^x,$$

and

$$y_p'' = \left[C_3 x^3 + (6C_3 + C_2)x^2 + (6C_3 + 4C_2 + C_1)x + (2C_2 + 2C_1)\right] e^x.$$

Substitution into equation (1.27) gives

$$\begin{aligned} &\{\left[C_3 x^3 + (6C_3 + C_2)x^2 + (6C_3 + 4C_2 + C_1)x + (2C_2 + 2C_1)\right] \\ &\quad +2\left[C_3 x^3 + (3C_3 + C_2)x^2 + (2C_2 + C_1)x + C_1\right] \\ &\quad -3\left[C_3 x^3 + C_2 x^2 + C_1 x\right]\} e^x = x^2 e^x \end{aligned}$$

Further simplification leads to

$$\left[12C_3 x^2 + (6C_3 + 8C_2)x + (2C_2 + 4C_1)\right] = x^2,$$

from which we obtain

$$C_3 = \tfrac{1}{12}, \quad C_2 = -\tfrac{1}{16} \quad \text{and} \quad C_1 = \tfrac{1}{32}.$$

The particular solution can now be written as

$$y_p = \left(\tfrac{1}{12}x^3 - \tfrac{1}{16}x^2 + \tfrac{1}{32}x\right) e^x.$$

□

In situations in which the forcing function is neither an exponential, a polynomial nor a product of the two, a more general method needs to be used. The method is called *Variation of Parameters* and is also applicable to equations with non-constant coefficients. We shall learn about this method in the next section.

EXERCISES 1.2 *Find the general solution for the following non-homogeneous differential equations by the method of undetermined coefficients.*

1.	$y'' + 2y' + y = \mathrm{e}^{-2x}$	2.	$y'' + 6y' + 8y = \mathrm{e}^{-4x}$
3.	$y'' - 5y' - 6y = \mathrm{e}^{x}$	4.	$y'' + 7y' + 12y = \mathrm{e}^{-3x}$
5.	$y'' - 3y' - 10y = \mathrm{e}^{-2x}$	6.	$y'' + 2y' - 15y = \mathrm{e}^{5x}$
7.	$y'' + y' - 12y = \sin 2x$	8.	$y'' - 7y' + 6y = \sin x + 2\cos 3x$
9.	$y'' + 6y' + 13y = \mathrm{e}^{-3x}$	10.	$y'' - 2y' + 10y = \mathrm{e}^{x}\cos 3x$
11.	$y'' + 4y' + 4y = \mathrm{e}^{-2x}$	12.	$y'' + 8y' + 12y = \mathrm{e}^{-2x} + \sin 2x$
13.	$y'' + 7y' + 10y = \mathrm{e}^{-5x}$	14.	$y'' + 2y' + 2y = 3\mathrm{e}^{-x}\sin x$
15.	$y'' - y' - 12y = x^3 + 4x$	16.	$y'' + 11y' - 12y = 3x^2 + 4 + \mathrm{e}^{-x}$
17.	$y'' - 9y' + 14y = x^2 + 4x$	18.	$y'' + 4y' + 5y = x^2\sin x$
19.	$y'' - 7y' + 10y = x^2\mathrm{e}^{2x}$	20.	$y'' - 6y' + 13y = x\mathrm{e}^{3x}\sin 2x$
21.	$y'' + 5y' - 6y = x^2 + x\sin x$	22.	$y'' + 9y = x^3 + x\cos 3x$

1.2.3 Variation of Parameters

Let us look at a general case of a second order linear differential equation,

$$a_2(x)y''(x) + a_1(x)y'(x) + a_0(x)y(x) = f(x). \tag{1.28}$$

The solution to the homogeneous equation

$$a_2(x)y''(x) + a_1(x)y'(x) + a_0(x)y(x) = 0 \tag{1.29}$$

can be written as

$$y_h = C_1y_1(x) + C_2y_2(x),$$

where y_1 and y_2 denote the two linearly independent solutions of equation (1.29). To obtain the solution satisfying the non-homogeneous equation (1.28), we replace the constants C_1 and C_2 with variable functions, v_1 and v_2, respectively. That is, we let

$$y(x) = v_1(x)y_1(x) + v_2(x)y_2(x). \tag{1.30}$$

The first derivative is

$$y' = v_1y_1' + v_2y_2' + v_1'y_1 + v_2'y_2. \tag{1.31}$$

It must be pointed out that the above expression (1.30) for $y(x)$ is excessively general. We may choose any relationship between v_1 and v_2. The most convenient one would be to set last two terms on the right-hand side of equation (1.31) equal to zero. Thus,

$$v_1'y_1 + v_2'y_2 = 0. \tag{1.32}$$

This leaves us with

$$y' = v_1 y_1' + v_2 y_2'.$$

The second derivative is

$$y'' = v_1 y_1'' + v_2 y_2'' + v_1' y_1' + v_2' y_2'.$$

The substitution of these derivatives into equation (1.28) gives

$$\begin{aligned} &a_2(x)\left[v_1 y_1'' + v_2 y_2'' + v_1' y_1' + v_2' y_2'\right] \\ &\quad + a_1(x)\left[v_1 y_1' + v_2 y_2'\right] + a_0(x)\left[v_1 y_1 + v_2 y_2\right] = f(x), \end{aligned}$$

which may be rearranged in the form

$$\begin{aligned} &a_2(x)\left(v_1' y_1' + v_2' y_2'\right) = f(x) \\ &\quad + v_1\left[a_2(x) y_1'' + a_1(x) y_1' + a_0(x) y_1\right] + v_2\left[a_2(x) y_2'' + a_1(x) y_2' + a_0(x) y_2\right]. \end{aligned}$$

Since y_1 and y_2 are solutions of the homogeneous equation (1.29), the terms within the square brackets [ ] above must vanish. Therefore, we have

$$a_2(x)\left(v_1' y_1' + v_2' y_2'\right) = f(x).$$

We also have the relationship (1.32) and together they may be written as a set of algebraic equations:

$$\begin{bmatrix} y_1(x) & y_2(x) \\ y_1'(x) & y_2'(x) \end{bmatrix} \begin{bmatrix} v_1'(x) \\ v_2'(x) \end{bmatrix} = \begin{bmatrix} 0 \\ F(x) \end{bmatrix}, \tag{1.33}$$

where $F(x) = f(x)/a_2(x)$. The solution for $v_1'(x)$ and $v_2'(x)$ is found by Cramer's rule and it is given by:

$$v_1'(x) = -\frac{y_2(x)F(x)}{W(x)} \tag{1.34}$$

and

$$v_2'(x) = \frac{y_1(x)F(x)}{W(x)}, \tag{1.35}$$

where $W(x)$ is the Wronskian determinant given by

$$W(x) = \begin{vmatrix} y_1(x) & y_2(x) \\ y_1'(x) & y_2'(x) \end{vmatrix}.$$

The equations (1.34) and (1.35) can now be integrated to give

$$v_1(x) = -\int_{c_1}^{x} \frac{y_2(\xi)F(\xi)}{W(\xi)} d\xi,$$

and

$$v_2(x) = \int_{c_2}^{x} \frac{y_1(\xi)F(\xi)}{W(\xi)} d\xi,$$

so that the complete solution can be written as

$$\begin{aligned} y(x) &= v_1(x)y_1(x) + v_2(x)y_2(x) \\ &= -y_1(x)\int_{c_1}^{x} \frac{y_2(\xi)F(\xi)}{W(\xi)}d\xi + y_2(x)\int_{c_2}^{x} \frac{y_1(\xi)F(\xi)}{W(\xi)}d\xi \end{aligned} \tag{1.36}$$

Example 1.18 As an example let us solve the differential equation,

$$y'' + 4y' + 4y = \frac{e^{-2x}}{x^2} \tag{1.37}$$

Here, the forcing function is not a simple product of an exponential and a polynomial. Therefore, the method of undetermined coefficients will not work. To apply the method of variation of parameters we first need to obtain the Wronskian which can be found from the solutions to the homogeneous equation. The characteristic equation for

$$y'' + 4y' + 4y = 0$$

is

$$m^2 + 4m + 4 = 0,$$

with roots $m_1 = m_2 = -2$. The two linearly independent solutions are, therefore, given by

$$y_1(x) = e^{-2x} \quad \text{and} \quad y_2(x) = xe^{-2x}.$$

The Wronskian may now be written as

$$\begin{aligned} W(x) &= \begin{vmatrix} y_1(x) & y_2(x) \\ y_1'(x) & y_2'(x) \end{vmatrix} = \begin{vmatrix} e^{-2x} & xe^{-2x} \\ -2e^{-2x} & (1-2x)e^{-2x} \end{vmatrix} \\ &= e^{-4x}. \end{aligned}$$

From equation (1.36) we obtain the complete solution as

$$\begin{aligned} y(x) &= -\int_{c_1}^{x} \frac{e^{-2x}\xi e^{-2\xi}e^{-2\xi}/\xi^2}{e^{-4\xi}}\,d\xi \;+\; \int_{c_2}^{x} \frac{xe^{-2x}e^{-2\xi}e^{-2\xi}/\xi^2}{e^{-4\xi}}\,d\xi \\ &= -e^{-2x}\int_{c_1}^{x} \frac{1}{\xi}\,d\xi \;+\; xe^{-2x}\int_{c_2}^{x} \frac{1}{\xi^2}\,d\xi \\ &= -e^{-2x}\ln x + \left[C_1^* e^{-2x} + C_2^* xe^{-2x}\right]. \end{aligned}$$

□

The next example is also a case for which the *Method of Undetermined Coefficients* does not work.

Example 1.19

$$y'' + 4y = \tan x.$$

The characteristic equation for the homogeneous equation is

$$m^2 + 4 = 0,$$

which has solutions, $m_1 = 2i$ and $m_2 = -2i$. The two linearly independent solutions, therefore, are:

$$y_1 = \sin 2x \qquad \text{and} \qquad y_2 = \cos 2x.$$

The Wronskian is

$$W(x) = \begin{vmatrix} \sin 2x & \cos 2x \\ 2\cos 2x & -2\sin 2x \end{vmatrix} = -2.$$

From equation (1.36), we obtain

$$\begin{aligned}
y(x) &= -\int_{c_1}^{x} \frac{\sin 2x \cos 2\xi \tan \xi}{-2}\, d\xi + \int_{c_2}^{x} \frac{\cos 2x \sin 2\xi \tan \xi}{-2}\, d\xi \\
&= \tfrac{1}{2}\sin 2x \int_{c_1}^{x} [2\cos\xi \sin\xi - \tan\xi]\; d\xi - \tfrac{1}{2}\cos 2x \int_{c_2}^{x} 2\sin^2\xi\; d\xi \\
&= \tfrac{1}{2}\sin 2x \left[-\tfrac{1}{2}\cos 2x + \ln\cos x\right] - \tfrac{1}{2}\cos 2x\left[x - \tfrac{1}{2}\sin 2x\right] + \left[C_1^* \sin 2x + C_2^* \cos 2x\right] \\
&= -\tfrac{1}{2}\left[x\cos 2x - (\ln\cos x)\sin 2x\right] + \left[C_1^* \sin 2x + C_2^* \cos 2x\right]
\end{aligned}$$

□

As stated earlier, the method of variation of parameters is very general and is also applicable to linear differential equations with variable coefficients. The general case of linear differential equations with variable coefficients require power series solutions. These will be treated in Chapter 4. In the next section we shall learn about the Euler equation which is a special case of variable coefficients.

EXERCISES 1.3 *Solve the following equations by the method of Variation of Parameters*

1. $y'' - 2y' + y = \dfrac{e^x}{x^2}.$
2. $y'' - y = \dfrac{1}{\sinh x}.$
3. $y'' + y = \cot x.$
4. $y'' + 4y = \sec 2x.$
5. $y'' + 2y' + 10y = e^{-x}\cot 3x.$
6. $y'' - 4y = \tanh x.$
7. $y'' + 4y' + 8y = 2\mathrm{e}^{-2x}\tan^2 x.$
8. $y'' - 6y' + 9y = xe^{3x}\ln x.$
9. $y'' - 4y' + 20y = \dfrac{e^{2x}}{\sin 4x}.$
10. $y'' + 3y' + \frac{9}{2}y = \dfrac{e^{-\frac{3}{2}x}}{\sin\frac{1}{2}x}.$

1.2.4 The Euler Equation

When the coefficients $a_n(x), a_{n-1}(x), \ldots, a_1(x), a_0(x)$ in equation(1.6) take on the following special form

$$\begin{aligned}
a_n(x) &= A_n x^n, \\
a_{n-1}(x) &= A_{n-1} x^{n-1}, \\
&\vdots \\
a_1(x) &= A_1 x, \\
a_0(x) &= A_0,
\end{aligned}$$

where the A_k's are constants, the differential equation can be transformed to the type with constant coefficients. This type of an equation is called the Euler equation[2]. The general form is

$$A_n x^n y^{(n)} + A_{n-1} x^{n-1} y^{(n-1)} + \cdots + A_3 x^3 y''' + A_2 x^2 y'' + A_1 x y' + A_0 y = f(x). \quad (1.38)$$

To solve this equation, we make the transformation

$$x = \mathrm{e}^z \quad \text{or} \quad z = \ln x,$$

where we can see that

$$\frac{dz}{dx} = \frac{1}{x}$$

Now we need to transform the derivatives with respect to x into derivatives with respect to z. We begin with $y'(x)$.

$$\begin{aligned} y'(x) &= \frac{dy}{dx} \\ &= \frac{dy}{dz} \cdot \frac{dz}{dx} \\ &= \frac{dy}{dz} \frac{1}{x}. \end{aligned}$$

Therefore, we can write

$$xy' = \frac{dy}{dz}.$$

We also have the operator equality

$$\frac{d}{dx} = \frac{1}{x} \frac{d}{dz}. \quad (1.39)$$

Let us now calculate $y''(x)$.

$$\begin{aligned} y''(x) &= \frac{d^2 y}{dx^2} \\ &= \frac{d}{dx}\left(\frac{1}{x}\frac{dy}{dz}\right) \\ &= -\frac{1}{x^2}\frac{dy}{dz} + \frac{1}{x}\left(\frac{d}{dx}\right)\frac{dy}{dz}. \end{aligned}$$

Here we replace

$$\frac{d}{dx} \quad \text{with} \quad \frac{1}{x}\frac{d}{dz}$$

[2]Also known as the Euler-Cauchy equation

and obtain

$$\begin{aligned} y''(x) &= -\frac{1}{x^2}\frac{dy}{dz} + \frac{1}{x}\left(\frac{1}{x}\frac{d}{dz}\right)\frac{dy}{dz}. \\ &= -\frac{1}{x^2}\frac{dy}{dz} + \frac{1}{x^2}\frac{d^2y}{dz^2}. \end{aligned}$$

If we now multiply each side by x^2 we obtain

$$x^2y''(x) = \frac{d^2y}{dz^2} - \frac{dy}{dz}.$$

For brevity we define the operator

$$D = \frac{d}{dz}.$$

With this notation, we can write

$$xy' = Dy,$$

and

$$x^2y'' = D(D-1)y.$$

By carrying on this process further to higher derivatives it is not difficult to show that

$$\begin{aligned} x^3y''' &= D(D-1)(D-2)y \\ &\vdots \\ x^n y^{(n)} &= [D(D-1)(D-2)\cdots(D-n+2)(D-n+1)]\, y. \end{aligned}$$

With this result the original differential equation (1.38) transforms into

$$\begin{aligned} A_n\,[D(D-1)(D-2)\cdots(D-n+2)(D-n+1)]\,y + \cdots \\ \cdots + A_2D(D-1)y + A_1Dy + A_0y = f(e^z), \end{aligned}$$

which can be written as

$$A_n^*D^ny(z) + A_{n-1}^*D^{n-1}y(z) + \ldots + A_2^*D^2y(z) + A_1^*Dy(z) + A_0^*y(z) = F(z). \quad (1.40)$$

Here the coefficients $A_n^*, \ldots, A_2^*, A_1^*, A_0^*$ are linear combinations of $A_n, \ldots, A_2, A_1, A_0$. Clearly, equation (1.40) is a differential equation with constant coefficients which has standard exponential solutions for the homogeneous case. Let us now consider some examples. We shall begin with homogeneous cases.

Homogeneous Euler Equation

Example 1.20

$$x^2y'' - xy' + y = 0. \quad (1.41)$$

With the transformation $x = e^z$, this differential equation becomes

$$D(D-1)y - Dy + y = 0,$$

or

$$D^2y - 2Dy + y = 0,$$

which may also be written as

$$\frac{d^2y(z)}{dz^2} - 2\frac{dy(z)}{dz} + y(z) = 0.$$

The characteristic equation for this equation is

$$m^2 - 2m + 1 = 0,$$

with roots $m_1 = m_2 = 1$. The general solution, therefore, is

$$y(z) = C_1e^z + C_2ze^z.$$

Using $z = \ln x$ changing back to x as the independent variable, we obtain

$$y(x) = C_1x + C_2(\ln x)x.$$

□

The next example is a third order equation.

Example 1.21

$$x^3y''' + 4x^2y'' - 2xy' - 4y = 0. \tag{1.42}$$

With the transformation $x = e^z$, the differential equation becomes

$$D(D-1)(D-2)y + 4D(D-1)y - 2Dy - 4y = 0,$$

or

$$\left[D^3 + D^2 - 4D - 4\right]y = 0.$$

This differential equation with constant coefficients has the characteristic equation

$$m^3 + m^2 - 4m - 4 = 0,$$

which factorizes into the form

$$(m+1)(m+2)(m-2) = 0.$$

The roots clearly are $m_1 = -1, \ \ m_2 = -2$ and $m_3 = 2$. The solution therefore is

$$y(z) = C_1e^{-z} + C_2e^{-2z} + C_3e^{2x},$$

or

$$y(x) = C_1\frac{1}{x} + C_2\frac{1}{x^2} + C_3x^2.$$

□

With the Euler equation we may by-pass the transformation $x = e^z$ and straight away assume a solution of the form $y = x^m$. For the purpose of illustration we consider a couple of examples.

Example 1.22

$$x^3y''' - 3xy' + 3y = 0. \tag{1.43}$$

Letting

$$y = x^m$$

we obtain

$$y' = mx^{m-1}, \quad y'' = m(m-1)x^{m-2}, \quad \text{and} \quad y''' = m(m-1)(m-2)x^{m-3}.$$

The substitution of y and its derivatives into equation (1.43) leads to

$$m(m-1)(m-2)x^m - 3mx^m + 3x^m = 0,$$

which yields the characteristic equation

$$m(m-1)(m-2) - 3m + 3 = 0.$$

This factorizes into the form

$$(m-1)(m+1)(m-3) = 0,$$

giving roots $m_1 = 1,\ m_2 = -1,$ and $m_3 = 3$. The solution therefore is

$$y(x) = C_1x + C_2\frac{1}{x} + C_3x^3.$$

□

In the next example we consider a case with repeated roots of the characteristic equation.

Example 1.23

$$x^2y'' + 5xy' + 4y = 0.$$

Letting $y = x^m$, we obtain the characteristic equation

$$m(m-1) + 5m + 4 = 0$$

or

$$m^2 + 4m + 4 = 0,$$

which has roots $m_1 = m_2 = -2$. The general solution in this case is

$$y = C_1\frac{1}{x^2} + C_2\frac{\ln x}{x^2}.$$

Here the $\ln x$ term appears in the same manner as it does in the solution of equation (1.41). □

In some instances we have complex roots of the characteristic equation. This leads to sin and cos terms as illustrated in the next two examples.

Example 1.24

$$x^3 y''' + 2x^2 y'' + xy' - y = 0. \tag{1.44}$$

With the transformation $x = e^z$, this differential equation may be written as

$$[D(D-1)(D-2) + 2D(D-1) + D - 1]\, y(z) = 0,$$

or

$$\left[D^3 - D^2 + D - 1\right] y(z) = 0.$$

This has the characteristic equation

$$m^3 - m^2 + m - 1 = 0, \tag{1.45}$$

which may be factorized to the form

$$(m-1)(m^2+1) = 0.$$

The roots are $m_1 = 1,\ m_2 = i$ and $m_3 = -i$. The solution therefore is

$$y(z) = C_1 e^z + C_2 \sin z + C_3 \cos z,$$

or in terms of x we have

$$y(x) = C_1 x + C_2 \sin(\ln x) + C_3 \cos(\ln x). \tag{1.46}$$

In this example if we choose to by-pass the transformation $x = e^z$, we assume a solution of the form $y = x^m$ and obtain the same characteristic equation as equation)1.45). With roots as given above, the general solution may be written as

$$y(x) = C_1^* x + C_2^* x^i + C_3^* x^{-i}.$$

Using the definition of the natural logarithm

$$x^m = e^{m \ln x},$$

we can write the above solution as

$$y(x) = C_1^* x + C_2^* e^{i \ln x} + C_3^* e^{-i \ln x}.$$

Now using Euler's formula, we obtain the solution as given in equation (1.46). □

Example 1.25 Let us consider the equation

$$x^2 y'' + 3xy' + 2y = 0.$$

Assuming a solution of the type $y = x^m$, the characteristic equation is

$$m(m-1) + 3m + 2 = 0,$$

or

$$m^2 + 2m + 2 = 0.$$

The roots are given by:

$$\begin{aligned} m &= \frac{-2 \pm \sqrt{(4-8)}}{2} \\ &= -1 \pm i. \end{aligned}$$

The solution takes the form

$$\begin{aligned} y &= C_1^* x^{-1+i} + C_2^* x^{-1-i} \\ &= \frac{1}{x}\left[C_1 \sin(\ln x) + C_2 \cos(\ln x)\right] \end{aligned}$$

□

Non-Homogeneous Euler Equation

For the non-homogeneous cases of the Euler equation, the method of undetermined coefficients will work for certain cases. We may recall that for linear equations with constant coefficients, if the forcing function contains terms exponentials, polynomials or products of these, then we can apply the method of undetermined coefficients. When written in terms of z as the independent variable, this includes terms such as

$$e^{-2z}, \quad z^2 + 3z + 5, \quad z^3 + z^2 e^{-z}, \quad \text{and} \quad (z^2 + 2z)z^{4z}.$$

Correspondingly, for the Euler equation, bearing in mind the transformation $x = e^z$ or $z = \ln x$, we can say that the method of undetermined coefficients is applicable to forcing functions such as

$$x^{-2}, \quad (\ln x)^2 + 3\ln x + 5, \quad (\ln x)^3 + (\ln x)^2 x^{-1}, \quad \text{and} \quad \left[(\ln x)^2 + 2\ln x\right] x^4.$$

The following examples will provide an illustration of the application of the method of undetermined coefficients.

Example 1.26

$$x^2 y'' - xy' + y = x^5. \tag{1.47}$$

Letting $x = e^z$, the differential equation transforms into

$$D^2 y - 2Dy + y = e^{5z}. \tag{1.48}$$

The characteristic equation is

$$m^2 - 2m + 1 = 0,$$

with solutions, $m_1 = m_2 = 1$. The solution of the homogeneous differential equation is

$$y_h = C_1 e^z + C_2 z e^z.$$

As we can see, the forcing function is not a secular term. Therefore, we may assume a particular solution of the form

$$y_p = C e^{5z}.$$

Substitution into equation (1.48) leads to

$$(25 - 10 + 1)Ce^{5z} = e^{5z},$$

which simplifies to

$$16C = 1, \qquad \text{or} \qquad C = \tfrac{1}{16}.$$

The particular solution can be written as

$$y_p = \tfrac{1}{16}e^{5z}.$$

The complete general solution is

$$y(z) = C_1e^z + C_2ze^z + \tfrac{1}{16}e^{5z},$$

or, in terms of x,

$$y(x) = C_1x + C_2(\ln x)x + \tfrac{1}{16}x^5.$$

We could have avoided the transformation into z altogether, by assuming a homogeneous solution of the type $y_h = x^m$ and a particular solution of the type $y_p = Cx^5$. This would have led to the same general solution. □

Example 1.27

$$x^2y'' - xy' + y = 3x. \tag{1.49}$$

Here we have the same differential equation as in the previous example, except for a different forcing function. This time the forcing function is a secular term because it is proportional to one of the solutions to the homogeneous solutions. In this case $y_p = C(\ln x)x$ will not work either because it is also a solution to the homogeneous equation. Therefore, we try

$$y_p = C(\ln x)^2x,$$

which in terms of $z = \ln x$ corresponds to $y_p = Cze^z$. The derivatives of y_p are:

$$y_p' = C\left[(\ln x)^2 + 2(\ln x)\right],$$

and

$$y_p'' = C\left[\frac{2\ln x}{x} + \frac{2}{x}\right].$$

Substitution into equation (1.49) gives

$$x^2C\left[\frac{2\ln x}{x} + \frac{2}{x}\right] - xC\left[(\ln x)^2 + 2(\ln x)\right] + C(\ln x)^2x = 3x,$$

which simplifies to

$$2Cx = 3x \qquad \text{or} \qquad C = \tfrac{3}{2},$$

and hence,

$$y_p = \tfrac{3}{2}x(\ln x)^2.$$

□

Example 1.28

$$x^2y'' + 3xy' + y = \frac{(\ln x)^2}{x}.$$

Assuming a homogeneous solution of the form $y_h = x^m$, we obtain the characteristic equation:

$$m(m-1) + 3m + 1 = 0, \quad \text{or} \quad m^2 + 2m + 1 = 0.$$

The roots are $m_1 = m_2 = -1$. The two linearly independent solutions of the homogeneous equation may be written as

$$y_1 = \frac{1}{x} \quad \text{and} \quad y_2 = \frac{\ln x}{x}.$$

Here, the forcing function is a product of a power in $\ln x$ and a simple power in x. Equivalently, in terms of $z = \ln x$, the forcing function is z^2e^{-z}. If we try a particular solution of the type,

$$y_p(x) = \frac{1}{x}\left[C_2(\ln x)^2 + C_1(\ln x) + C_0\right] = e^{-z}\left[C_2z^2 + C_1x + C_0\right],$$

and keep in mind that $y_1(x) = 1/x$ and $y_2(x) = (\ln x)/x$ are solutions of the homogeneous equation, the last two terms in $y_p(x)$ above will be redundant, and will vanish upon substitution of y_p into the left-hand side of the differenial equation. Additionally, terms on the left-hand side resulting from $(1/x)C_2(\ln x)^2$ will be reduced by two powers in $\ln x$. Therefore we need to consider

$$y_p(x) = \frac{1}{x}\left[C_4(\ln x)^4 + C_3(\ln x)^3 + C_2(\ln x)^2\right] = e^{-z}\left[C_4z^4 + C_1x^3 + C_2x^2\right].$$

Taking derivatives of y_p, we obtain

$$\begin{aligned}
y_p'(x) &= -\frac{1}{x^2}\left[C_4(\ln x)^4 + C_3(\ln x)^3 + C_2(\ln x)^2 - 4C_4(\ln x)^3 + 3C_3(\ln x)^2 + 2C_2(\ln x)\right] \\
&= \frac{1}{x^2}\left[-C_4(\ln x)^4 + (4C_4 - C_3)(\ln x)^3 + (3C_3 - C_2)(\ln x)^2 - 2C_2(\ln x)\right] \\
y_p''(x) &= \frac{1}{x^3}\left[2C_4(\ln x)^4 - 2(4C_4 - C_3)(\ln x)^3 - 2(3C_3 - C_2)(\ln x)^2 + 2C_2(\ln x)\right. \\
&\quad \left. -4C_4(\ln x)^3 + (12C_4 - 3C_3)(\ln x)^2 + (6C_3 - 2C_2)(\ln x) - 2C_2\right] \\
&= \frac{1}{x^3}\left[2C_4(\ln x)^4 - (12C_4 - 2C_3)(\ln x)^3 + (12C_4 - 9C_3 + 2C_2)(\ln x)^2\right. \\
&\quad \left. +6C_3(\ln x) - 2C_2\right]
\end{aligned}$$

Upon substituting the above expressions for $y_p(x)$ and its derivatives, we obtain

$$\begin{aligned}
&\frac{1}{x}\left[2C_4(\ln x)^4 - (12C_4 - 2C_3)(\ln x)^3 + (12C_4 - 9C_3 + 2C_2)(\ln x)^2 + 6C_3(\ln x) - 2C_2\right] \\
&+\frac{3}{x}\left[-C_4(\ln x)^4 + (4C_4 - C_3)(\ln x)^3 + (3C_3 - C_2)(\ln x)^2 - 2C_2(\ln x)\right] \\
&+\frac{1}{x}\left[C_4(\ln x)^4 + C_3(\ln x)^3 + C_2(\ln x)^2\right] = \frac{(\ln x)^2}{x}.
\end{aligned}$$

$$\frac{1}{x}\left[(12C_4 - 9C_3 + 2C_2)(\ln x)^2 \right.$$
$$\left. +6C_3(\ln x) - 2C_2 + (9C_3 - 3C_2)(\ln x)^2 - 2C_2(\ln x) + C_2(\ln x)^2\right] = \frac{(\ln x)^2}{x}.$$
$$\frac{1}{x}\left[(12C_4)(\ln x)^2 + [6C_3 - 2C_2](\ln x) - 2C_2\right] = \frac{(\ln x)^2}{x},$$

which yields

$$C_2 = 0, \qquad C_3 = 0, \qquad \text{and} \qquad C_4 = \tfrac{1}{12},$$

i.e.,

$$y_p = \frac{1}{12}\frac{(\ln x)^4}{x}.$$

The general solution is

$$y(x) = C_1\frac{1}{x} + C_2\frac{\ln x}{x} + \tfrac{1}{12}\frac{(\ln x)^4}{x}.$$

□

In cases in which the forcing functional is not a simple power in x, a polynomial in $\ln x$ or a product of these, the particular solution will have to be obtained by the method of variation of parameters. Let us now consider an example of such a case.

Example 1.29

$$x^2y'' + 3xy' + y = \frac{1}{x(\ln x)^2}.$$

Assuming a homogeneous solution of the form $y_h = x^m$, we obtain the characteristic equation:

$$m(m-1) + 3m + 1 = 0, \quad \text{or} \quad m^2 + 2m + 1 = 0.$$

The roots are $m_1 = m_2 = -1$. The two linearly independent solutions of the homogeneous equation may be written as

$$y_1 = \frac{1}{x} \quad \text{and} \quad y_2 = \frac{\ln x}{x}.$$

The Wronskian for this case is

$$\begin{aligned} W(x) &= \begin{vmatrix} y_1(x) & y_2(x) \\ y_1'(x) & y_2'(x) \end{vmatrix} = \begin{vmatrix} \frac{1}{x} & \frac{\ln x}{x} \\ -\frac{1}{x^2} & \frac{1-\ln x}{x^2} \end{vmatrix} \\ &= \frac{1}{x^3}. \end{aligned}$$

We can now apply the result given by equation (1.36) and write the solution as

$$\begin{aligned} y(x) &= -\frac{1}{x}\int_{c_1}^{x} \frac{\frac{\ln\xi}{\xi}\frac{1}{\xi(\ln\xi)^2\xi^2}}{\frac{1}{\xi^3}}d\xi + \frac{\ln x}{x}\int_{c_2}^{x}\frac{\frac{1}{\xi}\frac{1}{\xi(\ln\xi)^2\xi^2}}{\frac{1}{\xi^3}}d\xi \\ &= -\frac{\ln(\ln x)}{x} - \frac{1}{x} + C_1\frac{1}{x} + C_2\frac{\ln x}{x} \\ &= -\frac{\ln(\ln x)}{x} + C_1^*\frac{1}{x} + C_2\frac{\ln x}{x}. \end{aligned}$$

□

EXERCISES 1.4 *Obtain the general solution to the following differential equations:*

1. $x^2y'' + 14xy' + 40y = 0.$

2. $x^2y'' + 7xy' + 9y = 0.$

3. $x^2y'' - 3xy' + 5y = 0.$

4. $x^2y'' - 5xy' + 10y = 0.$

5. $x^2y'' + 5xy' + 13y = 0.$

6. $x^2y'' - 8xy' - 52y = 0.$

7. $x^2y'' + 2xy' - 12y = 0.$

8. $x^3y''' + 2xy' - 12y = 0.$

9. $x^3y''' - 3x^2y'' - 6xy' + 6y = 0.$

10. $x^2y'' - 2xy' - 10y = \dfrac{1}{x^2}.$

11. $x^2y'' + 8xy' + 10y = \dfrac{1}{x^5}.$

12. $x^2y'' - 12y = (\ln x)^3 + 4\ln x.$

13. $x^2y'' - 8xy' + 14y = (\ln x)^2 + 4\ln x.$

14. $x^2y'' - 6xy' + 10y = (x\ln x)^2.$

15. $x^2y'' + xy' + 16y = (\ln x)\left[\sin(4\ln x)\right].$

16. $x^2y'' + 3xy' + 10y = \dfrac{\cot(3\ln x)}{x}.$

Chapter 2

Power-Series Solutions to Differential Equations and Special Functions

2.1 Introduction and Basic Concepts

While a large class of engineering problems deal with differential equations having constant coefficients, there a numerous systems in which we have variable coefficients. We discuss the technique from a general perspective with a focus on special functions for the development of Fourier expansions of such functions. The procedure in such cases is to use a power series method. The technique is also referred to as the *Method of Frobenius*. Generally, we assume the solution to be a power series, substitute into the differential equation, and equate terms in like powers to obtain the coefficients of the power series. Let us begin with an example

Example 2.1

$$x^2 y''(x) + 2xy'(x) + (x^2 - 2)y(x) = 0. \tag{2.1}$$

Assume (and hope it works)

$$y(x) = C_0 + C_1 x + C_2 x^2 + \ldots = \sum_{n=0}^{\infty} C_n x^n. \tag{2.2}$$

The derivatives are:

$$y'(x) = \sum_{n=0}^{\infty} n C_n x^{n-1},$$

and

$$y''(x) = \sum_{n=0}^{\infty} n(n-1) C_n x^{n-2}.$$

Substitution of these series expansions into the differential equation (2.1) gives

$$x^2 \sum_{n=0}^{\infty} n(n-1) C_n x^{n-2} + 2x \sum_{n=0}^{\infty} n C_n x^{n-1} + (x^2 - 2) \sum_{n=0}^{\infty} C_n x^n = 0.$$

Further simplification yields

$$\sum_{n=0}^{\infty} n(n-1)C_n x^n + 2\sum_{n=0}^{\infty} nC_n x^n - 2\sum_{n=0}^{\infty} C_n x^n + \sum_{n=0}^{\infty} C_n x^{n+2} = 0,$$

or

$$\sum_{n=0}^{\infty} [n(n-1) + 2n - 2]\, C_n x^n + \sum_{n=0}^{\infty} C_n x^{n+2} = 0.$$

To get a better picture it is a good idea to open up the summations here. Doing so,

$$\begin{aligned} \left[-2C_0 + 0C_1 x + 4C_2 x^2 + 10C_3 x^3 + 18C_4 x^4 + 28C_5 x^5 + \ldots\right] & \\ + \left[C_0 x^2 + C_1 x^3 + C_2 x^4 + C_3 x^5 + \ldots\right] &= 0. \end{aligned}$$

By grouping terms in powers of x, we may write

$$\begin{aligned} -2C_0 + 0C_1 x + (4C_2 + C_0)x^2 + (10C_3 + C_1)x^3 & \\ +(18C_4 + C_2)x^4 + (28C_5 + C_3)x^5 + \ldots &= 0. \end{aligned}$$

Equating each side of the equation term-by-term, we find

$$\begin{aligned} -2C_0 &= 0, \\ C_1 &= \text{undetermined}, \\ 4C_2 + C_0 &= 0, \\ 10C_3 + C_1 &= 0, \\ 18C_4 + C_2 &= 0, \\ 28C_5 + C_3 &= 0, \\ &\vdots \end{aligned}$$

Is easy to see that $C_0 = C_2 = C_4 = C_6 = \ldots = 0$ and

$$\begin{aligned} C_3 &= -\tfrac{1}{10}C_1, \\ C_5 &= -\tfrac{1}{28}C_3, \\ &= \tfrac{1}{280}C_1, \\ &\vdots \end{aligned}$$

One solution can therefore be written as

$$y_1(x) = C_1\left[x - \tfrac{1}{10}x^3 + \tfrac{1}{280}x^5 \ldots\right].$$

The second solution does not come out in such a straightforward manner. The reason is that the assumed form in equation (2.2) is not sufficiently general to accommodate the second solution which actually contains some negative powers in x. In fact, in this example, the approach that we took has worked to the extent that it has, only because the power series expansion about $x = 0$ of one solution consists exclusively of positive integer powers. □

In general we need to admit negative as well as fractional powers. To that end, we assume a power series solution of the form

$$y(x) = C_0 x^s + C_1 x^{1+s} + C_2 x^{2+s} + \ldots = \sum_{n=0}^{\infty} C_n x^{n+s}, \tag{2.3}$$

where the values of s are determined by satisfying the differential equation. Taking derivatives, we obtain

$$y'(x) = \sum_{n=0}^{\infty} C_n (n+s) x^{n+s-1}, \tag{2.4}$$

and

$$y''(x) = \sum_{n=0}^{\infty} C_n (n+s)(n+s-1) x^{n+s-2}. \tag{2.5}$$

Substitution into the differential equation (2.1) yields

$$x^2 \sum_{n=0}^{\infty} C_n (n+s)(n+s-1) x^{n+s-2} + 2x \sum_{n=0}^{\infty} C_n (n+s) x^{n+s-1} + (x^2 - 2) \sum_{n=0}^{\infty} C_n x^{n+s} = 0,$$

which may be rewritten as

$$\sum_{n=0}^{\infty} C_n (n+s)(n+s-1) x^{n+s} + 2 \sum_{n=0}^{\infty} C_n (n+s) x^{n+s} - 2 \sum_{n=0}^{\infty} C_n x^{n+s} + \sum_{n=0}^{\infty} C_n x^{n+s+2} = 0.$$

With some rearrangement we obtain

$$\sum_{n=0}^{\infty} C_n \left[(n+s)(n+s-1) + 2(n+s) - 2\right] x^{n+s} + \sum_{n=0}^{\infty} C_n x^{n+s+2} = 0,$$

or

$$\sum_{n=0}^{\infty} C_n (n+s+2)(n+s-1) x^{n+s} + \sum_{n=0}^{\infty} C_n x^{n+s+2} = 0. \tag{2.6}$$

If we open up the summations, we have

$$\begin{aligned} (s+2)(s-1)C_0 x^s + (s+3)s C_1 x^{s+1} + (s+4)(s+1)C_2 x^{s+2} & \\ +(s+5)(s+2)C_3 x^{s+3} + & \;\cdots \\ +C_0 x^{s+2} + C_1 x^{s+3} + C_2 x^{s+4} + & \;\cdots \;= 0. \end{aligned}$$

Upon regrouping the powers in x, this becomes

$$\begin{aligned} &(s+2)(s-1)C_0 x^s + (s+3)s C_1 x^{s+1} \\ &\quad + \left[(s+4)(s+1)C_2 + C_0\right] x^{s+2} \\ &\quad + \left[(s+5)(s+2)C_3 + C_1\right] x^{s+3} \\ &\quad + \left[(s+6)(s+3)C_4 + C_2\right] x^{s+4} + \cdots = 0. \end{aligned}$$

Since this must be valid for all values of x, we can set the coefficient of every power of x to be equal to zero. Thus,

$$\begin{aligned}
(s+2)(s-1)C_0 &= 0, & (2.7)\\
(s+3)sC_1 &= 0, & (2.8)\\
[(s+4)(s+1)C_2 + C_0] &= 0, & (2.9)\\
[(s+5)(s+2)C_3 + C_1] &= 0, & (2.10)\\
[(s+6)(s+3)C_4 + C_2] &= 0, & (2.11)\\
&\vdots & (2.12)
\end{aligned}$$

We satisfy equation (2.7), by choosing a set of values of s,

$$s = -2, 1$$

and take $C_0 \neq 0$. Then, with these values of s in (2.8) we have $C_1 = 0$. By using equations (2.9–2.11) we can obtain the values of the coefficients. We can already see a pattern for general relationship. However, we can rigorously establish a general recurrence formula relating the C_n's by re-examining (2.6). Noting that the leading power in the first summation is x^s while in the second summation it is x^{s+2}, we extract the first two terms of the first summation. Therefore,

$$C_0(s+2)(s-1)x^s + C_1(s+3)sx^{s+1} + \sum_{n=2}^{\infty} C_n(n+s+2)(n+s-1)x^{n+s} + \sum_{n=0}^{\infty} C_n x^{n+s+2} = 0. \tag{2.13}$$

Now, the leading term in each summation has the power x^{s+2}. It would be better to shift the index of summation from 2 to 0 in the first sum.

2.1.1 Shifting the Index

Indices can be shifted as follows:
For a sum

$$S = \sum_{n=2}^{\infty} G_n x^{n+s}, \tag{2.14}$$

We write

$$S = \sum_{n=2}^{\infty} G_n x^{n-2+s+2} \tag{2.15}$$

We now define

$$n' = n - 2, \quad \text{or} \quad n = n' + 2.$$

Thus,

$$S = \sum_{n'+2=2}^{\infty} G_{n'+2} x^{n'+s+2} = \sum_{n'=0}^{\infty} G_{n'+2} x^{n'+s+2} \tag{2.16}$$

Since n' is a dummy index, it can be renamed as n. Therefore,

$$S = \sum_{n=0}^{\infty} G_{n+2} x^{n+s+2} \tag{2.17}$$

For the present case, we write equation (2.13) as:

$$C_0(s+2)(s-1)x^s + C_1(s+3)sx^{s+1} + \sum_{n=0}^{\infty} C_{n+2}(n+s+4)(n+s+1)x^{n+s+2} + \sum_{n=0}^{\infty} C_n x^{n+s+2} = 0, \tag{2.18}$$

i.e.,

$$C_0(s+2)(s-1)x^s + C_1(s+3)sx^{s+1} + \sum_{n=0}^{\infty} \left[C_{n+2}(n+s+4)(n+s+1) + C_n\right] x^{n+s+2} = 0 \tag{2.19}$$

Letting the coefficient of each power in x go to zero, we have

$$C_0(s+2)(s-1) = 0 \tag{2.20}$$

$$C_1(s+3)s = 0 \tag{2.21}$$

and

$$C_{n+2} = -\frac{C_n}{(n+s+4)(n+s+1)} \tag{2.22}$$

Equation (2.20) yields $s = 1, -2$ with $C_0 \neq 0$ and (2.21) gives $s = 0, -3$ with $C_1 \neq 0$. However, the pair of values of s must be consistent. Thus, if we select $s = 1, -2$, then we cannot have $s = 0, -3$. With the former pair of values of s, we conclude that in this example we must necessarily have $C_1 = 0$. This need not be the case if the second pair has one or both values of s coinciding with the first pair.

We begin with one value of s, say $s = 1$. We set

$$C_0 \neq 0 \tag{2.23}$$

As a result,

$$C_1 = 0$$

From (2.22) with $s = 1$, we see

$$\begin{aligned} C_{n+2} &= -\frac{C_n}{(n+5)(n+2)} \\ C_2 &= -\frac{C_0}{2(5)} \quad (n = 0) \\ C_3 &= -\frac{C_1}{3(6)} = 0 \quad (n = 1) \end{aligned}$$

$$\begin{aligned}
C_4 &= -\frac{C_2}{4(7)} = -\frac{C_0}{2(5)} \cdot \frac{-1}{4(7)} \\
&= \frac{C_0}{(2.4)(5.7)} = \frac{C_0}{280} \\
C_6 &= -\frac{C_4}{6(9)} = \frac{-C_0}{(2.4.6)(5.7.9)} \\
C_8 &= -\frac{C_6}{8(11)} = -\frac{C_0 3(8)}{9!} \cdot \frac{(-1)}{8(11)} \\
&= \frac{C_0}{(2.4.6.8)(5.7.9.11)} \\
&= \frac{C_0 3(10)}{1.2.3.4.5.6.7.8.9.10.11} = \frac{C_0 3(10)}{11!}
\end{aligned}$$

$$C_n = \frac{(-1)^{\frac{n}{2}} C_0 3(n+2)}{(n+3)!} \qquad n = 0, 2, 4, 6...$$

$$\begin{aligned}
y_1 &= C_0 x^1 \left\{ 1 - \frac{1}{10} x^2 + \frac{1}{280} x^4 + \cdots \right\} \\
&= C_0 x \sum_{n=0,2,4,\ldots}^{\infty} \frac{(-1)^{\frac{n}{2}} 3(n+2)}{(n+3)!} x^n \\
&= C_0 x \sum_{n=0}^{\infty} \frac{(-1)^n 3(2n+2)}{(2n+3)!} x^{2n}
\end{aligned}$$

Next, we try $s = -2$. Again, using (2.20–2.21) with $s = -2$, we see

$$\begin{aligned}
C_0 &\neq 0 \\
C_1 &= 0
\end{aligned}$$

From (2.22) with $s = -2$, we obtain the relationship

$$C_{n+2} = -\frac{C_n}{(n+2)(n-1)}$$

Following through with this formula, the coefficients turn out to be

$$\begin{aligned}
C_2 &= \frac{C_0}{2(-1)} = \frac{C_0}{2} \\
C_4 &= -\frac{C_2}{4(1)} = \frac{C_0}{2(4)(-1)(1)} \\
C_6 &= -\frac{C_4}{6(3)} = \frac{-C_0}{(2.4.6)(-1.1.3.5)} \\
&\vdots \\
C_n &= \frac{-(-1)^{\frac{n}{2}} C_0 (n-1)}{n!} \qquad n = 2, 4, 6, \ldots
\end{aligned}$$

While we have taken $C_1 = 0$, we find that the value of C_3 has to remain indeterminate because it is a zero-over-zero case. Therefore, we assume $C_3 \neq 0$. Following through, we find

$$\begin{aligned} C_5 &= \frac{-C_3}{5(2)} \\ C_7 &= \frac{-C_5}{7(4)} = \frac{C_3}{(5.7)(2.4)} \\ C_9 &= -\frac{C_7}{9(6)} = \frac{-C_3}{(5.7.9)(2.4.6)} \\ &= \frac{-C_3 3(8)}{9!} \\ &\vdots \\ C_n &= \frac{(-1)^{\frac{n+1}{2}} C_3 3(n-1)}{n!} \qquad n = 3, 5, 7, 9, \ldots \end{aligned}$$

The second solution can now be written as

$$y_2 = -\frac{C_0}{x^2} \sum_{n=0,2,4\ldots}^{\infty} \frac{(-1)^{\frac{n}{2}}(n-1)}{n!} x^n + \frac{C_3}{x^2} \sum_{n=3,5,7,\ldots}^{\infty} \frac{(-1)^{\frac{n+1}{2}} 3(n-1)}{n!} x^n$$

The second summation may be written as

$$\begin{aligned} & C_3 \sum_{n=3,5,7,\ldots}^{\infty} \frac{(-1)^{\frac{n+1}{2}} 3(n-1)}{n!} x^{n-2} \\ & \left[\text{let } n' = n-3, \quad \text{i.e., } n = n'+3\right] \\ &= C_3 \sum_{n'=0,2,4,\ldots,}^{\infty} \frac{(-1)^{\frac{n'+4}{2}} 3(n'+2)}{(n'+3)!} x^{n'+1} \\ &= C_3 x \sum_{n'=0,2,4,\ldots}^{\infty} \frac{(-1)^{\frac{n'}{2}} 3(n'+2)}{(n'+3)!} x^{n'} \end{aligned}$$

which is the first solution.

While we chose $s = 1, -2$, we may have chosen

$$s(s+3)C_1 = 0, \qquad \text{i.e.,} \quad s = 0, -3 \quad \text{with} \quad C_1 \neq 0.$$

That would have given us $C_0 = 0$ and we would have solutions:

$$\begin{aligned} y_1 &= C_1 x + C_2 x^2 + C_3 x^3 + \ldots \\ &= x(C_1 + C_2 x + C_3 x^2 + \ldots) \\ y_2 &= x^{-3}(C_1 x + C_2 x^2 + C_3 x^3 + \ldots) \\ &= \frac{1}{x^2}(C_1 + C_2 x + C_3 x^2 + \ldots) \end{aligned}$$

These two solutions would be the same as the ones obtained with

$$s = 1, -2, \quad C_0 \neq 0, \quad C_1 = 0$$

2.2 Solutions with Logarithmic Behavior

There are some situations in which the values of s are repeated. In such cases we cannot clearly identify two linearly independent solutions. In fact, the assumed power series of the type in equation (2.3) is not completely valid. To illustrate the point we consider the example below.

Example 2.2

$$x(2-x)y'' + 2(1-x)y' + 2y = 0 \tag{2.24}$$

As usual, we let

$$y = \sum_{n=0}^{\infty} C_n x^{n+s}, \quad y' = \sum_{n=0}^{\infty} (n+s) C_n x^{n+s-1},$$

and

$$y'' = \sum_{n=0}^{\infty} (n+s)(n+s-1) C_n x^{n+s-2} \tag{2.25}$$

Substitution into the differential equation gives

$$x(2-x) \sum_{n=0}^{\infty} (n+s)(n+s-1) C_n x^{n+s} + 2(1-x) \sum_{n=0}^{\infty} (n+s) 1 C_n x^{n+s} + 2 \sum_{n=0}^{\infty} C_n x^{n+s}$$

which adds up to

$$\begin{aligned}
-\sum_{n=0}^{\infty} C_n \left[(n+s)(n+s-1) + 2(n+s) - 2\right] x^{n+s} & \\
+ \sum_{n=0}^{\infty} C_n \left[2(n+s)(n+s-1) + 2(n+s)\right] x^{n+s-1} &= 0 \\
-\sum_{n=0}^{\infty} C_n \left[(n+s)^2 + (n+s) - 2\right] x^{n+s} + 2 \sum_{n=0}^{\infty} C_n \left[(n+s)(n+s)\right] x^{n+s-1} &= 0 \\
-\sum_{n=0}^{\infty} C_n (n+s+2)(n+s-1) x^{n+s} + 2 \sum_{n=0}^{\infty} C_n (n+s)^2 x^{n+s-1} &= 0
\end{aligned}$$

Now let us shift the index on the second summation

$$-\sum_{n=0}^{\infty} C_n (n+s+2)(n+s-1) x^{n+s} + 2 \sum_{n=-1}^{\infty} C_{n+1} (n+s+1)^2 x^{n+s} = 0$$

We may pull out the leading term from the second sum and combine the two summations to give

$$\begin{aligned} -\sum_{n=0}^{\infty} C_n(n+s+2)(n+s-1)x^{n+s} + 2C_0s^2x^{s-1} & \\ +2\sum_{n=0}^{\infty} C_{n+1}(n+s+1)^2x^{n+s} &= 0 \\ +2C_0s^2x^{s-1} + \sum_{n=0}^{\infty}\left[2C_{n+1}(n+s+1)^2\right. & \\ \left. - C_n(n+s+2)(n+s-1)\right]x^{n+s} &= 0 \end{aligned}$$

Equating powers in x, term by term, we find

$$\begin{aligned} x^{s-1}: &\quad (s)^2C_0 &&= 0 \quad \Rightarrow \quad s = 0, 0 \quad \text{taking } C_0 \neq 0 \\ x^{n+s}: &\quad 2C_{n+1}(n+s+1)^2 &&= C_n(n+s+2)(n+s-1) \end{aligned} \tag{2.26}$$

With $s = 0$,

$$C_{n+1} = C_n \frac{(n+2)(n-1)}{2(n+1)^2}$$

This gives

$$\begin{aligned} C_1 &= C_0\frac{(2)(-1)}{2(1)^2} = -C_0 \\ C_2 &= C_1\frac{(3)(0)}{2(2)^2} = 0 \\ C_3 &= C_2\frac{(4)(1)}{2(2)^2} = 0 \\ &\vdots \\ C_n &= 0, \quad n \geq 2 \end{aligned}$$

One solution is $y_1 = C_0(1-x)$. Since we have repeated values of s, a second linearly independent solution of the form given in (2.25) is not possible. To obtain the second solution, we can try the reduction of order method by assuming the second solution of the form

$$y_2(x) = f(x)(1-x). \tag{2.27}$$

This procedure will lead to an analytical solution that may not necessarily be a power-series solution. However, it will tell us more about the character of the solution so that we may construct a valid power-series form. Let us first take the derivatives of $y_2(x)$ as given in (2.27) so that

$$\begin{aligned} y_2'(x) &= -f(x) + (1-x)f'(x) \\ y_2''(x) &= -f'(x) - f'(x) + (1-x)f''(x) = (1-x)f''(x) - 2f'(x) \end{aligned}$$

The substitution of equation (2.27) and its derivatves into (2.24) gives

$$\begin{aligned} x(2-x)\left[(1-x)f''-2f'\right]+2(1-x)\left[(1-x)f'-f\right]+2(1-x)f &= 0 \\ x(2-x)(1-x)f''+2\left[(1-x)^2-x(2-x)\right]f'+[2(1-x)-2(1-x)]f &= 0 \\ x(2-x)(1-x)f''+2\left[2x^2-4x+1\right]f' &= 0 \end{aligned}$$

This may be reduced to

$$\frac{f''(x)}{f'(x)} = -\frac{2\left(2x^2-4x+1\right)}{x(2-x)(1-x)},$$

which may be further simplified in the form of partial fractions

$$\begin{aligned} \frac{f''(x)}{f'(x)} &= \frac{2\left(2x^2-4x+1\right)}{x(2-x)(1-x)} \\ &= \frac{A}{x}+\frac{B}{1-x}+\frac{C}{2-x} \\ &= -\frac{1}{x}+\frac{2}{1-x}+\frac{1}{2-x} \end{aligned}$$

Integration of both sides gives

$$\ln\left[f'(x)\right] = -\ln x - 2\ln(1-x) - \ln(2-x) + C_1',$$

or, by taking the inverse log of both sides, we have

$$f'(x) = \frac{C_1}{x(1-x)^2(2-x)}.$$

By decomposing, once again, into partial fractions, we have

$$f'(x) = C_1\left[\frac{D}{x}+\frac{E+Fx}{(1-x)^2}+\frac{G}{2-x}\right] = C_1\left[\frac{\frac{1}{2}}{x}+\frac{\frac{1}{2}}{2-x}+\frac{1}{(1-x)^2}\right]$$

Integrating once more, the expression for $f(x)$ is found to be

$$f(x) = C_1\left[\tfrac{1}{2}\ln x - \tfrac{1}{2}\ln(2-x) + \frac{1}{1-x}\right] + C_2.$$

The second solution may therefore be written as

$$\begin{aligned} y_2(x) = (1-x)f(x) &= C_1\left[\tfrac{1}{2}(1-x)\ln\left(\frac{x}{2-x}\right)+1\right]+C_2(1-x). \\ &= \tfrac{1}{2}C_1\left[(1-x)\ln\left(\frac{x}{2-x}\right)+2\right]+C_2(1-x). \end{aligned}$$

Here, we note that the term containing C_2 is the first solution $y_1(x)$. The second linearly independent solution may therefore be identified as just the term

$$y_2(x) = \left[(1-x)\ln\left(\frac{x}{2-x}\right)+2\right] = (1-x)\ln x - (1-x)\ln(2-x) + 2, \quad (2.28)$$

where we have dropped the integration constant $\frac{1}{2}C_1$. While the solution is complete here, we need to explore the possibilities for a power-series solution since that is the subject of the present discussion. With these two terms for $y_2(x)$, the second one may be expressed as a power series about $x = 0$. The first one, however, has a logarithmic singularity at $x = 0$, and therefore cannot be expressed as a power series in the form (2.25). To be able to apply the method of Frobenius, the assumed power-series solution needs to be modified to accommodate the logarithmic singularity. Therefore, we first assume a solution of the form

$$y_2(x) = y_1(x) \ln x + z(x), \tag{2.29}$$

where $z(x)$ may be expressed as a power series, corresponding to the term $(1 - x) \ln(2 - x)$. While we may substitute $y_1(x) = (1 - x)$, we shall retain the general form. The derivatives of $y_2(x)$ from this expression are

$$y_2'(x) = y_1'(x) \ln x + \frac{y_1(x)}{x} + z'(x),$$

and

$$y_2''(x) = y_1''(x) \ln x + \frac{2y_1'(x)}{x} - \frac{y_1(x)}{x^2} + z''(x).$$

The substitution of $y(x) = y_2(x)$ into the original differential equation (2.24) gives the following

$$\begin{aligned} x(2-x)&\left[y_1''(x) \ln x + \frac{2y_1'(x)}{x} - \frac{y_1(x)}{x^2} + z''(x)\right] \\ &+2(1-x)\left[y_1'(x) \ln x + \frac{y_1(x)}{x} + z'(x)\right] + 2\left[y_1(x) \ln x + z(x)\right] = 0. \end{aligned}$$

After some rearrangement

$$\begin{aligned} &\ln x \left\{x(2-x)y_1''(x) + 2(1-x)y_1'(x) + 2y_1(x)\right\} + 2(2-x)y_1'(x) \\ &+\left[\frac{2(1-x)}{x} - \frac{2-x}{x}\right] y_1(x) + \left\{x(2-x)z''(x) + 2(1-x)z'(x) + 2z(x)\right\} = 0. \end{aligned}$$

The term in the curly brackets {} multiplying $\ln x$ is exactly the left-hand side of the original differential equation (2.24) operating on a solution $y_1(x)$, and therefore vanishes. Next, by substituting $y_1(x) = (1 - x)$, we obtain

$$\begin{aligned} 2(2-x)(-1) + &\left[\frac{2(1-x)}{x} - \frac{2-x}{x}\right](1-x) \\ &+ \left\{x(2-x)z''(x) + 2(1-x)z'(x) + 2z(x)\right\} = 0, \end{aligned}$$

which simplifies to the following non-homogeneous equation for $z(x)$:

$$x(2-x)z''(x) + 2(1-x)z'(x) + 2(x) = 5 - 3x, \tag{2.30}$$

in which the left-hand side is the same as (2.24) with $z(x)$ in place of $y(x)$. Let us assume $z(x)$ to be the power series

$$z(x) = \sum_{n=0}^{\infty} D_n x^{n+s},$$

with derivatives

$$z'(x) = \sum_{n=0}^{\infty} D_n(n+s)x^{n-1}, \qquad \text{and} \qquad z''(x) = \sum_{n=0}^{\infty} D_n(n+s)(n-1)x^{n-2}$$

Since the homogeneous parts of (2.24) and (2.30) are the same, we can expect the values of s to be the same as well. That is, at this point, we can set $s = 0$, the value found in equation (2.26). For cases in which there are two distinct values of s, and logarithmic behavior at $x = 0$ persists, the treatment will be illustrated by appropriate examples. For now, we take $s = 0$, i.e.,

$$z(x) = \sum_{n=0}^{\infty} D_n x^n, \qquad z'(x) = \sum_{n=0}^{\infty} D_n n x^{n-1},$$

$$\text{and} \qquad z''(x) = \sum_{n=0}^{\infty} D_n n(n-1)x^{n-2}$$

The substitution into equation (2.30) leads to

$$x(2-x)\sum_{n=0}^{\infty} D_n n(n-1)x^{n-2}$$

$$+2(1-x)\sum_{n=0}^{\infty} D_n n x^{n-1} + 2\sum_{n=0}^{\infty} D_n x^n = 5 - 3x,$$

$$-\sum_{n=0}^{\infty} D_n\left[n(n-1) + 2n - 2\right]x^n + 2\sum_{n=0}^{\infty} D_n\left[n(n-1) + n\right]x^{n-1} = 5 - 3x,$$

$$-\sum_{n=0}^{\infty} D_n(n+2)(n-1)x^n + 2\sum_{n=0}^{\infty} D_n n^2 x^{n-1} = 5 - 3x,$$

$$-\sum_{n=0}^{\infty} D_n(n+2)(n-1)x^n + 2\sum_{n=-1}^{\infty} D_{n+1}(n+1)^2 x^n = 5 - 3x,$$

$$-\sum_{n=0}^{\infty} D_n(n+2)(n-1)x^n + 2D_0 0 + 2\sum_{n=0}^{\infty} D_{n+1}(n+1)^2 x^n = 5 - 3x$$

$$2D_0 0 + \sum_{n=0}^{\infty}\left[2D_{n+1}(n+1)^2 - D_n(n+2)(n-1)\right]x^n = 5 - 3x$$

We now equate both sides term by term to give the following expressions

$$x^0: \quad 2D_1(1) - D_0(2)(-1) = 5 \tag{2.31}$$

$$x^1: \quad 2D_2(4) - D_1(3)(0) = -3 \tag{2.32}$$

$$x^n: \quad 2D_{n+1}(n+1)^2 - D_n(n+2)(n-1) = 0, \quad n \geq 2 \tag{2.33}$$

To begin with, D_0 is indeterminate but we shall deal with that later. From (2.32), we obtain

$$D_2 = -\tfrac{3}{8},$$

and from (2.33),

$$D_{n+1} = D_n \frac{(n+2)(n-1)}{2(n+1)^2}, \quad n \geq 2.$$

Using this general expression, we can calculate the explicit numerical expressions for $D_3, D_4, D_5, \ldots$ in terms of D_2, i.e.,

$$\begin{aligned}
D_3 &= D_2 \frac{4(1)}{2(3)^2} \\
D_4 &= D_3 \frac{5(2)}{2(4)^2} = D_2 \frac{4(5)\ 1(2)}{2^2\ 3^2 4^2} = D_2 \frac{5!2!2^2}{3!2^2 4!^2} \\
D_5 &= D_4 \frac{6(3)}{2(5)^2} = D_2 \frac{4(5)6\ 1(2)3}{2^3\ 3^2 4^2 5^2} = D_2 \frac{6!3!}{3!5!^2} \\
D_6 &= D_5 \frac{7(4)}{5(6)^2} = D_2 \frac{4(5)6(7)\ 1(2)3(4)}{2^4\ 3^2 4^2 5^2 6^2} = D_2 \frac{7!4!2^2}{2^4 3!6!^2} \\
&\vdots \\
D_n &= D_2 \frac{(n+1)!(n-2)!}{3!n!^2 2^{n-4}} \\
&= D_2 \frac{(n+1)n!(n-2)!}{3!n!^2 2^{n-4}} \\
&= \frac{D_2}{3!} \frac{(n+1)(n-2)!}{n! 2^{n-4}} \\
&= \frac{D_2}{3!} \frac{(n+1)n!}{n! n(n-1) 2^{n-4}} \\
&= \frac{D_2}{3!} \frac{(n+1)}{n(n-1) 2^{n-4}} \\
&= \left(-\tfrac{3}{8}\right) \frac{2^4}{3!} \frac{(n+1)}{n(n-1)2^n} \qquad \left(\text{using} \quad D_2 = -\tfrac{3}{8}\right) \\
&= -\frac{(n+1)}{n(n-1)2^n}
\end{aligned}$$

Using equation (2.31), we may write

$$D_1 = \tfrac{5}{2} - D_0.$$

Using these expressions for the set of constants $\{D_n\}$, we may write the power-series expression for $z(x)$ as

$$\begin{aligned}
z(x) &= D_0 + \left(\tfrac{5}{2} + D_0\right) x - \sum_{n=2}^{\infty} \frac{(n+1)}{n(n-1)2^n} x^n \\
&= D_0(1-x) + \tfrac{5}{2}x - \sum_{n=2}^{\infty} \frac{(n+1)}{n(n-1)} \left(\frac{x}{2}\right)^n
\end{aligned}$$

We note that the term $D_0(1-x)$ is just the solution $y_1(x)$, and may be dropped. Using this expression for $z(x)$ in equation (2.29), we may write the expression for $y_2(x)$ as

$$y_2(x) = (1-x)\ln x + \tfrac{5}{2}x - \sum_{n=2}^{\infty} \frac{(n+1)}{n(n-1)} \left(\frac{x}{2}\right)^n.$$

It is not difficult to show that this is equivalent to the original solution given by equation (2.28). The general solution may be expressed as

$$y(x) = B_1(1-x) + B_2\left[(1-x)\ln x + \tfrac{5}{2}x - \sum_{n=2}^{\infty}\frac{(n+1)}{n(n-1)}\left(\frac{x}{2}\right)^n\right].$$

□

It should be noted that if a solution in a compact form such as (2.28) can be obtained, there is no need to construct a power series. However, this example was selected for illustrative purposes, and in particular, it was chosen because one of its solution $y_1(x) = (1-x)$ was particularly simple, and facilitated the straightforward (but perhaps lengthy) construction of $y_2(x)$ by the reduction of order method. For many other cases, the non-logarithmic solution may be an infinite power series, and there may not be an easy option such as one leading to (2.28). We shall consider such a case next.

Example 2.3

$$x^2y'' + 7xy' + (9+x)y = 0. \tag{2.34}$$

For the first solution we let

$$\begin{aligned} y &= \sum_{n=0}^{\infty} C_n x^{n+s} \\ y' &= \sum_{n=0}^{\infty}(n+s)C_n x^{n+s-1} \\ y'' &= \sum_{n=0}^{\infty}(n+s)(n+s-1)C_n x^{n+s-2} \end{aligned}$$

Substitution into the differential equation gives

$$\sum_{n=0}^{\infty}[(n+s)(n+s-1)+7(n+s)+9]\,C_n x^{n+s} + \sum_{n=0}^{\infty}C_n x^{n+s+1} = 0,$$

which may be rewritten as

$$\sum_{n=0}^{\infty}(n+s+3)^2 C_n x^{n+s} + \sum_{n=0}^{\infty}C_n x^{n+s+1} = 0.$$

Upon shifting the index for the first sum we have

$$\sum_{n=-1}^{\infty}(n+s+4)^2 C_{n+1}x^{n+s+1} + \sum_{n=0}^{\infty}C_n x^{n+s+1} = 0.$$

Let us extract the first term of the first sum, and then combine both the summations. This gives

$$C_0(s+3)^2x^s + \sum_{n=0}^{\infty}\left[(n+s+4)^2C_{n+1} + C_n\right]x^{n+s+1} = 0.$$

By equation like powers in x, we have

$$\begin{aligned} x^s: \quad C_0(s+3)^2 x^s &= 0 \\ x^{n+s+1}: \left[(n+s+4)^2 C_{n+1} + C_n\right] x^{n+s+1} &= 0, \quad n \geq 0 \end{aligned}$$

The first one of these gives us the indicial equation,

$$(s+3)^2 = 0, \quad \text{i.e.,} \quad s = -3, -3, \quad \text{with} \quad C_0 \neq 0. \tag{2.35}$$

The second one establishes the recurrence relationship

$$(n+s+4)^2 C_{n+1} + C_n = 0,$$

or with $s = -3$,

$$C_{n+1} = -\frac{C_n}{(n+1)^2}$$

from which we obtain

$$\begin{aligned} C_1 &= -\frac{C_0}{1^2} = -C_0 \\ C_2 &= -\frac{C_1}{2^2} = \frac{C_0}{1^2 2^2} \\ C_3 &= -\frac{C_2}{3^2} = -\frac{C_0}{1^2 2^2 3^2} = -\frac{C_0}{3!^2} \\ &\vdots \\ C_n &= C_0 \frac{(-1)^n}{n!^2} = \frac{(-1)^n}{n!^2}, \end{aligned} \tag{2.36, 2.37}$$

where we have set $C_0 = 1$ without any loss of generality. The first of the two linearly independent solutions can be written as

$$y_1 = \sum_{n=0}^{\infty} \frac{(-1)^n}{n!^2} x^{n-3} \tag{2.38}$$

This series can be further reduced to one of the well-known special functions called Bessel functions (see Section 2.4.3). In that notation, we may write

$$y_1 = \sum_{n=0}^{\infty} \frac{(-1)^n}{n!^2} x^{n-3} = \frac{1}{x^3} J_0\left(2x^{\frac{1}{2}}\right).$$

For the second solution, considering that the indicial equation (2.35) has repeated roots, we try:

$$\begin{aligned} y_2 &= y_1 \ln x + z \\ y_2' &= y_1' \ln x + \frac{y_1}{x} + z' \\ y_2'' &= y_1'' \ln x + \frac{2y_1'}{x} - \frac{y_1}{x^2} + z'' \end{aligned}$$

Substitution into equation (2.34) gives

$$x^2\left[y_1''\ln x+\frac{2y_1'}{x}-\frac{y_1}{x^2}+z''\right] \\ +7x\left[y_1'\ln x+\frac{y_1}{x}+z'\right]+(9+x)(y_1\ln x+z),$$

and with some rearrangement,

$$\ln x\left[x^2y_1''+7xy_1'+(9+x)y_1\right] \\ +\left[x^2z''+7xz'+(9+x)z\right]+\left[2xy_1'+6y_1(x)\right]=0,$$

where the first term vanishes. As a result,

$$x^2z''+7xz'+(9+x)z=-\left[2xy_1'+6y_1(x)\right], \tag{2.39}$$

where $y_1(x)$ is given by equation (2.38). Using this expression for $y_1(x)$, the right-hand side of (2.39) becomes

$$\begin{aligned} -\left[2xy_1'+4y_1(x)\right] &= -\sum_{n=0}^{\infty}\left[2x\frac{(-1)^n(n-3)}{n!2}x^{n-4}+6\frac{(-1)^n}{n!2}x^{n-3}\right] \\ &= -2\sum_{n=0}^{\infty}\left[\frac{(-1)^n n}{n!2}x^{n-3}\right]. \end{aligned}$$

Equation (2.39) now takes the form

$$x^2z''+7xz'+(9+x)z=-2\sum_{n=0}^{\infty}\left[\frac{(-1)^n n}{n!2}x^{n-3}\right]. \tag{2.40}$$

Now letting,

$$\begin{aligned} z &= \sum_{n=0}^{\infty}D_nx^{n+s} \qquad (s=-3) \\ &= \sum_{n=0}^{\infty}D_nx^{n-3} \\ z' &= \sum_{n=0}^{\infty}(n-3)D_nx^{n-4} \\ z'' &= \sum_{n=0}^{\infty}(n-3)(n-4)D_nx^{n-5} \end{aligned}$$

and substituting into (2.40) leads to

$$\sum_{n=0}^{\infty}D_n(n-3)(n-4)x^{n-3}+7\sum_{n=0}^{\infty}D_n(n-3)x^{n-3}$$

$$
\begin{aligned}
+9\sum_{n=0}^{\infty} D_n x^{n-3} + \sum_{n=0}^{\infty} D_n x^{n-2} &= -2\sum_{n=0}^{\infty}\left[\frac{(-1)^n n}{n!^2}x^{n-3}\right] \\
\sum_{n=0}^{\infty} D_n\left[(n-3)(n-4)\right. & \\
\left. +7(n-2)+9\right]x^{n-3} + \sum_{n=0}^{\infty} D_n x^{n-2} &= -2\sum_{n=0}^{\infty}\left[\frac{(-1)^n n}{n!^2}x^{n-3}\right] \\
\sum_{n=0}^{\infty} D_n n^2 x^{n-3} + \sum_{n=0}^{\infty} D_n x^{n-2} &= -2\sum_{n=0}^{\infty}\left[\frac{(-1)^n n}{n!^2}x^{n-3}\right]
\end{aligned}
$$

Let us shift the indices on the first sum on the left as well as the one on the right.

$$
\sum_{n=-1}^{\infty} D_{n+1}(n+1)^2 x^{n-2} + \sum_{n=0}^{\infty} D_n x^{n-2} = -2\sum_{n=-1}^{\infty}\left[\frac{(-1)^{n+1}(n+1)}{(n+1)!^2}x^{n-2}\right]
$$

Noting that the leading terms on the shifted sums vanish, we can start all summations from $n = 0$, and thus,

$$
\sum_{n=0}^{\infty}\left[D_{n+1}(n+1)^2 + D_n\right]x^{n-2} = -2\sum_{n=0}^{\infty}\left[\frac{(-1)^{n+1}(n+1)}{(n+1)!^2}x^{n-2}\right]
$$

We now equate powers of x, and obtain

$$
\left[D_{n+1}(n+1)^2 + D_n\right] = -2\left[\frac{(-1)^{n+1}(n+1)}{(n+1)!^2}\right], \qquad n \geq 0
$$

From here it follows that

$$
\begin{aligned}
n=0: \quad 1^2 D_1 + D_0 &= \frac{2(1)}{1!^2} \\
D_1 &= 2 - D_0, \qquad \text{where } D_0 \text{ is undefined} \\
n=1: \quad 2^2 D_2 + D_1 &= -\frac{2(2)}{2!^2} \\
D_2 &= \frac{-\frac{2(2)}{2!^2} - D_1}{2^2} = -2\left(\frac{1+\frac{1}{2}-D_0}{2^2}\right) \\
&= -2\left(\frac{1+\frac{1}{2}}{2!^2}\right) + \frac{D_0}{2!^2} \\
n=2: \quad 3^2 D_3 + D_2 &= \frac{2(3)}{3!^2} \\
D_3 &= \frac{\frac{2(3)}{3!^2} - D_2}{3^2} = 2\left(\frac{1+\frac{1}{2}+\frac{1}{3}-D_0}{3!^2}\right) \\
&= 2\left(\frac{1+\frac{1}{2}+\frac{1}{3}}{3!^2}\right) - \frac{D_0}{3!^2} \\
&\vdots \\
D_n &= 2(-1)^{n+1}\left(\frac{1+\frac{1}{2}+\frac{1}{3}+\cdots+\frac{1}{n}}{n!^2}\right) + \frac{D_0(-1)^n}{n!^2}, \qquad n \geq 1.
\end{aligned}
$$

While D_0 is undefined, we can see that it is just leading to the solution $y_1(x)$. We may therefore set $D_0 = 0$ without any loss of generality. The second solution is

$$\begin{aligned} y_2(x) &= y_1(x)\ln x + z(x) \\ &= \left[\sum_{n=0}^{\infty}\frac{(-1)^n}{n!^2}x^{n-3}\right]\ln x + 2\sum_{n=1}^{\infty}(-1)^{n+1}\left(\frac{1+\frac{1}{2}+\frac{1}{3}+\cdots+\frac{1}{n}}{n!^2}\right)x^{n-3} \end{aligned}$$

This result may also be expressed in terms of Bessel functions (discussed in Section 2.4.3). The general solution of equation (2.34) may be written as a linear combination of $y_1(x)$ and $y_2(x)$ so that

$$y(x) = Ay_1(x) + By_2(x).$$

□

In general, if the solution has logarithmic behavior which has not been incorporated into the assumed power series, then we allow for $\ln x$ behavior by assuming a solution of the form given by equation (2.29). For the two examples considered so far, we anticipated logarithmic behavior because the indicial equations (2.26) and (2.35) both had repeated roots. However, in some instances, even if the indicial equation has distinct roots, s_1 and s_2, $\ln x$ type behavior could be expected if $(s_1 - s_2)$ is an integer. This will be illustrated in the next example.

Example 2.4 Let us consider

$$xy''(x) - y = 0. \tag{2.41}$$

As in the previous examples, we let

$$\begin{aligned} y(x) &= \sum_{n=0}^{\infty} C_n x^{n+s} \qquad (2.42) \\ y'(x) &= \sum_{n=0}^{\infty}(n+s)C_n x^{n+s-1} \\ y''(x) &= \sum_{n=0}^{\infty}(n+s)(n+s-1)C_n x^{n+s-2} \end{aligned}$$

Substitution into the differential equation (2.41) gives

$$\sum_{n=0}^{\infty} C_n(n+s)(n+s-1)x^{n+s-1} - \sum_{n=0}^{\infty} C_n x^{n+s} = 0,$$

which, after shifting the index on the first sum, may be rewritten as

$$\sum_{n=-1}^{\infty} C_{n+1}(n+s+1)(n+s)x^{n+s} - \sum_{n=0}^{\infty} C_n x^{n+s} = 0,$$

Upon extracting the leading term ($n = -1$) in the first sum, and combining the remainder with the second sum, we have

$$C_0 s(s-1)x^{s-1} + \sum_{n=0}^{\infty} \left[C_{n+1}(n+s+1)(n+s) - C_n\right] x^{n+s} = 0.$$

The leading term gives the indicial equation

$$s(s-1) = 0, \quad \text{assuming } C_0 \neq 0. \tag{2.43}$$

The remaining terms under the summation lead us to the recurrence relationship

$$C_{n+1} = \frac{C_n}{(n+s+1)(n+s)}.$$

The roots of the indicial equation (2.43) are $s = 0, 1$. Let's start out with $s = 0$ for which the recurrence relationship is

$$C_{n+1} = \frac{C_n}{(n+1)n}.$$

Using this relationship,

$$C_1 = \frac{C_0}{(1)0} \to \infty. \tag{2.44}$$

This is an indication that the assumed solution (2.42) is not valid, and probably need a $\ln x$ term. However, we can attempt a solution with the second root, $s = 1$. With this value of s, the recurrence relationship takes the form

$$C_{n+1} = \frac{C_n}{(n+2)(n+1)},$$

resulting in

$$\begin{aligned}
C_1 &= \frac{C_0}{(2)1} = \tfrac{1}{2}C_0 \\
C_2 &= \frac{C_1}{(3)2} = \frac{C_0}{(3)2^2 1} \\
C_3 &= \frac{C_2}{(4)3} = \frac{C_0}{(4)3^2 2^2 1} = \frac{C_0}{(4)3!^2} \\
&\vdots \\
C_n &= \frac{C_0}{n!^2(n+1)} = \frac{C_0}{n!(n+1)!}
\end{aligned}$$

We may set $C_0 = 1$ without loss of generality, and write

$$C_n = \frac{1}{n!(n+1)!}.$$

The first solution therefore is

$$y_1(x) = \sum_{n=0}^{\infty} \frac{x^{n+1}}{n!(n+1)!}. \tag{2.45}$$

This result is also be represented in terms of Bessel functions (see Section 2.4.3) with the standard notation,

$$y_1(x) = \sum_{n=0}^{\infty} \frac{x^{n+1}}{n!(n+1)!} = x^{\frac{1}{2}} I_1\left(2x^{\frac{1}{2}}\right).$$

As in the previous two examples, let us now try a second solution of the form

$$y_2(x) = y_1(x)\ln x + z(x)$$

with derivatives

$$\begin{aligned} y_2'(x) &= y_1'(x)\ln x + \frac{y_1(x)}{x} + z'(x) \\ y_2''(x) &= y_1''(x)\ln x + \frac{2y_1'(x)}{x} - \frac{y_1(x)}{x^2} + z'' \end{aligned}$$

Once again, substitution into equation (2.41) gives

$$\ln x\left[xy_1''(x) - y_1(x)\right] + \left[xz''(x) - z(x)\right] + 2y_1'(x) - \frac{y_1(x)}{x} = 0,$$

Noting that the $\ln x$ term vanishes, and using the above expression (2.45) for $y_1(x)$ this equation for $z(x)$ simplifies to

$$\begin{aligned} xz''(x) - z(x) &= -2y_1'(x) + \frac{y_1(x)}{x} \\ &= \sum_{n=0}^{\infty}\left[-\frac{2(n+1)x^n}{n!(n+1)!} + \frac{x^n}{n!(n+1)!}\right] \\ &= -\sum_{n=0}^{\infty} \frac{(2n+1)x^n}{n!(n+1)!}. \end{aligned} \tag{2.46}$$

As in the previous examples, we assume

$$z(x) = \sum_{n=0}^{\infty} D_n x^{n+s},$$

where we take $s = s_2 = 0$. This is one of the two roots of the indicial equation (2.43) for which $s = s_1 = 1$ was used for constructing $y_1(x)$. With $s = s_2 = 0$, the second derivative of $z(x)$ is

$$z''(x) = \sum_{n=0}^{\infty} n(n-1)D_n x^{n-2}$$

Upon substituting these series expressions for $z(x)$ and $z''(x)$ into equation (2.45), we obtain

$$\sum_{n=0}^{\infty} n(n-1)D_n x^{n-1} - \sum_{n=0}^{\infty} D_n x^n = -\sum_{n=0}^{\infty} \frac{(2n+1)x^n}{n!(n+1)!}.$$

By shifting the index on the first sum (and noting that its leading term vanishes), we may combine the two summations to obtain

$$\sum_{n=0}^{\infty}[(n+1)nD_{n+1}-D_n]\,x^n \quad = \quad -\sum_{n=0}^{\infty}\frac{(2n+1)x^n}{n!(n+1)!}.$$

We may now equate like powes in x on each side term by term. This procedure yields the following:

$$
\begin{aligned}
x^0: \quad & (1)0D_1 - D_0 &=& -1, \qquad D_0 = 1, \ \text{ and } D_1 \text{ remains indeterminate}\\
x^1: \quad & (2)1D_2 - D_1 &=& -\frac{3}{1!2!} = -\tfrac{3}{2}, \qquad D_2 = \left(D_1 - \tfrac{3}{2}\right)\tfrac{1}{2}\\
x^2: \quad & (3)2D_3 - D_2 &=& -\frac{5}{2!3!} = -\tfrac{5}{12}, \qquad D_3 = \left(D_2 - \tfrac{5}{12}\right)\tfrac{1}{6}\\
& D_3 &=& -\left(\frac{3}{2(1)}\frac{1}{2!1!} + \frac{5}{2!3!}\right)\frac{1}{3(2)} + \frac{D_1}{3!2!}\\
& &=& -\left(\frac{3}{(3)2^2 1}\frac{1}{2!1!} + \frac{5}{3(2)2!3!}\right) + \frac{D_1}{3!2!}\\
x^3: \quad & (4)3D_4 - D_3 &=& -\frac{7}{4!3!},\\
& D_4 &=& \left(D_3 - \frac{7}{3!4!}\right)\frac{1}{(4)3}\\
& &=& -\left(\frac{3}{(4)3^2 2^2(1)2!1!} + \frac{5}{(4)3^2(2)2!3!} + \frac{7}{4(3)3!4!}\right) + \frac{D_1}{4!3!}\\
& &=& -\left(\frac{3}{(2)1} + \frac{5}{3(2)} + \frac{7}{(4)3}\right)\frac{1}{4!3!} + \frac{D_1}{4!3!}\\
x^4: \quad & (5)4D_5 - D_4 &=& \frac{9}{5!4!},\\
& D_5 &=& \left(D_4 - \frac{9}{5!4!}\right)\frac{1}{(5)4}\\
& &=& -\left(\frac{3}{(5)4^2 3^2 2^2(1)2!1!} + \frac{5}{(5)4^2 3^2(2)2!3!}\right.\\
& & & \left. + \frac{7}{(5)4^2(3)3!4!} + \frac{9}{5(4)5!4!}\right) + \frac{D_1}{5!4!}\\
& &=& -\left(\frac{3}{(2)1} + \frac{5}{(3)2} + \frac{7}{(4)3} + \frac{9}{(5)4}\right)\frac{1}{5!4!} + \frac{D_1}{5!4!}\\
& &\vdots& \\
& D_n &=& -\left(\sum_{k=1}^{n-1}\frac{(2k+1)}{(k+1)k}\right)\frac{1}{n!(n-1)!} + \frac{D_1}{n!(n-1)!},\\
& &=& -\left[\sum_{k=1}^{n-1}\left(\frac{1}{k} + \frac{1}{k+1}\right)\right]\frac{1}{n!(n-1)!} + \frac{D_1}{n!(n-1)!},
\end{aligned}
$$

$n \geq 2.$

With this result, we may write

$$y_2(x) = y_1(x)\ln x + z(x)$$
$$= \left(\sum_{n=0}^{\infty} \frac{x^{n+1}}{n!(n+1)!}\right)\ln x - \sum_{n=2}^{\infty}\left[\sum_{k=1}^{n-1}\left(\frac{1}{k}+\frac{1}{k+1}\right)\right]\frac{x^n}{n!(n-1)!}$$
$$+D_1\sum_{n=1}^{\infty}\frac{x^n}{n!(n-1)!},$$

where the last term may be identified as $D_1 y_1(x)$ and is therefore redundant. Upon removing this term, and shifting the index on the second summation, we obtain

$$y_2(x) = y_1(x)\ln x + z(x)$$
$$= \left(\sum_{n=0}^{\infty} \frac{x^{n+1}}{n!(n+1)!}\right)\ln x - \sum_{n=1}^{\infty}\left[\sum_{k=1}^{n}\left(\frac{1}{k}+\frac{1}{k+1}\right)\right]\frac{x^{n+1}}{n!(n+1)!}.$$

This solution can also be related to the modified Bessel functions. □

At this point we can summarize the general procedure for the method of Frobenius when applied to second-order linear ordinary differential equations.

2.3 General Procedure for Second-Order Equations

For a linear second-order equation of the type,

$$a_2(x)y''(x) + a_1(x)y'(x) + a_0(x)y(x) = 0, \tag{2.47}$$

we may assume a power-series solution about $x = 0$,

$$y(x) = \sum_{n=0}^{\infty} C_n x^{n+s}.$$

For the second-order indicial equation [such as (2.20), (2.26), (2.35) and (2.43)], there will be two roots, s_1 and s_2. There are three different cases with the possible values of these roots.

Case 1: If $(s_1 - s_2)$ is not an integer, then there are two linerly independent solutions,

$$y_1(x) = \sum_{n=0}^{\infty} C_n^{(1)} x^{n+s_1} \quad \text{and} \quad y_2(x) = \sum_{n=0}^{\infty} C_n^{(2)} x^{n+s_2}. \tag{2.48}$$

Case 2: If s_1 and s_2 are equal (say s_0), there is one solution,

$$y_1(x) = \sum_{n=0}^{\infty} C_n x^{n+s_0},$$

and a second solution of the form

$$y_2(x) = y_1(x)\ln x + z(x),$$

where $z(x)$ satisfies a non-homogeneous differential with left-hand side the same as (2.47), i.e.,

$$\begin{aligned} &a_2(x)z''(x) + a_1(x)z'(x) + a_0(x)z(x) \\ &\quad = a_2(x)\left(\frac{y_1(x)}{x^2} - 2\frac{y_1'(x)}{x}\right) - a_1(x)\left(\frac{y_1(x)}{x}\right). \end{aligned} \tag{2.49}$$

A power-series solution for $z(x)$ may be assumed to be of the form

$$z(x) = \sum_{n=0}^{\infty} D_n x^{n+s_0},$$

where s_0 has the same value as determined from the indicial equation. The set of coefficients D_n is determined by substituting this series expansion into (2.49) and following the procedure outlined in the Examples 2.2 and 2.3.

Case 3: If $(s_1 - s_2)$ is an integer (take s_1 as the larger of the two), then there is one solution of the type

$$y_1(x) = \sum_{n=0}^{\infty} C_n x^{n+s_1}.$$

The second solution may have a $\ln x$ term as in Case 2 above and Example 2.4, or may have a simple power series such as equation (2.48). To cover both possibilities, the second solution is assumed to have the form

$$y_2(x) = Ay_1(x)\ln x + z(x),$$

where

$$z(x) = \sum_{n=0}^{\infty} D_n x^{n+s_2}.$$

The constant A may turn out to be zero if the $\ln x$ term is not required. However, if a second solution of the type (2.48) is assumed, and if a $\ln x$ term is required, the latter will become apparent by some of the coefficients becoming infinite [see equation (2.44) in Example 2.4].

2.4 Special Functions

For many mathematical problems in engineering and physics, one encounters various classes of functions. Among the most common ones are Legendre polynomials (as well as Legendre functions) and Bessel functions. These come up frequently in the solutions of various partial differential equations in spherical and cylindrical geometry.

2.4.1 Legendre Polynomials and Functions

For various problems in spherical geometry, we come across Legendre's equation

$$(1-x^2)\frac{d^2y}{dx^2} - 2x\frac{dy}{dx} + n(n+1)y = 0. \tag{2.50}$$

Let us consider here $n = 3$ so that

$$(1-x^2)y'' - 2xy' + 12y = 0.$$

Let

$$y(x) = \sum_{n=0}^{\infty} C_n x^{n+s}.$$

Taking derivatives, we obtain

$$\begin{aligned} y'(x) &= \sum_{n=0}^{\infty} C_n (n+s) x^{n+s-1}, \\ y''(x) &= \sum_{n=0}^{\infty} C_n (n+s)(n+s-1) x^{n+s-2}. \end{aligned}$$

Substitution into the differential equation leads to

$$\begin{aligned} &(1-x^2)\sum_{n=0}^{\infty} C_n (n+s)(n+s-1) x^{n+s-2} \\ &\quad -2\sum_{n=0}^{\infty} C_n (n+s) x^{n+s} + 12\sum_{n=0}^{\infty} C_n x^{n+s}. = 0, \end{aligned}$$

$$\begin{aligned} &\sum_{n=0}^{\infty} C_n (n+s)(n+s-1) x^{n+s-2} - \sum_{n=0}^{\infty} C_n (n+s)(n+s-1) x^{n+s} \\ &\qquad - 2\sum_{n=0}^{\infty} C_n (n+s) x^{n+s} + 12\sum_{n=0}^{\infty} C_n x^{n+s} = 0. \end{aligned}$$

By grouping together the second, the third and the fourth summations, we find

$$\begin{aligned} &\sum_{n=0}^{\infty} C_n (n+s)(n+s-1) x^{n+s-2} \\ &\quad -\sum_{n=0}^{\infty} C_n \left[(n+s)(n+s-1) + 2(n+s) - 12\right] x^{n+s} = 0, \end{aligned}$$

which may also be written as

$$\sum_{n=0}^{\infty} C_n(n+s)(n+s-1)x^{n+s-2} - \sum_{n=0}^{\infty} C_n(n+s+4)(n+s-3)x^{n+s} = 0.$$

Let us pull out the first two terms of the first summation. These terms are those with the lowest powers in x and have no corresponding terms with like powers in the other summation. Thus,

$$\begin{aligned} & C_0 s(s-1)x^{s-2} + C_1(s+1)sx^{s-1} \\ & \quad \sum_{n=2}^{\infty} C_n(n+s)(n+s-1)x^{n+s-2} - \sum_{n=0}^{\infty} C_n(n+s+4)(n+s-3)x^{n+s} = 0. \end{aligned}$$

Now, shift the index on the first summation.

$$\begin{aligned} & C_0 s(s-1)x^{s-2} + C_1(s+1)sx^{s-1} \\ & \quad \sum_{n=0}^{\infty} \left[C_{n+2}(n+s+2)(n+s+1) - C_n(n+s+4)(n+s-3)\right] x^{n+s} = 0. \end{aligned}$$

Setting the coefficients of each power in x equal to zero,

$$\begin{aligned} C_0 s(s-1) &= 0 \quad \text{resulting in} \quad C_0 \neq 0, \quad s = 0, 1 \\ C_1(s+1)s &= 0 \quad \Rightarrow C_1 \text{ is indeterminate for } s = 0 \\ & \qquad C_1 = 0 \text{ for } s = 1, \\ C_{n+2} &= \frac{(n+s+4)(n+s-3)}{(n+s+2)(n+s+1)} C_n, \qquad n = 0, 1, 2, 3, 4, 5, \ldots \end{aligned}$$

Let us set $s = 0$. Therefore,

$$C_{n+2} = \frac{(n+4)(n-3)}{(n+2)(n+1)} C_n \qquad n = 0, 1, 2, 3, 4, 5, \ldots \tag{2.51}$$

This leads to

$$\begin{aligned} C_2 &= \frac{4(-3)}{2(1)} C_0 = -6C_0 \\ C_4 &= \frac{6(-1)}{4(3)} C_2 = \frac{6(4)(-3)(-1)}{4!} C_0 = 3C_0 \\ C_6 &= \frac{8(1)}{6(5)} C_4 = \frac{(8)6(4)(-3)(-1)1}{6!} C_0 = \frac{4}{5} C_0 \\ C_8 &= \frac{(10)3}{8(7)} C_6 = \frac{10(8)6(4)(-3)(-1)1(3)}{8!} C_0 = \frac{3}{7} C_0 \\ C_{10} &= \frac{(12)5}{10(9)} C_6 = \frac{(12)10(8)6(4)(-3)(-1)1(3)5}{10!} C_0 = \frac{2}{7} C_0 \\ & \vdots \end{aligned}$$

Alternately, we could write from Equation(2.51)

$$\begin{aligned}
C_{n+2} &= \frac{(n+4)!!(n-3)!!}{(n+2)!!(n+1)!!}C_0^* \quad n=0,2,4,6,\ldots \\
&= \frac{(n+4)!!(n+3)!!C_0^*}{(n+2)!!(n+1)!!(n+3)(n+1)(n-1)} \\
&= \frac{(n+4)!C_0^*}{(n+2)!(n+3)(n+1)(n-1)} \\
&= \frac{(n+4)(n+3)C_0^*}{(n+3)(n+1)(n-1)} \\
&= \frac{(n+4)C_0^*}{(n+1)(n-1)} \\
C_n &= \frac{(n+2)C_0^*}{(n-1)(n-3)}, \quad n=0,2,4,6,\ldots
\end{aligned} \tag{2.52}$$

To obtain the relationship between C_0 and C_0^*, we set $n = 0$ and obtain

$$C_0 = \tfrac{2}{3}C_0^*,$$

or

$$C_0^* = \tfrac{3}{2}C_0,$$

from where we may write

$$C_n = \frac{\frac{3}{2}(n+2)C_0}{(n-1)(n-3)}, \quad n=0,2,4,6,\ldots$$

Since with $s = 0$, we have $C_1 \neq 0$. We obtain the rest of the odd coefficients as follows

$$\begin{aligned}
C_3 &= \frac{5(-2)}{3(2)}C_1 = -\frac{5}{3}C_1 \\
C_5 &= \frac{7(0)}{5(4)}C_3 = 0
\end{aligned}$$

It is now easy to see that $C_7 = C_9 = C_{11} = \ldots = 0$.

We may now write down the solution as

$$\begin{aligned}
y(x) &= C_0\left[1 - 6x^2 + 3x^4 + \tfrac{4}{5}x^6 + \tfrac{3}{7}x^8 + \tfrac{2}{7}x^{10} + \ldots\right] + C_1\left(x - \tfrac{5}{3}x^3\right) \\
&= C_0 \sum_{n=0,2,4,\ldots}^{\infty} \frac{\frac{3}{2}(n+2)}{(n-1)(n-3)}x^n - \tfrac{1}{3}C_1(5x^3 - 3x)
\end{aligned}$$

The infinite series can be written in a compact form give

$$y(x) = \left(-\tfrac{2}{3}C_1\right)\tfrac{1}{2}(5x^3 - 3x) + \left(\tfrac{3}{2}C_0\right)\left[\tfrac{1}{2}(5x^3 - 3x)\tfrac{1}{2}\ln\left(\frac{1+x}{1-x}\right) + \left(\tfrac{2}{3} - \tfrac{5}{2}x^2\right)\right]$$

The two linearly independent solutions are

$$\begin{aligned} y_1(x) &= P_3(x) = \tfrac{1}{2}(5x^3 - 3x) \\ y_2(x) &= Q_3(x) = \tfrac{1}{2}P_3(x)\ln\left(\frac{1+x}{1-x}\right) + \left(\tfrac{2}{3} - \tfrac{5}{2}x^2\right). \end{aligned}$$

The polynomial solutions are denoted by $P_0(x), P_1(x), \ldots, P_n(x), \ldots$, and the other set of linearly independent solutions are denoted by $Q_0(x), Q_1(x), \ldots, Q_n(x), \ldots$. A few of these solutions are given here.

$$\begin{aligned} P_0(x) &= 1 \\ P_1(x) &= x \\ P_2(x) &= \tfrac{1}{2}\left(3x^2 - 1\right) \\ P_3(x) &= \tfrac{1}{2}\left(5x^3 - 3x\right) \\ P_4(x) &= \tfrac{1}{8}\left(35x^4 - 30x^2 + 3\right) \\ P_5(x) &= \tfrac{1}{8}\left(63x^5 - 70x^3 + 15x\right) \\ P_6(x) &= \tfrac{1}{16}\left(231x^6 - 315x^4 + 105x^2 - 5\right) \\ P_7(x) &= \tfrac{1}{16}\left(429x^7 - 693x^5 + 315x^5 - 35x\right) \end{aligned} \tag{2.53}$$

$$\begin{aligned} Q_0(x) &= \tfrac{1}{2}\ln\left(\frac{1+x}{1-x}\right) \\ Q_1(x) &= \tfrac{1}{2}x\ln\left(\frac{1+x}{1-x}\right) - 1 \\ Q_2(x) &= \tfrac{1}{2}P_2(x)\ln\left(\frac{1+x}{1-x}\right) - \tfrac{3}{2}x \\ Q_3(x) &= \tfrac{1}{2}P_3(x)\ln\left(\frac{1+x}{1-x}\right) + \left(\tfrac{2}{3} - \tfrac{5}{2}x^2\right) \end{aligned} \tag{2.54}$$

The polynomials are normalized so that in the range $-1 \leq x \leq 1$, the integral of the square of each polynomial integrates to the following value

$$\int_{-1}^{1} [P_n(x)]^2 \, dx = \frac{2}{2n+1}.$$

The polynomials have other important properties that are useful for various applications. These are discussed later in Chapter 3.

EXERCISES 2.1 Obtain two linearly independent solutions of Legendre's equation for the cases, $n = 0, 1, 2, 4$.

1. $(1 - x^2)y'' - 2xy' = 0.$ $(n = 0)$
2. $(1 - x^2)y'' - 2xy' + 2y = 0.$ $(n = 1)$
3. $(1 - x^2)y'' - 2xy' + 6y = 0.$ $(n = 2)$
4. $(1 - x^2)y'' - 2xy' + 20y = 0.$ $(n = 4)$

5. Obtain one polynomial solution for Legendre's equation in the general case for any integer values of n. [Do not use n as an index of summation; use k instead.]

2.4.2 Laguerre Polynomials

This class of polynomials appears in quantum physics, and we shall briefly address the pertaining differential equation,

$$xy'' + (1-x)y' + ny = 0. \tag{2.55}$$

The polynomial solutions are given by

$$L_n(x) = \sum_{k=0}^{n} \frac{n!(-1)^k}{k!^2(n-k)!} x^k, \qquad , n = 0, 1, 2, 3, \ldots \tag{2.56}$$

A few of the solutions for specific n values are as follows:

$$\begin{aligned} L_0 &= 1 \\ L_1 &= 1 - x \\ L_2 &= \tfrac{1}{2}\left[x^2 - 4x + 2\right] \\ L_3 &= \tfrac{1}{6}\left[6 - 18x + 9x^2 - x^3\right] \end{aligned}$$

The second solution has logarithmic behavior about $x = 0$ and is left as a set of exercises. Additional properties are discussed in the next chapter in connection with Fourier expansions.

EXERCISES 2.2 Obtain two linearly independent solutions of equation (2.55) for

1. $n = 0$
2. $n = 1$
3. $n = 2$

2.4.3 Bessel Functions

For a large class of problems in spherical and cylindrical geometry, we need to deal with Bessel's equation, or some variations of it. While the following form is what is commonly known as Bessel's equation,

$$x^2y'' + xy' + (x^2 - \nu^2)y = 0, \tag{2.57}$$

other forms come up. Among these are, for example, the modified Bessel's equation, the spherical Bessel's equation, and the modified spherical Bessel's equation. For the sake of clear reference, we shall refer to (2.57) as the cylindrical Bessel's equation.

Cylindrical Bessel Functions

For many cases in cylindrical geometry, we have $\nu = n$, as integer values. For these cases, the two linearly independent solutions consist of one integer power series and the other containing a $\ln x$ term. The former set is denoted by $J_n(x)$ and the latter by $Y_n(x)$.

Let us consider the case $n = 1$ as an example

Example 2.5

$$x^2 y''(x) + xy'(x) + (x^2 - 1)y(x) = 0 \tag{2.58}$$

Let

$$y(x) = \sum_{n=0}^{\infty} C_n x^{n+s}.$$

Taking derivatives, we obtain

$$y'(x) = \sum_{n=0}^{\infty} C_n (n+s)x^{n+s-1}, \tag{2.59}$$

$$y''(x) = \sum_{n=0}^{\infty} C_n (n+s)(n+s-1)x^{n+s-2}. \tag{2.60}$$

Substitution into the differential equation leads to

$$x^2 \sum_{n=0}^{\infty} C_n (n+s)(n+s-1)x^{n+s-2} + x \sum_{n=0}^{\infty} C_n (n+s)x^{n+s} + (x^2-1) \sum_{n=0}^{\infty} C_n x^{n+s} = 0, \tag{2.61}$$

$$\sum_{n=0}^{\infty} C_n (n+s)(n+s-1)x^{n+s} + \sum_{n=0}^{\infty} C_n (n+s)x^{n+s} + \sum_{n=0}^{\infty} C_n x^{n+s+2} - \sum_{n=0}^{\infty} C_n x^{n+s} = 0.$$

By grouping together the first, the second and the fourth summation, we find

$$\sum_{n=0}^{\infty} C_n \left[(n+s)(n+s-1) + (n+s) - 1\right] x^{n+s} + \sum_{n=0}^{\infty} C_n x^{n+s+2} = 0, \tag{2.62}$$

which may also be written as

$$\sum_{n=0}^{\infty} C_n \left[(n+s)^2 - 1\right] x^{n+s} + \sum_{n=0}^{\infty} C_n x^{n+s+2} = 0. \tag{2.63}$$

The powers in x in the first summation start matching with the second after $n = 2$ in the former. Therefore, let us pull out the first two terms of the first summation. These terms are those with the lowest powers in x and have no corresponding terms with like powers in the other summation. Thus,

$$C_0(s^2 - 1)x^s + C_1\left[(s+1)^2 - 1\right] x^{s+1}$$
$$+ \sum_{n=2}^{\infty} C_n \left[(n+s)^2 - 1\right] x^{n+s} + \sum_{n=0}^{\infty} C_n x^{n+s+2} = 0.$$

Now, shift the index on the first summation,

$$C_0(s+1)(s-1)x^s + C_1(s+2)sx^{s+1} + \sum_{n=0}^{\infty} C_{n+2}\left[(n+s+2)^2-1\right]x^{n+s+2} + \sum_{n=0}^{\infty} C_n x^{n+s+2} = 0,$$

and rewrite as

$$C_0(s+1)(s-1)x^s + C_1(s+2)sx^{s+1} + \sum_{n=0}^{\infty}\left[C_{n+2}(n+s+3)(n+s+1)+C_n\right]x^{n+s+2} = 0.$$

Setting the coefficients of each power in x equal to zero,

$$\begin{aligned} C_0(s+1)(s-1) &= 0 \quad \text{resulting in} \quad C_0 \neq 0, \quad s_1 = 1, s_2 = -1, \\ C_1(s+2)s &= \quad C_1 = 0 \quad \text{for} \quad s = 1, -1 \\ C_{n+2} &= -\frac{C_n}{(n+s+3)(n+s+1)}, \qquad n = 0, 1, 2, 3, \ldots \end{aligned} \tag{2.64}$$

With $s = s_1 = 1$,

$$C_{n+2} = -\frac{C_n}{(n+4)(n+2)} \qquad n = 0, 1, 2, 3, 4, 5, \ldots \tag{2.65}$$

This leads to

$$\begin{aligned} C_2 &= -\frac{1}{4(2)}C_0 = -\frac{1}{8}C_0 = -\frac{4(2)}{4^2 2^2} \\ C_4 &= -\frac{1}{6(4)}C_2 = +\frac{6(2)}{6^2 4^2 2^2}C_0 = +\frac{1}{192}C_0 \\ C_6 &= -\frac{1}{8(6)}C_2 = -\frac{8(2)}{8^2 6^2 4^2 2^2}C_0 = -\frac{1}{9216}C_0 \\ &\vdots \\ C_n &= (-1)^{\frac{1}{2}n}\frac{2(n+2)}{(n+2)^2 n^2 \ldots 6^2 4^2 2^2}C_0, \\ &= (-1)^{\frac{1}{2}n}\frac{2(n+2)}{[(n+2)/2]!^2 2^{n+2}}C_0 \qquad n = 0, 2, 4, 6, 8, \ldots \end{aligned}$$

Since $C_1 = 0$, we can easily show that $C_3 = C_5 = C_7 = \ldots = 0$.

The first solution, therefore, is

$$\begin{aligned} y_1(x) &= C_0\left[x - \frac{1}{8}x^3 + \frac{1}{192}x^5 - \frac{1}{9216}x^7 \ldots\right] \\ &= 2C_0 \sum_{n=0,2,4,\ldots}^{\infty} (-1)^{\frac{1}{2}n}\frac{(n+2)}{[(n+2)/2]!^2 2^{n+2}}x^{n+1} \\ &= 2C_0 \sum_{n=0}^{\infty}(-1)^n \frac{1}{n!(n+1)!}\left(\tfrac{1}{2}x\right)^{2n+1} = 2C_0 J_1(x), \end{aligned}$$

where $J_1(x)$ is known as the Bessel function of the first kind and order one. We may set $2C_0 = 1$, and write

$$y_1(x) = J_1(x).$$

For the second solution, we take $s = s_2 = -1$. From the recurrence relationship (2.64), we obtain

$$C_{n+2} = -\frac{C_n}{(n+2)n} \qquad n = 0, 1, 2, 3, 4, 5, \ldots$$

When we attempt to calculate C_2, we end up with $C_2 \to \infty$. This signals the possibility of a $\ln x$ term. Let us therefore use

$$y_2(x) = y_1(x)\ln x + z(x),$$

and substitute into the differential equation (2.58). Upon taking derivatives, we obtain

$$\begin{aligned} y_2'(x) &= y_1'(x)\ln x + \frac{y_1(x)}{x} + z'(x) \\ y_2''(x) &= y_1''(x)\ln x + 2\frac{y_1'(x)}{x} - \frac{y_1(x)}{x^2} + z''(x). \end{aligned}$$

Using these in (2.58), we find

$$\begin{aligned} &\ln x\left[x^2 y_1''(x) + x y_1'(x) + (x^2-1)y_1(x)\right] \\ &\quad \left[2xy_1'(x) - y_1(x) + y_1(x)\right] + \left[x^2 z''(x) + xz'(x) + (x^2-1)z(x)\right] = 0. \end{aligned}$$

The first set of terms containing $\ln x$ vanish because the quantities in brackets represent the left side on the original differential equation (2.58). We are therefore left with

$$\left[x^2 z''(x) + xz'(x) + (x^2-1)z(x)\right] = -2xy_1'(x),$$

for which we assume the solution

$$z(x) = \sum_{n=0}^{\infty} D_n x^{n+s_2}.$$

Taking derivatives, and following the procedure similar to equations (2.60-2.63), we obtain

$$\sum_{n=0}^{\infty} D_n\left[(n+s_2)^2 - 1\right]x^{n+s_2} + \sum_{n=0}^{\infty} D_n x^{n+s_2+2} = -2\sum_{n=0}^{\infty} C_n(n+1)x^{n+1} \quad (2.66)$$

With $s_2 = -1$,

$$\sum_{n=0}^{\infty} D_n\left[(n-1)^2 - 1\right]x^{n-1} + \sum_{n=0}^{\infty} D_n x^{n+1} = -2\sum_{n=0}^{\infty} C_n(n+1)x^{n+1}$$

which can be simplified slightly to give

$$\sum_{n=0}^{\infty} D_n n(n-2)x^{n-1} + \sum_{n=0}^{\infty} D_n x^{n+1} = -2\sum_{n=0}^{\infty} C_n(n+1)x^{n+1}$$

Now, we extract the first two terms from the first sum on the left and shift its index by two places. Therefore,

$$D_0 0x^{-1} - D_1 x^0 + \sum_{n=0}^{\infty} D_{n+2}(n+2)nx^{n+1} + \sum_{n=0}^{\infty} D_n x^{n+1} = -2\sum_{n=0}^{\infty} C_n(n+1)x^{n+1},$$

or by grouping the two summations,

$$D_0 0x^{-1} - D_1 x^0 + \sum_{n=0}^{\infty} [D_{n+2}(n+2)n + D_n]\, x^{n+1} = -2\sum_{n=0}^{\infty} C_n(n+1)x^{n+1},$$

Equating powers in x gives,

$$\begin{aligned}
x^{-1} &: \quad 0D_0 = 0 \quad \Rightarrow D_0 \text{ is indeterminate (for the moment)} \\
x^{0} &: \quad -D_1 = 0 \\
x^{1} &: \quad 0D_2 + D_0 = -2C_0 \quad \Rightarrow D_0 = -2C_0 \\
x^{2} &: \quad 3D_3 + D_1 = -4C_1 = 0 \quad \Rightarrow D_3 = 0 \\
x^{3} &: \quad 8D_4 + D_2 = -6C_2 = \tfrac{3}{4}C_0 \\
&\quad \vdots \\
x^{n} &: \quad D_{n+2}(n+2)n + D_n = -2(n+1)C_n.
\end{aligned}$$

It is not difficult to see that

$$D_{n+2} = -\frac{2(n+1)C_n + D_n}{(n+2)n}.$$

Here, D_2 remains undefined, and carrying on with an unspecified value for D_2 will lead to the $y_1(x)$ solution which is also the solution for the homogeneous equation for $z(x)$. Nonetheless, we shall retain D_2 for the moment, and see how it plays out.

It is also easy to see that $D_1 = D_3 = D_5 = \ldots = 0$. The values of the remaining coefficients are determined as follows:

$$\begin{aligned}
D_4 &= -\frac{2(3)C_2 + D_2}{4(2)} \\
&= -\frac{2(3)}{4(2)}\left(-\frac{4(2)}{4^2 2^2}\right) + \frac{D_2}{4(2)} \\
&= \frac{\frac{3}{2}}{(4)2^2} - \frac{D_2}{4(2)} \\
D_6 &= -\frac{2(5)C_4 + D_4}{6(4)} \\
&= -\frac{2(5)6(2)}{6(4)6^2 4^2 2^2} - \frac{\frac{3}{2}}{(6)4^2 2^2} + \frac{D_2}{(6)4^2 2} \\
&= -\frac{\frac{5}{6}}{(6)4^2 2^2} - \frac{\frac{3}{2}}{(6)4^2 2^2} + \frac{D_2}{(6)4^2 2}
\end{aligned}$$

$$
\begin{aligned}
&= -\frac{\frac{5}{6}+\frac{3}{2}}{(6)4^2 2^2} + \frac{D_2}{(6)4^2 2} \\
D_8 &= -\frac{2(7)C_6 + D_6}{8(6)} \\
&= \frac{\frac{7}{12}+\frac{5}{6}+\frac{3}{2}}{(8)6^2 4^2 2^2} - \frac{D_2}{(8)6^2 4^2 2} \\
D_{10} &= -\frac{2(9)C_8 + D_8}{10(8)} \\
&= -\frac{\frac{9}{20}+\frac{7}{12}+\frac{5}{6}+\frac{3}{2}}{(10)8^2 6^2 4^2 2^2} + \frac{D_2}{(10)8^2 6^2 4^2 2} \\
&= -\frac{\left(\frac{1}{5}+\frac{1}{4}\right)+\left(\frac{1}{4}+\frac{1}{3}\right)+\left(\frac{1}{3}+\frac{1}{2}\right)+\left(\frac{1}{2}+1\right)}{(10)8^2 6^2 4^2 2^2} + \frac{D_2}{(10)8^2 6^2 4^2 2} \\
&= -\frac{\left(\frac{1}{2}+\frac{1}{3}+\frac{1}{4}+\frac{1}{5}\right)+\left(1+\frac{1}{2}+\frac{1}{3}+\frac{1}{4}\right)}{(10)8^2 6^2 4^2 2^2} + \frac{D_2}{(10)8^2 6^2 4^2 2} \\
&= -\frac{\left(1+\frac{1}{2}+\frac{1}{3}+\frac{1}{4}+\frac{1}{5}\right)+\left(1+\frac{1}{2}+\frac{1}{3}+\frac{1}{4}\right)}{(10)8^2 6^2 4^2 2^2} + \frac{\left(D_2+\frac{1}{2}\right)}{(10)8^2 6^2 4^2 2} \\
&= -\frac{\Phi(5)+\Phi(4)}{(10)8^2 6^2 4^2 2^2} + \frac{\left(D_2+\frac{1}{2}\right)}{(10)8^2 6^2 4^2 2} \\
&= -\frac{\Phi(5)+\Phi(4)}{5!4!2^9} + \frac{(2D_2+1)}{5!4!2^9} \\
&\ \vdots \\
D_n &= \left[(-1)^{\frac{1}{2}n-1}\right]\left\{\frac{\Phi\left(\frac{1}{2}n\right)+\Phi\left(\frac{1}{2}n-1\right)}{\left(\frac{1}{2}n\right)!\left(\frac{1}{2}n-1\right)!2^{n-1}} + \frac{(2D_2+1)}{\left(\frac{1}{2}n\right)!\left(\frac{1}{2}n-1\right)!2^{n-1}}\right\}, \qquad n = 2,4,6,8,\ldots
\end{aligned}
$$

where

$$\Phi(k) = 1 + \tfrac{1}{2} + \tfrac{1}{3} + \tfrac{1}{4} + \cdots + \frac{1}{k}.$$

With D_n defined above, and $D_0 = -2$ we may write the following expression for $z(x)$:

$$
\begin{aligned}
z(x) &= -\frac{2}{x} - \sum_{n=2,4,6,\ldots}^{\infty} \left[(-1)^{\frac{1}{2}n-1}\right]\left\{\frac{\Phi\left(\frac{1}{2}n\right)+\Phi\left(\frac{1}{2}n-1\right)}{\left(\frac{1}{2}n\right)!\left(\frac{1}{2}n-1\right)!2^{n-1}} - \frac{(2D_2+1)}{\left(\frac{1}{2}n\right)!\left(\frac{1}{2}n-1\right)!2^{n-1}}\right\} x^{n-1} \\
&= -\frac{2}{x} - \sum_{n=1}^{\infty}\left[(-1)^{n-1}\right]\left\{\frac{\Phi(n)+\Phi(n-1)}{n!\,(n-1)!2^{2n-1}} - \frac{(2D_2+1)}{n!(n-1)!2^{2n-1}}\right\} x^{2n-1}, \\
&= -\frac{2}{x} - \sum_{n=0}^{\infty}\left[(-1)^{n}\right]\left\{\frac{\Phi(n+1)+\Phi(n)}{(n+1)!n!} - \frac{(2D_2+1)}{(n+1)!n!}\right\}\left(\tfrac{1}{2}x\right)^{2n+1},
\end{aligned}
$$

Here, the second term [containing $(2D_2+1)$] is proportional to $y_1(x) = J_1(x)$, and may be removed without any loss of generality. The second solution is therefore,

$$y_2(x) = \left[\sum_{n=0}^{\infty} \frac{(-1)^n}{(n+1)!n!}\left(\tfrac{1}{2}x\right)^{2n+1}\right] \ln x - \frac{2}{x} - \sum_{n=0}^{\infty} \frac{(-1)^n\left[\Phi(n+1)+\Phi(n)\right]}{(n+1)!n!}\left(\tfrac{1}{2}x\right)^{2n+1}$$

This is the second linearly-independent solution of Bessel's equation (2.58). However, it is not the standard form which is represented as follows

$$Y_1(x) = \frac{2}{\pi}\left\{[\ln\left(\tfrac{1}{2}x\right) + \gamma]\left[\sum_{n=0}^{\infty}\frac{(-1)^n}{(n+1)!n!}\left(\tfrac{1}{2}x\right)^{2n+1}\right] - \frac{1}{x} - \tfrac{1}{2}\sum_{n=0}^{\infty}\frac{(-1)^n\left[\Phi(n+1)+\Phi(n)\right]}{(n+1)!n!}\left(\tfrac{1}{2}x\right)^{2n+1}\right\},$$

where $\gamma = 0.5772157\ldots$ is Euler's constant defined by

$$\gamma = \lim_{k\to\infty}\left[\Phi(k) - \ln k\right].$$

□

We have just derived the two linearly-independent solutions for Bessels equation of order one (i.e., $n = 1$). For the general case of the nth order equation (2.57), the two solutions may be similarly derived with power-series expressions,

$$J_n(x) = \sum_{k=0}^{\infty}\frac{(-1)^k}{(k+n)!k!}\left(\tfrac{1}{2}x\right)^{2k+n}, \quad n = 0, 1, 2, 3\ldots \tag{2.67}$$

$$Y_n(x) = \frac{2}{\pi}\left\{[\ln\left(\tfrac{1}{2}x\right) + \gamma]J_n(x) - \tfrac{1}{2}\sum_{k=0}^{n-1}\frac{(n-k-1)!}{k!}\left(\tfrac{1}{2}x\right)^{2k-n} - \tfrac{1}{2}\sum_{k=0}^{\infty}\frac{(-1)^k\left[\Phi(n+k)+\Phi(k)\right]}{(n+k)!k!}\left(\tfrac{1}{2}x\right)^{2k+n}\right\}, \quad n = 1, 2, 3.. \tag{2.68}$$

For $n = 0$, the expressions are

$$J_0(x) = \sum_{k=0}^{\infty}\frac{(-1)^k}{k!^2}\left(\tfrac{1}{2}x\right)^{2k} \tag{2.69}$$

$$Y_0(x) = \frac{2}{\pi}\left\{[\ln\left(\tfrac{1}{2}x\right) + \gamma]J_0(x) - \sum_{k=0}^{\infty}\frac{(-1)^k\Phi(k)}{k!^2}\left(\tfrac{1}{2}x\right)^{2k+n}\right\}. \tag{2.70}$$

Plots of $J_n(x)$ and $Y_n(x)$ for $n = 0, 1, 2, 3$ are given in Figure 2.1.

EXERCISES 2.3 Derive the above solutions to Bessel's equation (2.57)

1. $\nu = 0$
2. $\nu = 2$.

We next go on to the modified cylindrical Bessel functions.

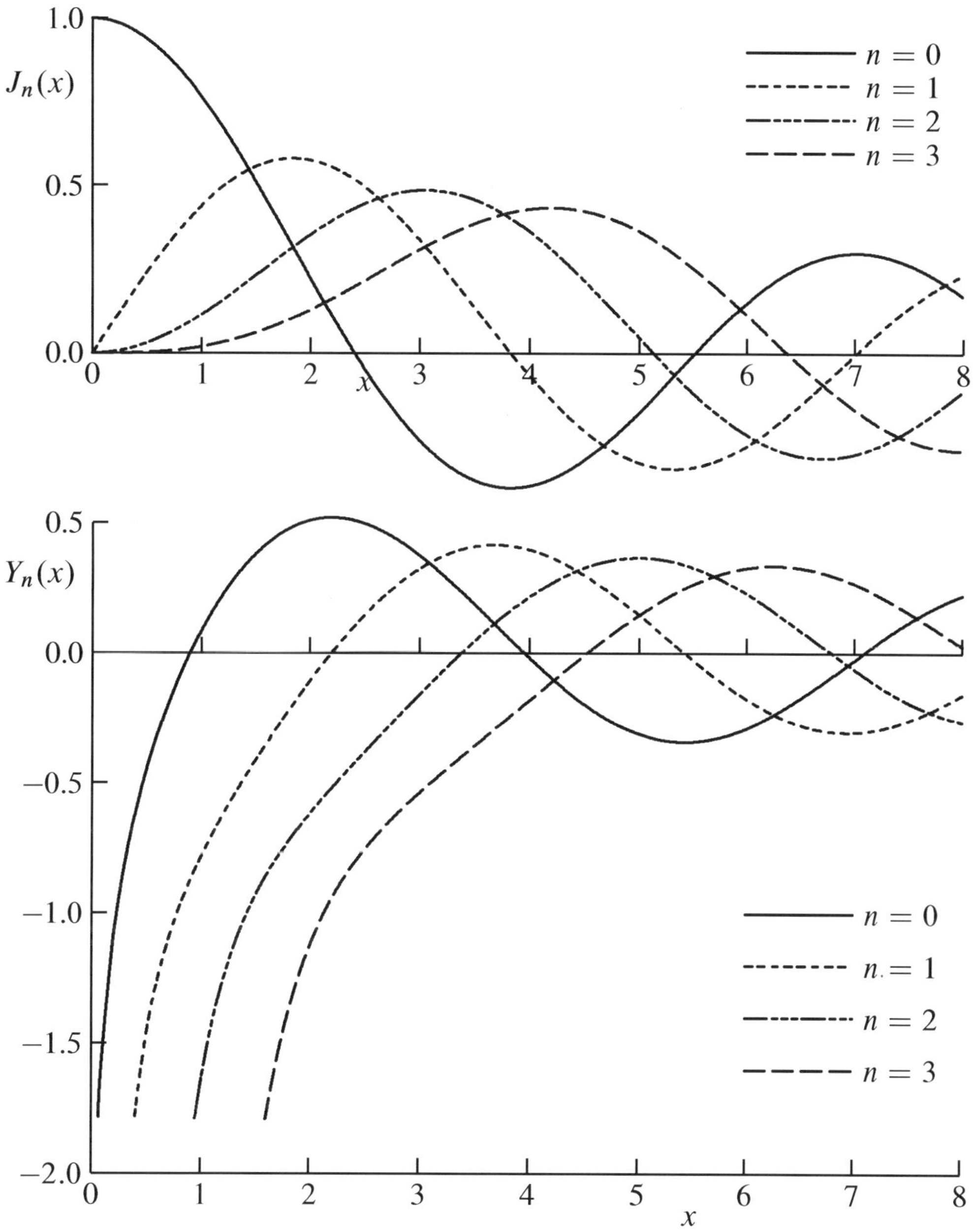

Figure 2.1: Plots for $J_n(x)$ and $Y_n(x)$ for $n = 0, 1, 2, 3$

Modified Cylindrical Bessel Functions

In equation (2.57), if we replace the $x^2y(x)$ term with $-x^2y(x)$, the equation becomes

$$x^2y'' + xy' - (x^2 + \nu^2)y = 0. \tag{2.71}$$

Again, for integer values of ν, this equation comes up in cylindrical geometry, and the solutions may be constructed by the power-series method. In principle, the modified Bessel's equation is equivalent to replacing x^2 with $-x^2$ (i.e., x with ix) throughout (2.57). The solution is therefore a modification of equations (2.67) and (2.68), i.e.,

$$I_n(x) = i^{-n}J_n(ix) = \left[\sum_{k=0}^{\infty} \frac{1}{(k+n)!k!}\left(\tfrac{1}{2}x\right)^{2k+n}\right] \tag{2.72}$$

$$K_n(x) = \left\{(-1)^{n+1}\left[\ln\left(\tfrac{1}{2}x\right) + \gamma\right]I_n(x) + \tfrac{1}{2}\sum_{k=0}^{n-1}\frac{(-1)^k(n-k-1)!}{k!}\left(\tfrac{1}{2}x\right)^{2k-n}\right.$$
$$\left. + \tfrac{1}{2}(-1)^n\sum_{k=0}^{\infty}\frac{[\Phi(n+k)+\Phi(k)]}{(n+k)!k!}\left(\tfrac{1}{2}x\right)^{2k+n}\right\}. \tag{2.73}$$

Plots of these are given in Figure 2.2.

In spherical geometry, a slightly different form of equation (2.57) comes up frequently. We shall discuss this equation and its solutions next.

Spherical Bessel Functions

The spherical Bessel's equation has the form

$$x^2y'' + 2xy' + \left[x^2 - n(n+1)\right]y = 0. \tag{2.74}$$

The two linearly independent solutions to this equation may be constructed in the same way as the cylindrical Bessel's equation. However, in this case, the $\ln x$ terms do not come up. The derivation by the power-series method is quite straightforward and left as exercises (see problems 3 and 4 in Exercises 2.4 on page 74). The two linearly independent solutions [denoted by $j_n(x)$ and $y_n(x)$] are related to those of the cylindrical Bessel's equation through the relationships,

$$j_n(x) = \left(\frac{\pi}{2x}\right)^{\frac{1}{2}} J_{n+\frac{1}{2}}(x) \quad \text{and} \quad y_n(x) = \left(\frac{\pi}{2x}\right)^{\frac{1}{2}} Y_{n+\frac{1}{2}}(x). \tag{2.75}$$

The power-series solution will yield

$$j_n(x) = \sum_{k=0}^{\infty}\frac{(-1)^k(k+n)!2^n}{k!(2k+2n+1)!}x^{2k+n} = x^n\sum_{k=0}^{\infty}\frac{(-1)^k(k+n)!2^{k+n}}{k!(2k+2n+1)!}\left(\tfrac{1}{2}x^2\right)^k \tag{2.76}$$

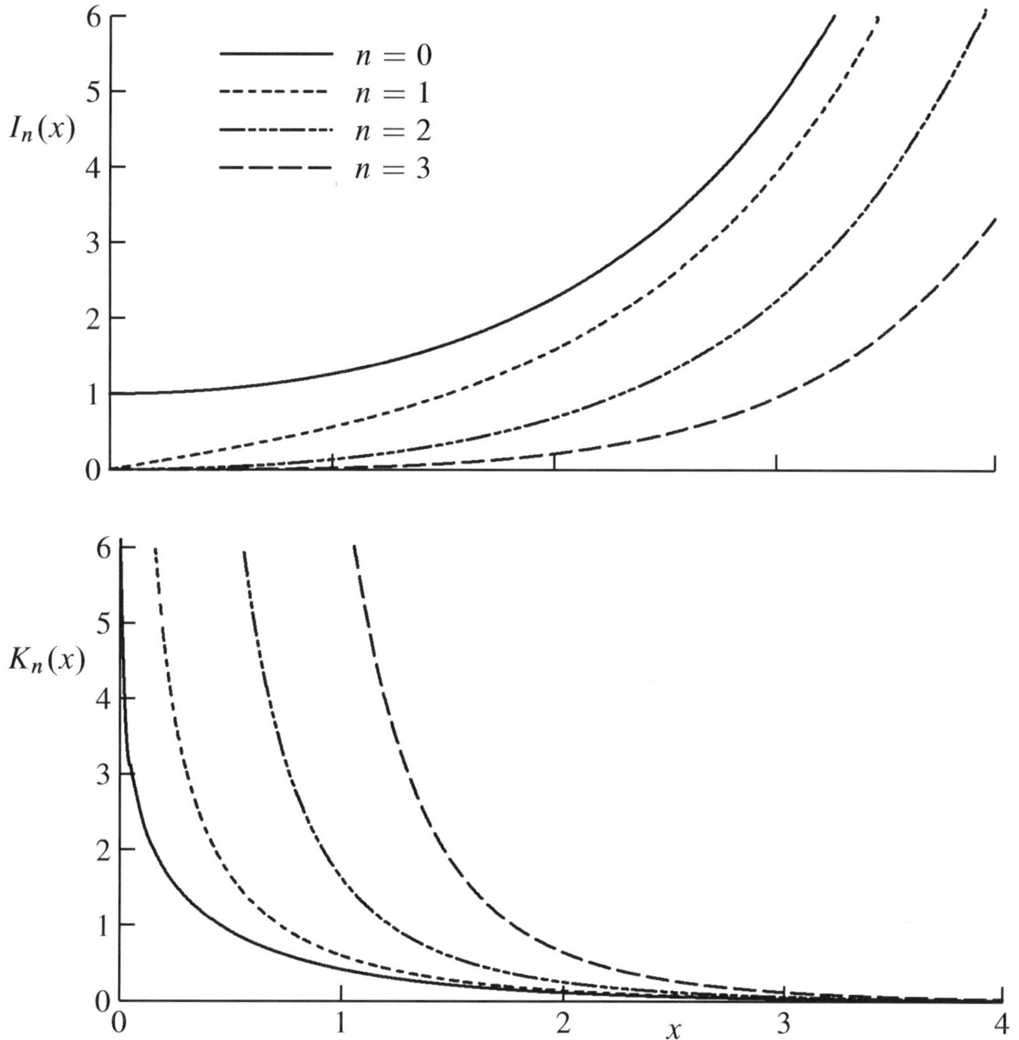

Figure 2.2: Plots for $I_n(x)$ and $K_n(x)$ for $n = 0, 1, 2, 3$

and

$$
\begin{aligned}
y_n(x) &= -\frac{1}{2^{n-1}x^{n+1}} \sum_{k=1}^{n} \frac{(2n-2k+1)!}{(n-k)!} x^{2k-2} - \frac{x^{n-1}}{2^n n!} \\
&\qquad + \frac{x^{n+1}}{2^{n+1}} \sum_{k=0}^{\infty} \frac{(-1)^k k!}{(n+k+1)!(2k+1)!} x^{2k} \\
&= -\frac{1}{x^{n+1}} \sum_{k=1}^{n} \frac{(2n-2k+1)!}{(n-k)!2^{n-k}} \left(\tfrac{1}{2}x^2\right)^{k-1} - \frac{x^{n-1}}{2^n n!} \\
&\qquad + \frac{1}{x^{n+1}} \sum_{k=0}^{\infty} \frac{(-1)^k k! 2^k}{(n+k+1)!(2k+1)!} \left(\tfrac{1}{2}x^2\right)^{k+n+1}
\end{aligned} \tag{2.77}
$$

Other convenient forms of these solutions may be expressed as [1]

$$\begin{aligned} j_n(x) &= \frac{x^n}{1\cdot 3\cdot 5\ldots(2n+1)}\left\{1-\frac{\frac{1}{2}x^2}{1!(2n+3)}+\frac{\left(\frac{1}{2}x^2\right)^2}{2!(2n+3)(2n+5)}-\cdots\right\} \\ y_n(x) &= -\frac{1\cdot 3\cdot 5\ldots(2n-1)}{x^{n+1}}\left\{1-\frac{\frac{1}{2}x^2}{1!(1-2n)}+\frac{\left(\frac{1}{2}x^2\right)^2}{2!(1-2n)(3-2n)}-\cdots\right\} \end{aligned}$$

Besides these power-series solutions, the spherical Bessel functions may be written as combinations of sine and cosine functions. This form turns out to be very convenient for calculations since it forms finite series expressions. In general,

$$h_n^{(1)}(x) = j_n(x) + iy_n(x) = (-i)^{n+1}\frac{e^{ix}}{x}\sum_{k=0}^{n}\frac{i^k}{k!(2x)^k}\frac{(n+k)!}{(n-k)!}. \tag{2.78}$$

Specifically, for $n = 0, 1, 2$, we have

$$j_0(x) = \frac{\sin x}{x} \tag{2.79}$$

$$j_1(x) = \frac{\sin x}{x^2} - \frac{\cos x}{x} \tag{2.80}$$

$$j_2(x) = \left(\frac{3}{x^3}-\frac{1}{x}\right)\sin x - \left(\frac{3}{x^2}\right)\cos x \tag{2.81}$$

and

$$y_0(x) = -\frac{\cos x}{x} \tag{2.82}$$

$$y_1(x) = -\frac{\cos x}{x^2} - \frac{\sin x}{x} \tag{2.83}$$

$$y_2(x) = \left(-\frac{3}{x^3}+\frac{1}{x}\right)\cos x - \left(\frac{3}{x^2}\right)\sin x \tag{2.84}$$

These functions have application in various spherical-geometry problems such as in heat conduction, electromagnetic wave theory, and acoustics.

Modified Spherical Bessel Functions

A slight variation of equation (2.74) is

$$x^2y'' + 2xy' - \left[x^2 + n(n+1)\right]y = 0. \tag{2.85}$$

This is the equivalent of replacing x with ix in (2.74), and the two linearly independent solutions are denoted by $i_n(x)$ and $k_n(x)$. The relationships between the corresponding modified cylindrical Bessel functions are as follows:

$$i_n(x) = \left(\frac{\pi}{2x}\right)^{\frac{1}{2}} I_{n+\frac{1}{2}}(x) \quad \text{and} \quad k_n(x) = \left(\frac{\pi}{2x}\right)^{\frac{1}{2}} K_{n+\frac{1}{2}}(x). \tag{2.86}$$

Also used is the pair $i_n(x)$ and $i_{-n-1}(x)$ where

$$i_{-n-1}(x) = \left(\frac{\pi}{2x}\right)^{\frac{1}{2}} I_{-n-\frac{1}{2}}(x)\,. \tag{2.87}$$

The power-series expressions for these functions are

$$\begin{aligned}
i_n(x) &= \frac{x^n}{1\cdot 3\cdot 5\ldots(2n+1)}\left\{1+\frac{\frac{1}{2}x^2}{1!(2n+3)}+\frac{\left(\frac{1}{2}x^2\right)^2}{2!(2n+3)(2n+5)}+\cdots\right\}\\
&= \sum_{k=0}^{\infty}\frac{(k+n)!2^n}{k!(2k+2n+1)!}x^{2k+n} = x^n\sum_{k=0}^{\infty}\frac{(k+n)!2^{k+n}}{k!(2k+2n+1)!}\left(\tfrac{1}{2}x^2\right)^k\\
i_{-n-1}(x) &= \frac{1\cdot 3\cdot 5\ldots(2n-1)}{(-1)^n x^{n+1}}\left\{1+\frac{\frac{1}{2}x^2}{1!(1-2n)}+\frac{\left(\frac{1}{2}x^2\right)^2}{2!(1-2n)(3-2n)}+\cdots\right\}\\
&= \frac{(-1)^n}{2^{n-1}x^{n+1}}\sum_{k=1}^{n}\frac{(2n-2k+1)!(-1)^k}{(n-k)!}x^{2k-2}+\frac{x^{n-1}}{2^n n!}\\
&\quad +\frac{x^{n+1}}{2^{n+1}}\sum_{k=0}^{\infty}\frac{k!}{(n+k+1)!(2k+1)!}x^{2k}\\
&= \frac{(-1)^n}{x^{n+1}}\sum_{k=1}^{n}\frac{(2n-2k+1)!(-1)^k}{(n-k)!2^{n-k}}\left(\tfrac{1}{2}x^2\right)^{k-1}+\frac{x^{n-1}}{2^n n!}\\
&\quad +\frac{1}{x^{n+1}}\sum_{k=0}^{\infty}\frac{k!2^k}{(n+k+1)!(2k+1)!}\left(\tfrac{1}{2}x^2\right)^{k+n+1}
\end{aligned}$$

In terms of hyperbolic functions, we may write for $n = 0, 1, 2$

$$\begin{aligned}
i_0(x) &= \frac{\sinh x}{x}\\
i_1(x) &= -\frac{\sinh x}{x^2}+\frac{\cosh x}{x}\\
i_2(x) &= \left(\frac{3}{x^3}+\frac{1}{x}\right)\sinh x-\left(\frac{3}{x^2}\right)\cosh x
\end{aligned}$$

and

$$\begin{aligned}
i_{-1}(x) &= \frac{\cosh x}{x}\\
i_{-2}(x) &= -\frac{\cosh x}{x^2}+\frac{\sinh x}{x}\\
i_{-3}(x) &= \left(\frac{3}{x^3}+\frac{1}{x}\right)\cosh x-\left(\frac{3}{x^2}\right)\sinh x.
\end{aligned}$$

These may be combined to give the exponential forms

$$\begin{aligned}
k_0(x) &= \tfrac{1}{2}\pi\left(\frac{1}{x}\right)e^{-x} \\
k_1(x) &= \tfrac{1}{2}\pi\left(\frac{1}{x}+\frac{1}{x^2}\right)e^{-x} \\
k_2(x) &= \tfrac{1}{2}\pi\left(\frac{1}{x}+\frac{3}{x^2}+\frac{3}{x^3}\right)e^{-x} \\
&\vdots \\
k_n(x) &= \tfrac{1}{2}\pi\left(\frac{e^{-x}}{x}\right)\sum_{k=0}^{n}\frac{1}{k!(2x)^k}\frac{(n+k)!}{(n-k)!} \\
&= \tfrac{1}{2}\pi\left(\frac{e^{-x}}{x}\right)R_n\left(\frac{1}{2x}\right),
\end{aligned}$$

where

$$R_n(u) = \sum_{k=0}^{n}\frac{(n+k)!u^k}{(n-k)!k!} \tag{2.88}$$

is a class of polynomials whose properties have been analyzed by Sadhal [3] in relation to problems involving spherical particles with large radial flow.

EXERCISES 2.4 Obtain two linearly independent solutions for each of the following differential equations

1. $x^2y'' + x(2-x)y' - 2y = 0$
2. $x^2y'' + x(3+x)y' + y = 0$
3. $x^2y'' + 2xy' + (x-6)y = 0$
4. $x^2y'' + 2xy' - (x+12)y = 0$
5. $x^2y'' + x(x-\frac{1}{4})y' + \frac{1}{4}y = 0$
6. $xy'' + (1+x)y' + 2y = 0$
7. $x^2y'' + 4xy' + xy = 0$
8. $y'' - 2xy' + 2ny = 0, \quad n = 0, 1, 2.$ *(Hermite's differential equation)*

Chapter 3

Fourier Series and Fourier Integrals

3.1 Introduction

Fourier series and Fourier integrals form a powerful mathematical tool for solving linear partial differential equations. While there are numerous other applications, the focus here is mainly towards developing the groundwork for solving partial differential equations.

Simply put, Fourier series consists of a representation of an arbitrary piecewise continuous function as a series of sine and cosine functions. The theoretical basis of Fourier series is easier to understand from the perspective of periodic functions and we shall initially concentrate on such functions.

3.2 Periodic Functions

A periodic function is one that repeats itself. While the term periodic implies repetition in time, it is generically applied to spatially repetitive functions. An example is given below.

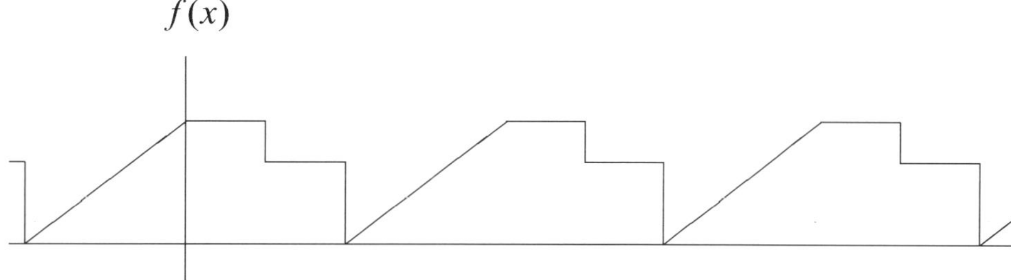

3.3 Even and Odd Functions

The knowledge about symmetry or antisymmetry of a function can be useful in defining what type of expansion is applicable. For this purpose we define even and odd functions. An even function is one that is symmetric, i.e., $f(x) = f(-x)$. An odd one is antisymmetric with the property that $g(x) = -g(-x)$. Some examples are given here.

Even Functions $[f(x) = f(-x)]$

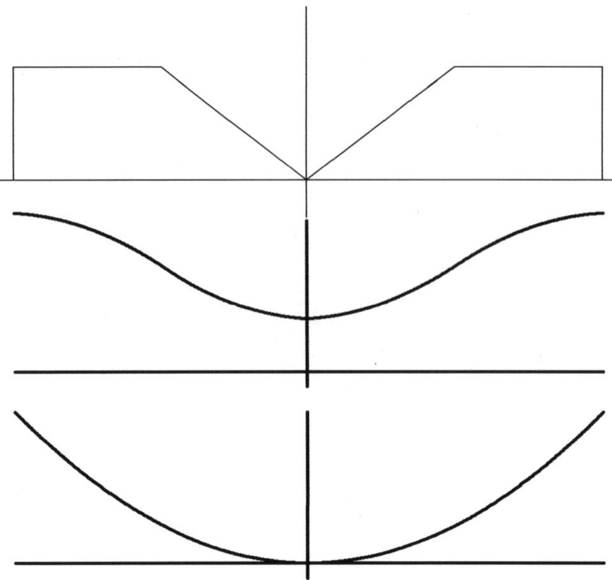

Odd Functions $[g(x) = -g(-x)]$

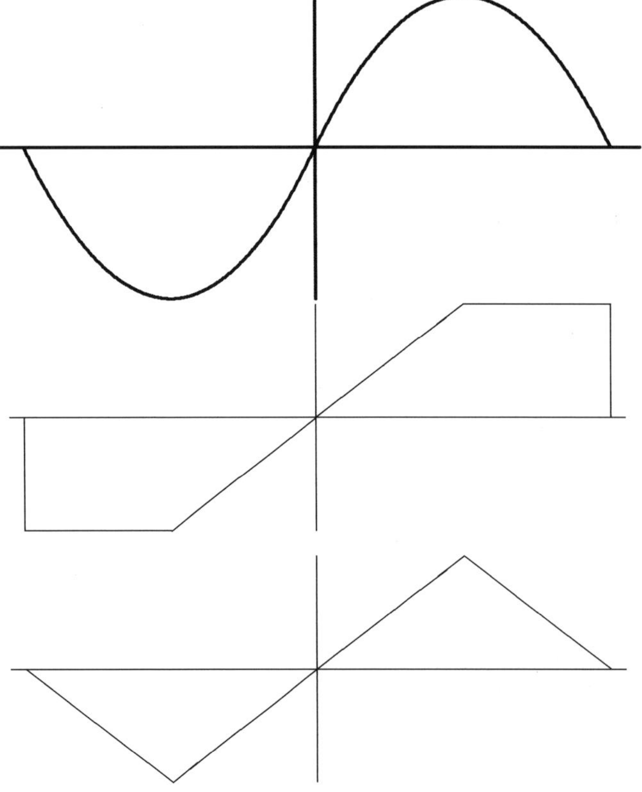

Every function can be represented as a sum of even and odd functions. Consider

$$\begin{aligned} F(x) &= \frac{F(x)+F(-x)}{2} + \frac{F(x)-F(-x)}{2} \\ &= f(x)+g(x), \end{aligned}$$

where we define

$$f(x) = \frac{F(x)+F(-x)}{2} \tag{3.1}$$

and

$$g(x) = \frac{F(x)-F(-x)}{2}. \tag{3.2}$$

By replacing x with $-x$ in equation (3.1), we see that

$$\begin{aligned} f(-x) &= \frac{F(-x)+F(-(-x))}{2} \\ &= \frac{F(-x)+F(x)}{2} \\ &= f(x) \end{aligned}$$

Since $f(x) = f(-x)$, we conclude that $f(x)$ is an even function.

Similarly, by replacing x with $-x$ in equation (3.2), we have

$$\begin{aligned} g(-x) &= \frac{F(-x)-F(-(-x))}{2} \\ &= \frac{F(-x)-F(x)}{2} \\ &= -\left(\frac{F(x)-F(-x)}{2}\right) \\ &= -g(x), \end{aligned}$$

which proves that $g(x)$ is an odd function. Thus, every function, regardless of (anti)symmetry can be expressed as a sum of an even and an odd function.

Example 3.1

Consider

$$F(x) = e^{-x} + x^2 + 2x = f(x) + g(x).$$

The decomposition would give the even part as

$$\begin{aligned} f(x) &= \frac{1}{2}[F(x)+F(-x)] \\ &= \frac{1}{2}\left[(e^{-x}+x^2+2x)+(e^x+x^2-2x)\right] \\ &= \cosh x + x^2, \end{aligned}$$

and the odd part as

$$\begin{aligned} g(x) &= \frac{1}{2}[F(x) - F(-x)] \\ &= \frac{1}{2}\left[(e^{-x} + x^2 + 2x) - (e^x + x^2 - 2x)\right] \\ &= \sinh x + 2x. \end{aligned}$$

□

3.4 Periodic Functions as Series of Periodic Functions

3.4.1 Odd Functions

Consider an odd function, $g(x)$, which has a period of 2π.

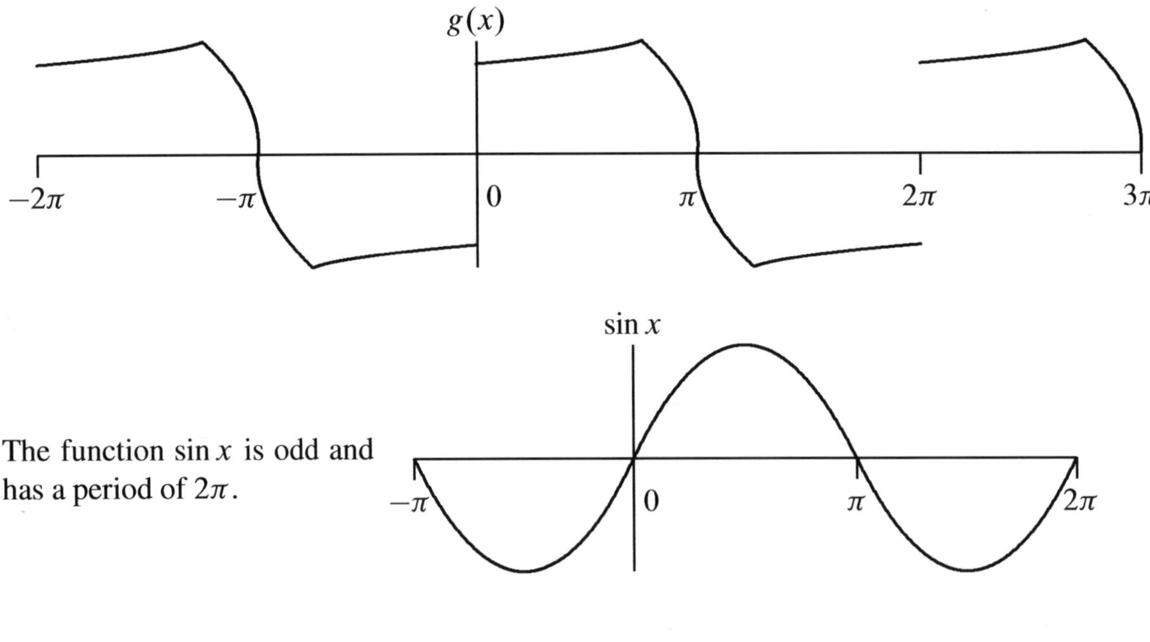

The function $\sin x$ is odd and has a period of 2π.

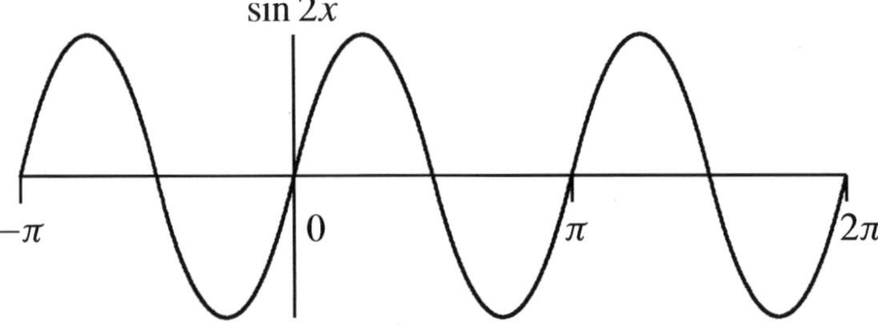

The function $\sin 2x$ is odd and has a period of π as well as 2π.

The function $\sin 3x$ is odd and has a period of π as well as $\frac{2}{3}\pi$.

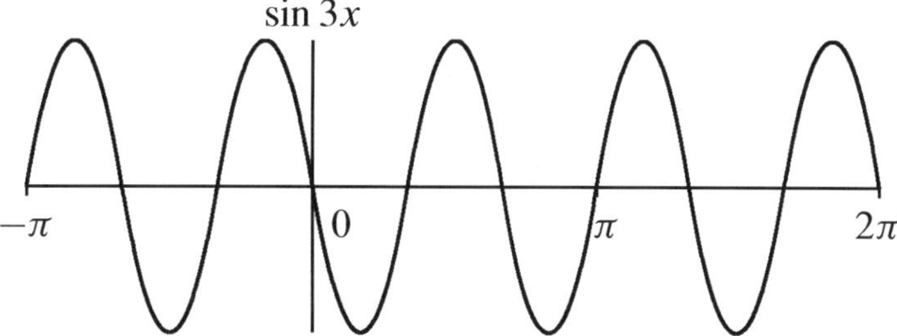

The common property among all these functions in the set $\{\sin nx\}$ is that they all are odd and have a period of 2π. This set is therefore a suitable candidate for expansion of an arbitrary function that is also odd with a period of 2π.

We may therefore write

$$\begin{aligned} g(x) &= a_1 \sin x + a_2 \sin 2x + a_3 \sin 3x + \ldots \\ &= \sum_{n=1}^{\infty} a_n \sin nx. \end{aligned} \tag{3.3}$$

Let us multiple each side of this equation by $\sin mx$ and integrate over the range $[-\pi, \pi]$. Following through with this, we obtain

$$\int_{-\pi}^{\pi} g(x) \sin mx \, dx = \sum_{n=1}^{\infty} a_n \int_{-\pi}^{\pi} \sin mx \sin nx \, dx. \tag{3.4}$$

An expression for the integral on the right can be obtained in a straightforwatd manner. Denoting that integral by I_{mn}, we may write

$$\begin{aligned} I_{mn} &- \int_{-\pi}^{\pi} \sin mx \sin nx \, dx \\ &= \int_{-\pi}^{\pi} \tfrac{1}{2} [\cos(m-n)x - \cos(m+n)x] \, dx \\ &= \tfrac{1}{2} \left[\frac{\sin(m-n)x}{(m-n)} - \frac{\sin(m+n)x}{(m+n)} \right]_{-\pi}^{\pi} \\ &= \begin{cases} 0, & (m \neq n) \\ \int_{-\pi}^{\pi} \sin^2 mx \, dx, & (m = n) \end{cases} \end{aligned}$$

The integral can be easily calculated as follows:

$$\begin{aligned} I_{mm} &= \int_{-\pi}^{\pi} \sin^2 mx \, dx \\ &= \int_{-\pi}^{\pi} \frac{1 - \cos 2mx}{2} \, dx \\ &= \left[\tfrac{1}{2} x - \frac{\sin 2mx}{4m} \right]_{-\pi}^{\pi} \\ &= \pi. \end{aligned}$$

Thus, we can write

$$I_{mn} = \begin{cases} 0, & (m \neq n) \\ \pi, & (m = n) \end{cases} \tag{3.5}$$

The fact that the integral has a zero value for $m \neq n$ and nonzero for $m = n$ makes $\{\sin nx\}$ an orthogonal set. This is analogous to a set of perpendicular vectors.

Vector Analogy

A set of N perpendicular vectors,

$$\boldsymbol{x}_1, \boldsymbol{x}_2, \boldsymbol{x}_3, \ldots, \boldsymbol{x}_n, \ldots, \boldsymbol{x}_N$$

has the property that

$$\begin{aligned}
\boldsymbol{x}_1 \cdot \boldsymbol{x}_2 &= 0 \\
\boldsymbol{x}_1 \cdot \boldsymbol{x}_3 &= 0 \\
&\vdots \\
\boldsymbol{x}_1 \cdot \boldsymbol{x}_n &= 0 \\
&\vdots \\
\boldsymbol{x}_1 \cdot \boldsymbol{x}_N &= 0 \\
\\
\boldsymbol{x}_2 \cdot \boldsymbol{x}_3 &= 0 \\
&\vdots \\
\boldsymbol{x}_2 \cdot \boldsymbol{x}_n &= 0 \\
&\vdots \\
\boldsymbol{x}_2 \cdot \boldsymbol{x}_N &= 0
\end{aligned}$$

and

$$\boldsymbol{x}_n \cdot \boldsymbol{x}_n \neq 0, \qquad n = 1, 2, 3, \ldots, N.$$

Here we are comparing

$$\int_{-\pi}^{\pi} \sin mx \sin nx \, dx \qquad \text{with} \qquad \boldsymbol{x}_m \cdot \boldsymbol{x}_n$$

both of which are

$$\begin{cases} = 0, & (m \neq n) \\ \neq 0, & (m = n) \end{cases}$$

The series expansion

$$\begin{aligned}
g(x) &= \sum_{n=1}^{\infty} a_n \sin nx \\
&= a_1 \sin x + a_2 \sin 2x + a_3 \sin 3x + \ldots
\end{aligned}$$

can be compared with the vector representation

$$\begin{aligned} \boldsymbol{f} &= a_1\boldsymbol{x}_1 + a_2\boldsymbol{x}_2 + a_3\boldsymbol{x}_3 + \ldots + a_N\boldsymbol{x}_N \qquad (3.6)\\ &= \sum_{n=1}^{N} a_n\boldsymbol{x}_n. \qquad (3.7) \end{aligned}$$

If we take the dot product of each side with $\boldsymbol{x}_m$, we obtain

$$\begin{aligned} \boldsymbol{f}\cdot\boldsymbol{x}_m &= a_1\boldsymbol{x}_1\cdot\boldsymbol{x}_m + a_2\boldsymbol{x}_2\cdot\boldsymbol{x}_m + \ldots + a_m\boldsymbol{x}_m\cdot\boldsymbol{x}_m + \ldots + a_N\boldsymbol{x}_N\cdot\boldsymbol{x}_m \\ &= a_m(\boldsymbol{x}_m\cdot\boldsymbol{x}_m). \end{aligned}$$

Therefore,

$$a_m = \frac{\boldsymbol{f}\cdot\boldsymbol{x}_m}{(\boldsymbol{x}_m\cdot\boldsymbol{x}_m)}$$

Correspondingly,

$$\begin{aligned} \int_{-\pi}^{\pi} g(x)\sin mx\,dx &= a_m\int_{-\pi}^{\pi}\sin^2 mx\,dx \\ &= a_m\pi \\ a_m &= \frac{\int_{-\pi}^{\pi} g(x)\sin mx\,dx}{\int_{-\pi}^{\pi}\sin^2 mx\,dx} \\ &= \frac{1}{\pi}\int_{-\pi}^{\pi}\underbrace{g(x)}_{\text{odd}}\ \underbrace{\sin mx}_{\text{odd}}\,dx \\ &= \frac{2}{\pi}\int_{0}^{\pi} g(x)\sin mx\,dx. \end{aligned}$$

Since function $g(x)$ exhibits repetitive behavior together with its antisymmetry, the information is required over only half a period. The integral spans over the range $[0,\pi]$ and takes information about the function $g(x)$ only over that range.

3.4.2 Even Functions

In a manner similar to the above expansion, we may carry out an expansion for an even function, $f(x)$, in a series of even functions given by the set $\{\cos nx\}$,

$$\begin{aligned} f(x) &= \sum_{n=0}^{\infty} b_n\cos nx \\ &= b_0 + \sum_{n=1}^{\infty} b_n\cos nx \end{aligned}$$

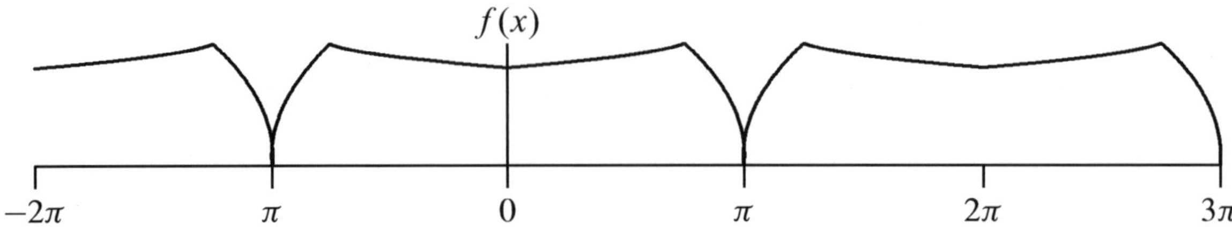

For an even function, the average value is nonzero in general, and the leading constant term often survives. That is not the case for odd functions which always have the average value equal to zero. By integrating both sides of the above expansion over one period, we obtain

$$\begin{aligned}\int_{-\pi}^{\pi} f(x)\,dx &= b_0 \int_{-\pi}^{\pi} dx + \sum_{n=1}^{\infty} \int_{-\pi}^{\pi} \cos nx\,dx \\ &= b_0 2\pi\end{aligned}$$

Therefore,

$$b_0 = \frac{1}{2\pi} \int_{-\pi}^{\pi} f(x)\,dx,$$

which is in fact the average value of $f(x)$ over the range $-\pi \le x \le \pi$. Since $f(x)$ is an even function, we may integrate over only half the period and multiply the result by two. Thus,

$$b_0 = \frac{1}{\pi} \int_{0}^{\pi} f(x)\,dx.$$

The application of the orthogonality principle, similar to the sine expansion gives

$$b_n = \frac{2}{\pi} \int_{0}^{\pi} f(x) \cos nx\,dx, \quad n \ge 1.$$

3.4.3 Expansion of Non-Symmetric Functions

As we have discussed before, a function that does not exhibit any sort of symmetry, may be decomposed as a sum of an even and an odd function in the following manner

$$\begin{aligned}F(x) &= \underbrace{g(x)}_{\text{odd}} + \underbrace{f(x)}_{\text{even}} \\ &= \left(\sum_{n=1}^{\infty} a_n \sin nx\right) + \left(b_0 + \sum_{n=1}^{\infty} b_n \cos nx\right).\end{aligned}$$

Here we independently expand the decomposed parts, $g(x)$ and $f(x)$, in terms of their respective (anti)symmetric Fourier series. Therefore, for

$$g(x) = \left(\sum_{n=1}^{\infty} a_n \sin nx\right),$$

$$
\begin{aligned}
a_n &= \frac{2}{\pi}\int_0^{\pi} g(x)\sin nx\,dx \\
&= \frac{2}{\pi}\int_0^{\pi} \frac{F(x)-F(-x)}{2}\sin nx\,dx \\
&= \frac{1}{\pi}\int_0^{\pi} F(x)\sin nx\,dx + -\frac{1}{\pi}\int_0^{\pi} F(-x)\sin nx\,dx
\end{aligned}
$$

Let $x' = -x$ in the second integral. With this substitution, we have

$$
\begin{aligned}
a_n &= \frac{1}{\pi}\int_0^{\pi} F(x)\sin nx\,dx - \frac{1}{\pi}\int_{x'=0}^{x'=-\pi} F(-(-x'))\sin(-nx')\,(-dx') \\
&= \frac{1}{\pi}\int_0^{\pi} F(x)\sin nx\,dx + \frac{1}{\pi}\int_{x'=-\pi}^{x'=0} F(x')\sin(nx')\,dx' \\
&= \frac{1}{\pi}\int_{-\pi}^{\pi} F(x)\sin nx\,dx
\end{aligned} \tag{3.8}
$$

Similarly, we expand the even part of $F(x)$, i.e.,

$$
f(x) = \left(\sum_{n=0}^{\infty} b_n \cos nx\right),
$$

$$
\begin{aligned}
b_n &= \frac{2}{\pi}\int_0^{\pi} f(x)\cos nx\,dx \\
&= \frac{2}{\pi}\int_0^{\pi} \frac{F(x)+F(-x)}{2}\cos nx\,dx \\
&= \frac{1}{\pi}\int_0^{\pi} F(x)\cos nx\,dx + +\frac{1}{\pi}\int_0^{\pi} F(-x)\cos nx\,dx \\
&= \frac{1}{\pi}\int_0^{\pi} F(x)\cos nx\,dx + \frac{1}{\pi}\int_{x'=0}^{x'=-\pi} F(-(-x'))\cos(-nx')\,(-dx') \\
&= \frac{1}{\pi}\int_0^{\pi} F(x)\cos nx\,dx + \frac{1}{\pi}\int_{x'=-\pi}^{x'=0} F(x')\cos(nx')\,dx' \\
&= \frac{1}{\pi}\int_{-\pi}^{\pi} F(x)\cos nx\,dx, \quad n = 1, 2, 3, 4, \ldots
\end{aligned} \tag{3.9}
$$

$$
\begin{aligned}
b_0 &= \frac{1}{\pi}\int_0^{\pi} f(x)\,dx \\
&= \frac{1}{\pi}\int_0^{\pi} \frac{F(x)+F(-x)}{2}\,dx \\
&= \frac{1}{2\pi}\int_0^{\pi} F(x)\,dx + +\frac{1}{2\pi}\int_0^{\pi} F(-x)\,dx \\
&= \frac{1}{2\pi}\int_0^{\pi} F(x)\,dx + \frac{1}{2\pi}\int_{x'=0}^{x'=-\pi} F(-(-x'))\,(-dx')
\end{aligned}
$$

$$\begin{aligned} &= \frac{1}{2\pi}\int_0^{\pi} F(x)\,dx + \frac{1}{2\pi}\int_{x'=-\pi}^{x'=0} F(x')\,dx' \\ &= \frac{1}{2\pi}\int_{-\pi}^{\pi} F(x)\,dx, \qquad \text{(average value)} \end{aligned} \tag{3.10}$$

Example 3.2 Consider the function,

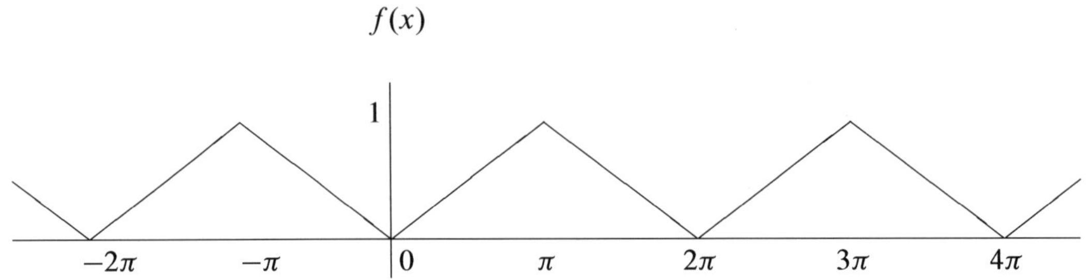

This function may be expressed as:

$$f(x) = \begin{cases} \left(\dfrac{1}{\pi}\right)x, & 0 < x < \pi, \\ 2 - \left(\dfrac{1}{\pi}\right)x, & \pi < x < 2\pi. \end{cases}$$

Since this is an even function, we may use a cosine expansion. Therefore.

$$f(x) = \sum_{n=0}^{\infty} b_n \cos nx. \tag{3.11}$$

The leading coefficient is given by

$$\begin{aligned} b_0 &= \frac{1}{\pi}\int_0^{\pi}\left(\frac{1}{\pi}\right)x\,dx \\ &= \frac{1}{\pi}\left(\frac{1}{\pi}\right)\tfrac{1}{2}x^2\Big|_0^{\pi} = \tfrac{1}{2}. \end{aligned}$$

The remaining coefficients are

$$\begin{aligned} b_n &= \frac{2}{\pi}\int_0^{\pi}\left(\frac{1}{\pi}\right)x\,\cos nx\,dx \\ &= \frac{2}{\pi}\left[\frac{x}{\pi}\frac{\sin nx}{n}\Big|_0^{\pi} - \frac{1}{\pi}\int_0^{\pi}\frac{\sin nx}{n}\,dx\right] \\ &= \frac{2}{\pi}\left[0 + \frac{\cos nx}{\pi n^2}\Big|_0^{\pi}\right] \\ &= \frac{2}{\pi}\left[\frac{(-1)^n - 1}{\pi n^2}\right] \\ &= -2\frac{[1-(-1)^n]}{n^2\pi^2} \end{aligned}$$

With these coefficients evaluated, $f(x)$ may be written as

$$f(x) = \tfrac{1}{2} - \frac{2}{\pi^2}\sum_{n=1}^{\infty}\frac{1-(-1)^n}{n^2}\cos nx.$$

Example 3.3 Next, let us consider an odd function.

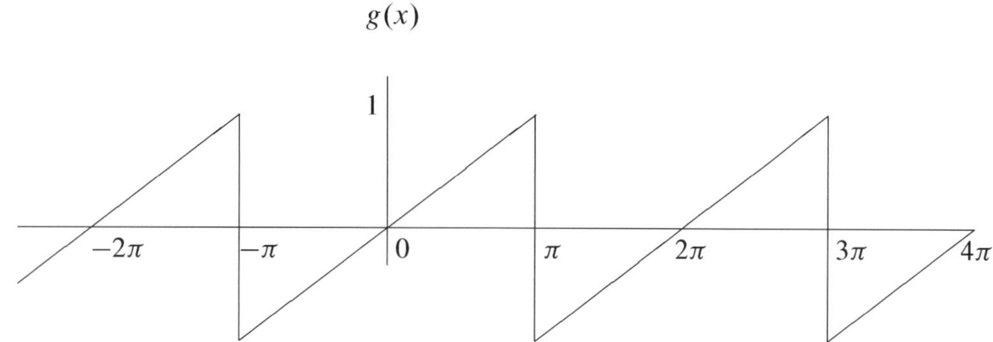

This function may be expressed as:

$$g(x) = \left(\frac{1}{\pi}\right)x, \qquad -\pi < x < \pi.$$

For an odd function such as this one, a sine expansion is appropriate. Therefore.

$$g(x) = \sum_{n=1}^{\infty} a_n \sin nx. \tag{3.12}$$

The coefficients are

$$\begin{aligned}
a_n &= \frac{2}{\pi}\int_0^{\pi}\left(\frac{1}{\pi}\right)x \,\sin nx\,dx \\
&= \frac{2}{\pi}\left[\frac{x}{\pi}\left(-\frac{\cos nx}{n}\right)\Big|_0^{\pi} - \frac{1}{\pi}\int_0^{\pi} -\frac{\cos nx}{n}\,dx\right] \\
&= \frac{2}{\pi}\left[-\frac{(-1)^n}{n} + \frac{\sin nx}{\pi n^2}\Big|_0^{\pi}\right] \\
&= \frac{2}{\pi}\left[\frac{(-1)^n}{n} + 0\right] \\
&= -\frac{2(-1)^n}{n\pi^2}
\end{aligned}$$

With these coefficients determined, $g(x)$ may be expressed as

$$g(x) = -2\sum_{n=1}^{\infty}\frac{(-1)^n}{n\pi}\sin nx.$$

Example 3.4 We now consider a function that is a combination of the functions in the previous two examples, i.e.,

$$F(x) = f(x) + \tfrac{1}{2}g(x)$$

Careful addition of the graphic representation of these two functions yields the following:

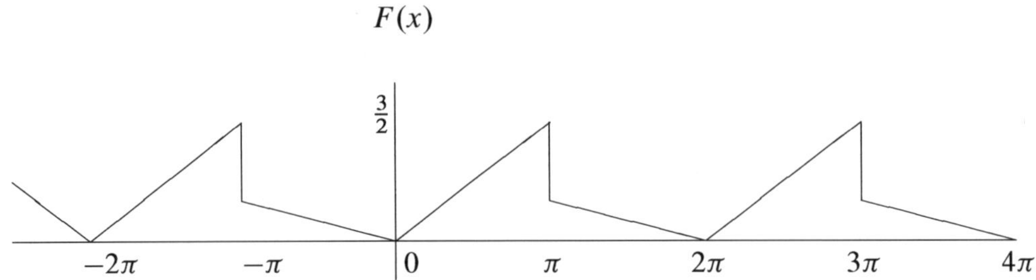

This example does not exhibit any symmetry characteristics and should, therefore, require a full sine and cosine expansion. Considering that the even and odd components have been expanded as cosine and sine series, respectively, both with the same period, the corresponding combination of the expansions will suffice. Therefore, we may write:

$$\begin{aligned} F(x) &= f(x) + \tfrac{1}{2}g(x) \\ &= \tfrac{1}{2} - \frac{2}{\pi^2}\sum_{n=1}^{\infty}\frac{1-(-1)^n}{n^2}\cos nx - \sum_{n=1}^{\infty}\frac{(-1)^n}{n\pi}\sin nx. \end{aligned} \tag{3.13}$$

The mathematical expression for this function is This function may be expressed as:

$$F(x) = \begin{cases} -\left(\dfrac{1}{2\pi}\right)x, & -\pi < x < 0, \\ \left(\dfrac{3}{2\pi}\right)x, & 0 < x < \pi. \end{cases}$$

A formal full Fourier series expansion of the type

$$F(x) = \sum_{n=1}^{\infty} a_n \sin nx + \sum_{n=0}^{\infty} b_n \cos nx,$$

may be carried out, and the coefficients are given by equations (3.8), (3.9), and (3.10). By carrying out the integration for the coefficients, the above expression (3.13) would be obtained.

3.4.4 Expansion for Arbitrary Period

So far we have carried out expansions with a period of 2π. This is generally applicable to circular geometry geometry. A function of the angular coordinate is unconditionally 2π-periodic, unless there is a cut in the circle.

The idea of periodicity can be readily extended to an arbitrary dimension. In general, we define the period as $2l$. The expansion procedures may be easily extended to the general case by stretching the variable x. Consider a function $f(x)$ that has a period of $2l$ in the variable x. We redefine the independent variable as:

$$x' = \frac{\pi x}{l}.$$

Therefore,

$$dx' = \left(\frac{\pi}{l}\right) dx.$$

Then, in the variable x', the function

$$f(x) = F(x') = f(x'l/\pi)$$

would have a period of 2π.

Let us now take the general case of a function $F(x')$ that has a period of 2π. Then, from the previously-developed procedures for calculating the coefficients, we shall obtain the expressions for the coefficients for a function that has a period of $2l$. The function $F(x')$ may be expressed as series of functions that have a period of 2π as follows:

$$F(x') = \sum_{n=1}^{\infty} a_n \sin nx' + \sum_{n=0}^{\infty} b_n \cos nx'.$$

In terms of $x = \pi x'/l$, we have

$$F(x') = f(x) = \sum_{n=1}^{\infty} a_n \sin\left(\frac{n\pi x}{l}\right) + \sum_{n=0}^{\infty} b_n \cos\left(\frac{n\pi x}{l}\right).$$

The coefficients remain unchanged, and are given by

$$\begin{aligned} a_n &= \frac{1}{\pi}\int_{x'=-\pi}^{x'=\pi} F(x')\sin(nx')\,dx' \\ &= \frac{1}{\pi}\int_{x=-l}^{x=l} f(x)\sin\left(\frac{n\pi x}{l}\right)\left(\frac{\pi}{l}\right)dx \\ &= \frac{1}{l}\int_{x=-l}^{x=l} f(x)\sin\left(\frac{n\pi x}{l}\right)\,dx \end{aligned}$$

Similarly,

$$\begin{aligned} b_n &= \frac{1}{\pi}\int_{x'=-\pi}^{x'=\pi} F(x')\cos(nx')\,dx' \\ &= \frac{1}{\pi}\int_{x=-l}^{x=l} f(x)\cos\left(\frac{n\pi x}{l}\right)\left(\frac{\pi}{l}\right)dx \\ &= \frac{1}{l}\int_{x=-l}^{x=l} f(x)\cos\left(\frac{n\pi x}{l}\right)\,dx, \end{aligned}$$

and

$$b_0 = \frac{1}{2\pi}\int_{x'=-\pi}^{x'=\pi} F(x')\,dx' = \frac{1}{2l}\int_{x=-l}^{x=l} f(x)\,dx.$$

EXERCISES 3.1

Expand each of the following functions, first as a Fourier sine series, and then as a cosine series. In each case define the half period as the region over which the information is given.

Infer from the periodicity and the (anti)symmetery the behavior of each series and make approximate plots (without numerical calculation of the series) over a couple of periods.

1.
$$f(x) = 2\ell x + x^2, \qquad 0 \le x \le \ell$$

2.
$$f(x) = x^2 + 1, \qquad 0 \le x \le 2.$$

3.
$$f(x) = \begin{cases} e^{-x}, & 0 \le x < 1 \\ e^{-1}, & 1 < x \le 2 \end{cases}$$

4.
$$f(x) = \begin{cases} \frac{1}{2}, & 0 \le x < 1 \\ 1 & 1 < x \le 2 \end{cases}$$

5.
$$f(x) = \begin{cases} x, & 0 \le x < a \\ \frac{1}{2}(a - x), & a < x \le 2a \end{cases}$$

6.
$$f(x) = \left[\pi^2 - (\pi - x)^2\right], \qquad 0 \le x \le \pi$$

7.
$$f(x) = \begin{cases} -x, & 0 \le x < 1 \\ x, & 1 < x \le \pi \end{cases}$$

8.
$$f(x) = 1 - x, \qquad 0 \le x \le 1.$$

9.
$$f(x) = 2x^2 + 1, \qquad 0 \le x \le 2.$$

10.
$$f(x) = \begin{cases} 2, & 0 \le x < 1 \\ x, & 1 < x \le 2 \end{cases}$$

EXERCISES 3.2

For Problems 1-5, expand the given functions as Fourier series:

1. Period $= 2L$

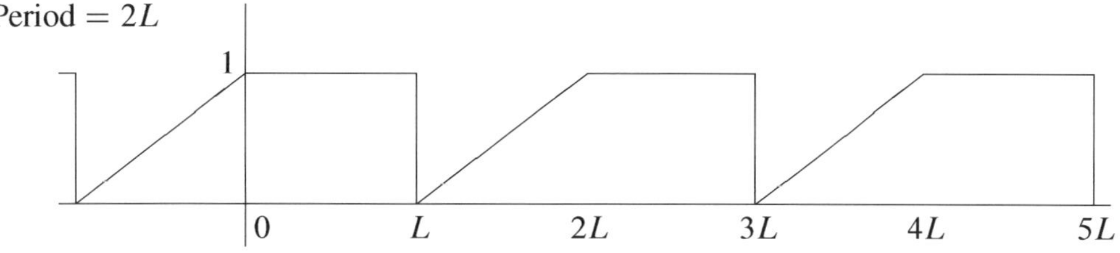

2. period $= 2L$

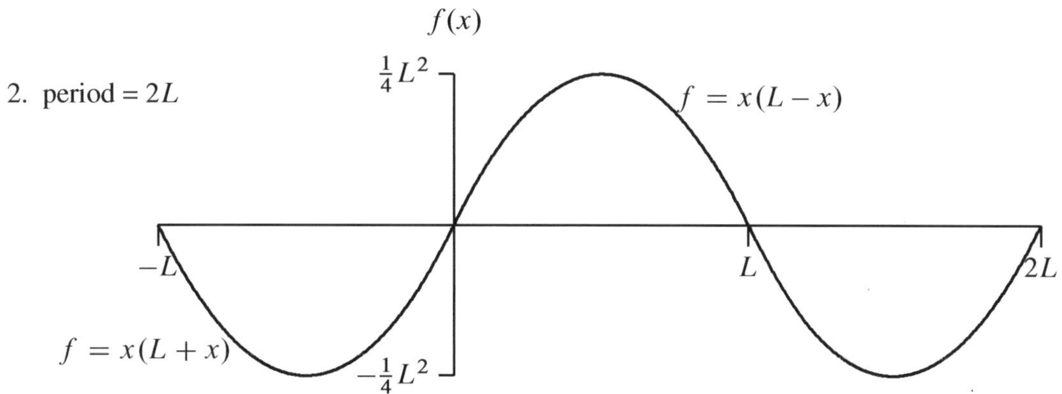

3. Period $= 2L$

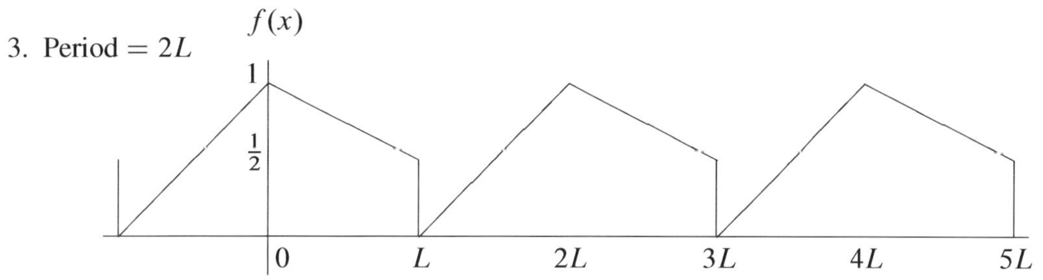

4. period $= 2L$

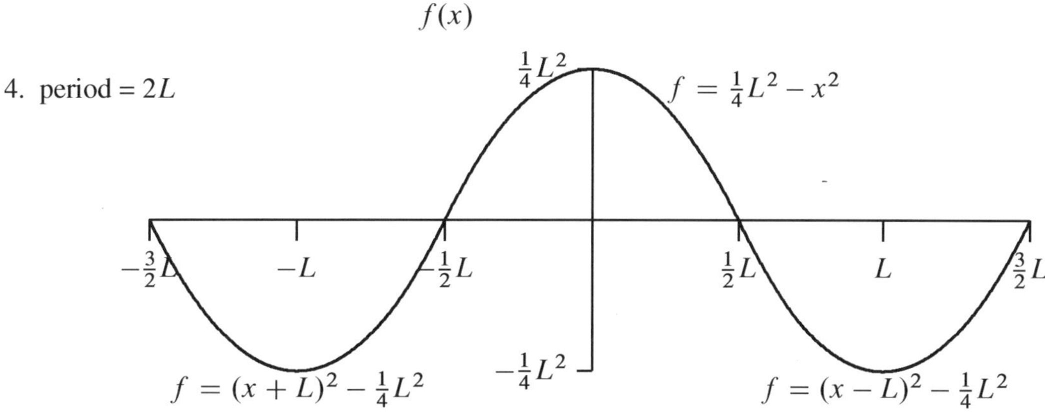

5. Period $= 2\ell$

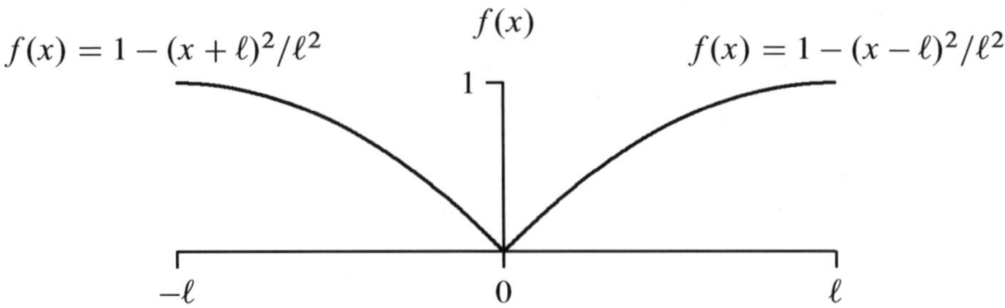

6. For the periodic function with period $2l$,

$$f(x) = \sum_{n=1}^{\infty} A_n \sin\left(\frac{n\pi x}{l}\right) + \sum_{n=0}^{\infty} B_n \cos\left(\frac{n\pi x}{l}\right),$$

show that

$$A_n = \frac{1}{l}\int_{-l+x_0}^{l+x_0} f(x)\sin\left(\frac{n\pi x}{l}\right)\,dx, \qquad n = 1, 2, 3, \ldots,$$

$$B_n = \frac{1}{l}\int_{-l+x_0}^{l+x_0} f(x)\cos\left(\frac{n\pi x}{l}\right)\,dx, \qquad n = 1, 2, 3, \ldots,$$

and

$$B_0 = \frac{1}{2l}\int_{-l+x_0}^{l+x_0} f(x)\,dx, \qquad n = 1, 2, 3, \ldots,$$

where x_0 is a arbitrary constant, and $f(x)$ is defined as a periodic function so that $f(x) = f(x + 2ml)$ where m is an integer.

3.5 Double Fourier Series

In situations when f is a function of two variables, the procedures developed earlier are easily applicable on each variable in turn. For example, if we have a function $f(x, y)$ that may be required to be expanded as a sine series in x over $0 < x < a$ and a sine series in y over $0 < y < b$, the we may express the function as follows:

$$f(x, y) = \sum_{m=1}^{\infty} F_m(y)\sin\left(\frac{m\pi x}{a}\right), \tag{3.14}$$

where the coefficient $F_m(y)$ is a function of y, and is given by

$$F_m(y) = \frac{2}{a}\int_0^a f(x, y)\sin\left(\frac{m\pi x}{a}\right)\,dx.$$

We may further expand $F_m(y)$ as a Fourier sine series in y over $0 < y < b$. Thus, we express

$$F_m(y) = \sum_{n=1}^{\infty} F_{mn} \sin\left(\frac{n\pi y}{b}\right), \tag{3.15}$$

where

$$\begin{aligned} F_{mn} &= \frac{2}{b}\int_0^b F_m(y) \sin\left(\frac{n\pi y}{b}\right)\, dy \\ &= \frac{2}{b}\int_0^b \left[\frac{2}{a}\int_0^a f(x,y)\sin\left(\frac{m\pi x}{a}\right)\, dx\right] \sin\left(\frac{n\pi y}{b}\right)\, dy \\ &= \frac{2}{a}\frac{2}{b}\int_0^a\int_0^b f(x,y)\sin\left(\frac{m\pi x}{a}\right)\sin\left(\frac{n\pi y}{b}\right)\, dy\, dx \end{aligned}$$

With these coefficients known, and by substituting the expression for $F_m(y)$ in equation (3.15) into 3.14, the function $f(x, y)$ may be expressed as

$$\begin{aligned} f(x,y) &= \sum_{m=1}^{\infty}\left[\sum_{n=1}^{\infty} F_{mn}\sin\left(\frac{n\pi y}{b}\right)\right]\sin\left(\frac{m\pi x}{a}\right) \\ &= \sum_{m=1}^{\infty}\sum_{n=1}^{\infty} F_{mn}\sin\left(\frac{n\pi y}{b}\right)\sin\left(\frac{m\pi x}{a}\right) \end{aligned}$$

Example 3.5 As an example, we expand the function $f(x, y)$ shown in the figure. It is required that the function should be expanded as a sine series in x and a full series in y. The periods need to be specified; for x the half period is l, and in y the full period is h.

First, we need a mathematical description of the function. In terms of x and y coordinates, we may express $f(x, y)$ as

$$f(x,y) = \begin{cases} 0, & 0 < x < \frac{1}{2}l, \quad 0 < y < h \\ H(y), & \frac{1}{2}l < x < l, \quad 0 < y < h \end{cases}, \tag{3.16}$$

where $H(y)$ is

$$H(y) = \begin{cases} 1, & 0 < y < \frac{1}{2}h \\ (2 - \frac{2}{h}y), & \frac{1}{2}h < y < h \end{cases} \tag{3.17}$$

We begin with a sine expansion in x so that

$$f(x,y) = \sum_{m=1}^{\infty} F_m(y)\sin\left(\frac{m\pi x}{l}\right).$$

The set of coefficients, $F_m(y)$ is given by

$$\begin{aligned} F_m(y) &= \frac{2}{l}\int_0^l f(x,y)\sin\left(\frac{m\pi x}{l}\right)\, dx \\ &= \frac{2}{l}\left\{\int_0^{\frac{1}{2}l}(0)\sin\left(\frac{m\pi x}{l}\right)\, dx + \int_{\frac{1}{2}l}^{l} H(y)\sin\left(\frac{m\pi x}{l}\right)\, dx\right\} \end{aligned}$$

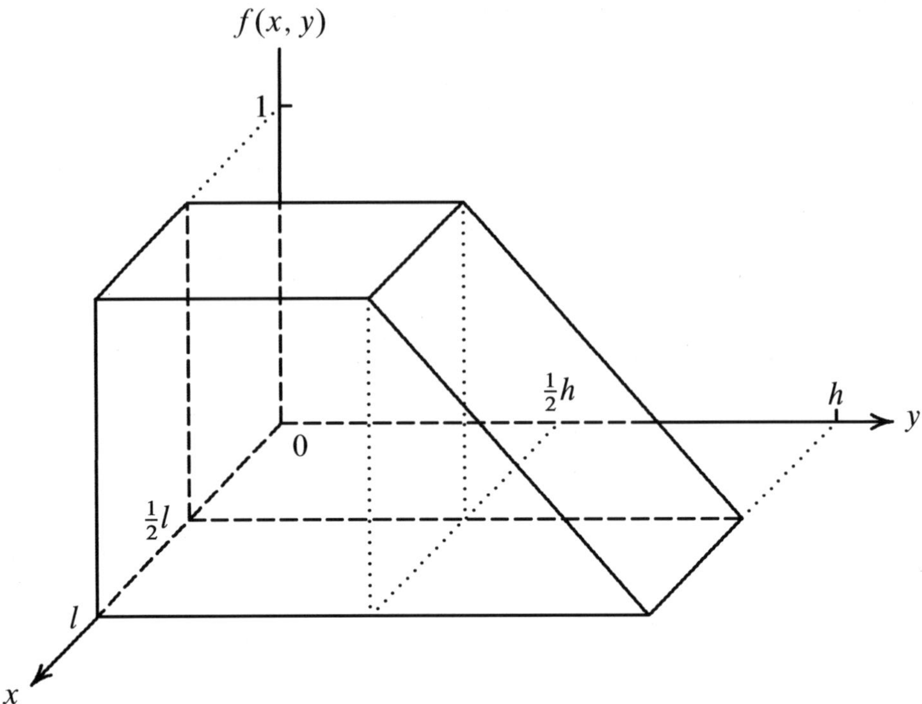

Figure 3.1: Graphic expression of a two-variable function defined by equations (3.16)-(3.17)

$$\begin{aligned}
&= \frac{2}{l}H(y)\left[-\frac{\cos\left(\frac{m\pi x}{l}\right)}{\frac{m\pi}{l}}\right]_{\frac{1}{2}l}^{l} \\
&= \frac{2}{l}H(y)\left[\frac{-(-1)^m+\cos(\frac{1}{2}m\pi)}{m\pi/l}\right] \\
F_m(y) &= -\frac{2}{m\pi}H(y)\left[(-1)^m-\cos\left(\tfrac{1}{2}m\pi\right)\right] \\
&= a_m H(y),
\end{aligned}$$

where

$$a_m = -\frac{2}{m\pi}\left[(-1)^m-\cos\left(\tfrac{1}{2}m\pi\right)\right]$$

At this point, the coefficient $F_m(y)$ can be expanded as a full series in y. The period has been defined as h, and therefore, the half period is $\frac{1}{2}h$. Thus, $F_m(y)$ may be expressed as:

$$F_m(y) = \sum_{n=1}^{\infty} F_{mn}\sin\left(\frac{n\pi y}{\frac{1}{2}h}\right) + G_{m0} + \sum_{n=1}^{\infty} G_{mn}\cos\left(\frac{n\pi y}{\frac{1}{2}h}\right).$$

The coefficients, F_{mn} and G_{mn} are given by

$$
\begin{aligned}
G_{m0} &= \frac{1}{h}\int_0^h a_m H(y)\,dy \\
&= a_m\left(\frac{1}{h}\right)\left[\int_0^{\frac{1}{2}h} 1\,dy + \int_{\frac{1}{2}h}^{h}\left(2-\frac{2}{h}y\right)dy\right] \\
&= \tfrac{3}{4}a_m. \\
G_{mn} &= \frac{1}{\frac{1}{2}h}\int_0^h a_m H(y)\cos\left(\frac{n\pi y}{\frac{1}{2}h}\right)dy \\
&= a_m\left(\frac{1}{h}\right)\left[\int_0^{\frac{1}{2}h} 1\cos\left(\frac{n\pi y}{\frac{1}{2}h}\right)dy + \int_{\frac{1}{2}h}^{h}\left(2-\frac{2}{h}y\right)\cos\left(\frac{n\pi y}{\frac{1}{2}h}\right)dy\right] \\
&= \frac{2a_m}{h}\left\{\frac{\sin(2n\pi y/h)}{2n\pi/h}\Bigg|_0^{\frac{1}{2}h} + \left(2-\frac{2y}{h}\right)\frac{\sin(2n\pi y/h)}{2n\pi/h}\Bigg|_{\frac{1}{2}h}^{h} - \left(\frac{2}{h}\right)\frac{\cos(2n\pi y/h)}{(2n\pi/h)^2}\Bigg|_{\frac{1}{2}h}^{h}\right\} \\
&= \frac{2a_m}{h}\left\{-\left(\frac{2}{h}\right)\frac{[1-(-1)^m]}{(2n\pi/h)^2}\right\} \\
G_{mn} &= \frac{2[1-(-1)^n]}{n^2\pi^2}\cdot\frac{\left[(-1)^m-\cos\left(\frac{1}{2}m\pi\right)\right]}{m\pi}, \qquad m\geq 1
\end{aligned}
$$

Similarly, the set of coefficients, F_{mn}, is found to be

$$
\begin{aligned}
F_{mn} &= \frac{1}{\frac{1}{2}h}\int_0^h a_m H(y)\sin\left(\frac{n\pi y}{\frac{1}{2}h}\right)dy \\
&= a_m\left(\frac{1}{h}\right)\left[\int_0^{\frac{1}{2}h} 1\sin\left(\frac{n\pi y}{\frac{1}{2}h}\right)dy + \int_{\frac{1}{2}h}^{h}\left(2-\frac{2}{h}y\right)\sin\left(\frac{n\pi y}{\frac{1}{2}h}\right)dy\right] \\
&= \frac{2a_m}{h}\left\{-\frac{\cos(2n\pi y/h)}{2n\pi/h}\Bigg|_0^{\frac{1}{2}h} - \left(2-\frac{2y}{h}\right)\frac{\cos(2n\pi y/h)}{2n\pi/h}\Bigg|_{\frac{1}{2}h}^{h} + \left(\frac{2}{h}\right)\frac{\sin(2n\pi y/h)}{(2n\pi/h)^2}\Bigg|_{\frac{1}{2}h}^{h}\right\} \\
&= \frac{2a_m}{h}\left(\frac{1}{2n\pi/h}\right) \\
F_{mn} &= -\frac{2\left[(-1)^m-\cos\left(\frac{1}{2}m\pi\right)\right]}{mn\pi^2}, \qquad m\geq 1
\end{aligned}
$$

$$
\begin{aligned}
f(x,y) &= \sum_{m=1}^{\infty} G_{m0}\sin\left(\frac{m\pi x}{l}\right) \\
&\quad + \sum_{n=1}^{\infty}\sum_{m=1}^{\infty}\left[G_{mn}\cos\left(\frac{n\pi y}{\frac{1}{2}h}\right) + F_{mn}\sin\left(\frac{n\pi y}{\frac{1}{2}h}\right)\right]\sin\left(\frac{m\pi x}{l}\right)
\end{aligned}
$$

□

The idea of a double series can be applied to various combinations of expansions in x and y, depending on what the respective situation calls for. The most general case would be full sine and cosine series in both x and y.

3.6 Expansions Over Infinite Domains

We now discuss the expansion procedures for functions for which the information is derived from the range $0 < x < \infty$ or $-\infty < x < \infty$. This may be a function that does not have periodic behavior.

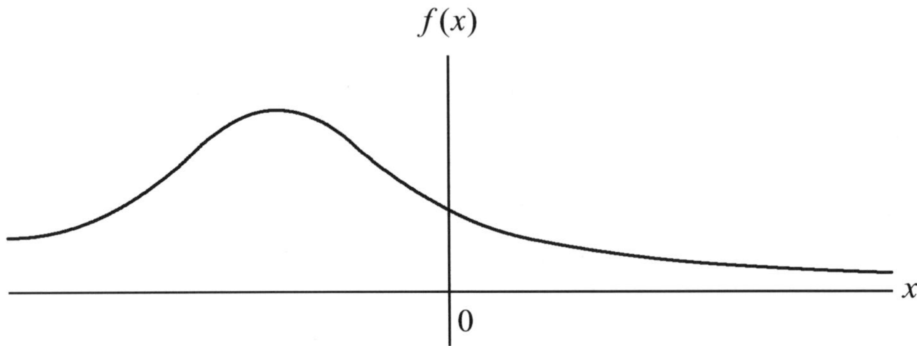

We consider an expansion with a half-period l, and then let $l \to \infty$. Let us take a sine expansion in x, i.e.,

$$g(x) = \sum_{n=0}^{\infty} a_n \sin\left(\frac{n\pi x}{l}\right)$$

where

$$a_n = \frac{2}{l}\int_0^l g(\xi) \sin\left(\frac{n\pi \xi}{l}\right) d\xi.$$

Let us define

$$\lambda_n = \frac{n\pi}{l},$$

so that

$$g(x) = \sum_{n=0}^{\infty} a_n \sin(\lambda_n x) \tag{3.18}$$

and

$$a_n = \frac{2}{l}\int_0^l g(\xi) \sin(\lambda_n \xi)\, d\xi. \tag{3.19}$$

Considering the sequence, $\{\lambda_n, n = 1, 2, 3, \ldots\}$, or

$$\begin{aligned}
\lambda_1 &= \frac{\pi}{l} \\
\lambda_2 &= \frac{2\pi}{l} \\
\lambda_3 &= \frac{3\pi}{l} \\
\lambda_4 &= \frac{4\pi}{l} \\
&\vdots
\end{aligned}$$

It is clear that the difference between any successive values of λ_n is π/l. We write

$$\Delta\lambda_n = \lambda_n - \lambda_{n-1} = \frac{\pi}{l}. \tag{3.20}$$

Now substituting equation (3.19) into (3.18), we obtain

$$g(x) = \sum_{n=0}^{\infty}\left[\frac{2}{l}\int_0^l g(\xi)\sin(\lambda_n\xi)\,d\xi\right]\sin(\lambda_n x).$$

Now, using equation (3.20), we may express the above as

$$\begin{aligned} g(x) &= \sum_{n=0}^{\infty}\left[\frac{2}{\pi}\Delta\lambda_n\int_0^l g(\xi)\sin(\lambda_n\xi)\,d\xi\right]\sin(\lambda_n x), \\ &= \sum_{n=0}^{\infty}\left[\frac{2}{\pi}\int_0^l g(\xi)\sin(\lambda_n\xi)\,d\xi\right]\sin(\lambda_n x)\,\Delta\lambda_n, \\ &= \sum_{n=0}^{\infty}[G(\lambda_n)]\sin(\lambda_n x)\,\Delta\lambda_n, \end{aligned} \tag{3.21}$$

where

$$G(\lambda_n) = \frac{2}{\pi}\int_0^l g(\xi)\sin(\lambda_n\xi)\,d\xi.$$

In the figure below, the function $[G(\lambda_n)\sin(\lambda_n x)]$, is schematically plotted on a histogram for different values of λ_n, $n = 1, 2, 3, \ldots$.

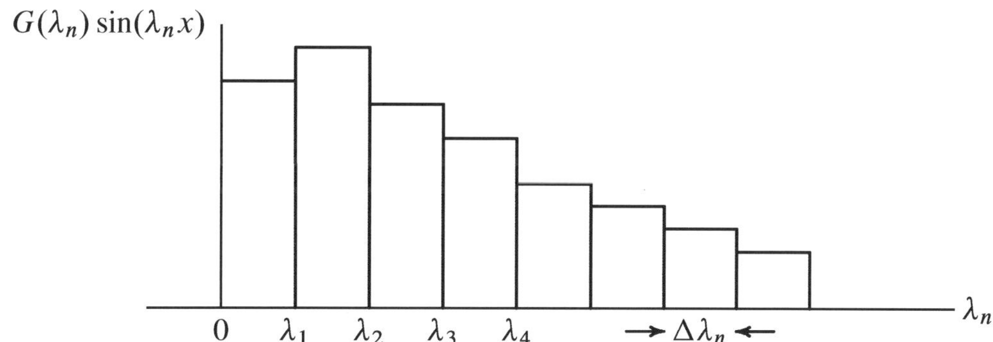

The sum given by $g(x)$ in equation (3.21) is the area under the histogram. In the limit of $l \to \infty$, the difference $\Delta\lambda_n$ between successive values of λ_n becomes infinitesimally small. Thus,

$$\Delta\lambda_n \to d\lambda.$$

In this limit, equation (3.21) becomes

$$g(x) = \int_0^\infty G(\lambda) \sin \lambda x \, d\lambda,$$

where

$$G(\lambda) = \frac{2}{\pi} \int_0^{l \to \infty} g(\xi) \sin(\lambda \xi) \, d\xi = \frac{2}{\pi} \int_0^\infty g(\xi) \sin(\lambda \xi) \, d\xi. \tag{3.22}$$

The function $G(\lambda)$ is known as the Fourier sine transform of $g(x)$. The information about $g(x)$ over $0 < x < \infty$ may be recovered by the inverse transform,

$$g(x) = \int_0^\infty G(\lambda) \sin(\lambda x) \, d\lambda. \tag{3.23}$$

Similarly, we may obtain a transform and inverse pair for the cosine integral. That is,

$$\begin{aligned} f(x) &= \int_0^\infty F(\lambda) \cos(\lambda x) \, d\lambda, \\ F(\lambda) &= \frac{2}{\pi} \int_0^\infty f(\xi) \cos(\lambda \xi) \, d\xi. \end{aligned}$$

Example 3.6 Let us consider the function,

$$g(x) = \mathrm{e}^{-x}, \qquad 0 < x < \infty,$$

and express it as a sine integral,

$$g(x) = \int_0^\infty G(\lambda) \sin \lambda x \, d\lambda.$$

The transform $G(\lambda)$ is given by

$$\begin{aligned} G(\lambda) &= \frac{2}{\pi} \int_0^\infty g(x) \sin(\lambda x) \, dx \\ &= \frac{2}{\pi} \int_0^\infty \mathrm{e}^{-x} \sin(\lambda x) \, dx \\ &= \frac{2}{\pi} \left(\frac{\lambda}{1 + \lambda^2} \right) \end{aligned}$$

The result is

$$g(x) = \frac{2}{\pi} \int_0^\infty \left(\frac{\lambda}{1 + \lambda^2} \right) \sin \lambda x \, d\lambda.$$

A plot of this integral expansion would give us e^{-x} over the range $0 < x < \infty$. The projection of the expansion into the negative x side will not give the original function e^{-x} because the information from that side was never taken. Instead, the antisymmetry of the sine function takes over, as shown here.

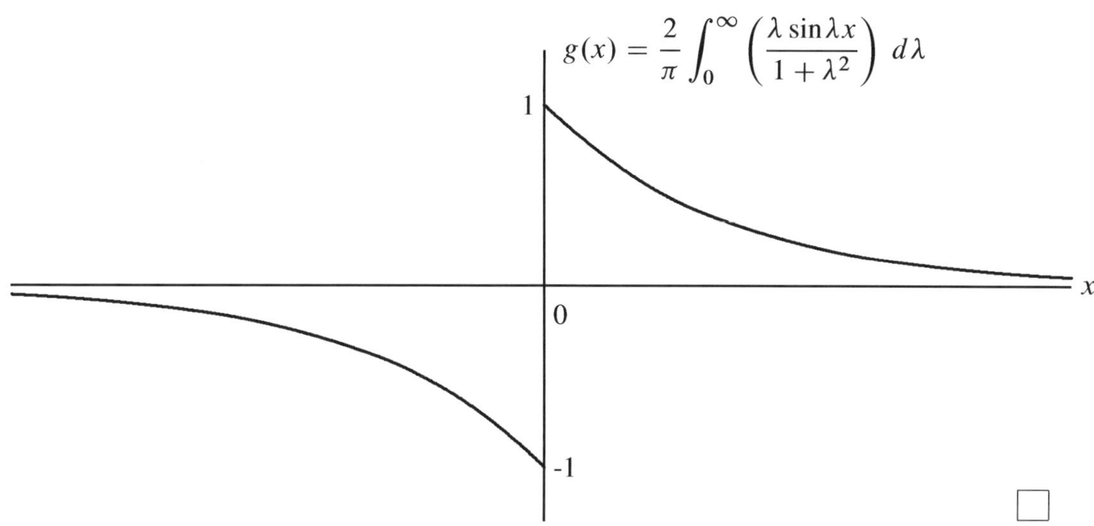

□

Example 3.7 Let us consider the same function as in the previous example and express it as a cosine integral,

$$f(x) = \mathrm{e}^{-x} = \int_0^\infty F(\lambda)\cos\lambda x \, d\lambda, \quad (0 < x < \infty)$$

The Fourier cosine transform $F(\lambda)$ is given by

$$\begin{aligned} F(\lambda) &= \frac{2}{\pi}\int_0^\infty f(x)\cos(\lambda x)\, dx \\ &= \frac{2}{\pi}\int_0^\infty \mathrm{e}^{-x}\cos(\lambda x)\, dx \\ &= \frac{2}{\pi}\left(\frac{1}{1+\lambda^2}\right) \end{aligned}$$

The result is

$$f(x) = \frac{2}{\pi}\int_0^\infty \left(\frac{1}{1+\lambda^2}\right)\cos\lambda x \, d\lambda.$$

Again, a plot of this integral expansion would give us e^{-x} over the range $0 < x < \infty$ but on the negative side, symmetry takes over, as shown here.

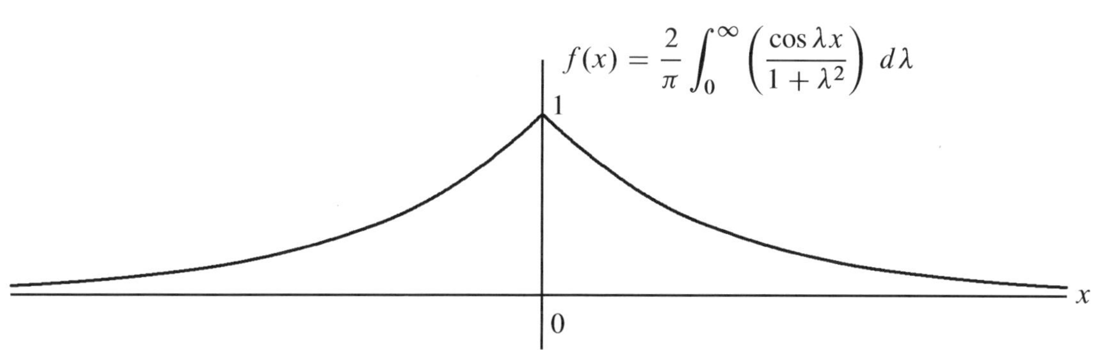

□

3.6.1 Non-Symmetric Expansions Over $-\infty < x < \infty$

As discussed earlier, a function may be decomposed into even and odd components. We now consider a function $f(x)$ described over the region, $-\infty < x < \infty$. We decompose this as

$$f(x) = f_{even}(x) + f_{odd}(x).$$

We may expand $f_{even}(x)$ as a cosine expansion,

$$f_{even}(x) = \int_0^\infty F(\lambda)\cos\lambda x \, d\lambda,$$

and $f_{odd}(x)$ may be expanded as a sine integral,

$$f_{odd}(x) = \int_0^\infty G(\lambda)\sin\lambda x \, d\lambda.$$

The transform $F(\lambda)$ is given by

$$\begin{aligned} F(\lambda) &= \frac{2}{\pi}\int_0^\infty f_{even}(x)\cos\lambda x \, dx && (3.24) \\ &= \frac{1}{\pi}\int_{-\infty}^\infty f_{even}(x)\cos\lambda x \, dx \\ &= \frac{1}{\pi}\int_{-\infty}^\infty [f_{even}(x) + f_{odd}(x)]\cos\lambda x \, dx \\ &= \frac{1}{\pi}\int_{-\infty}^\infty f(x)\cos\lambda x \, dx && (3.25) \end{aligned}$$

Similarly,

$$\begin{aligned} G(\lambda) &= \frac{2}{\pi}\int_0^\infty f_{odd}(x)\sin\lambda x \, dx \\ &= \frac{1}{\pi}\int_{-\infty}^\infty f_{odd}(x)\sin\lambda x \, dx \\ &= \frac{1}{\pi}\int_{-\infty}^\infty [f_{even}(x) + f_{odd}(x)]\sin\lambda x \, dx \\ &= \frac{1}{\pi}\int_{-\infty}^\infty f(x)\sin\lambda x \, dx && (3.26) \end{aligned}$$

The function $f(x)$ may be expressed as a complete integral expansion in the form

$$f(x) = \int_0^\infty F(\lambda)\cos\lambda x \, d\lambda + \int_0^\infty G(\lambda)\sin\lambda x \, d\lambda. \quad (3.27)$$

Example 3.8 Let us consider the function described in the figure below.

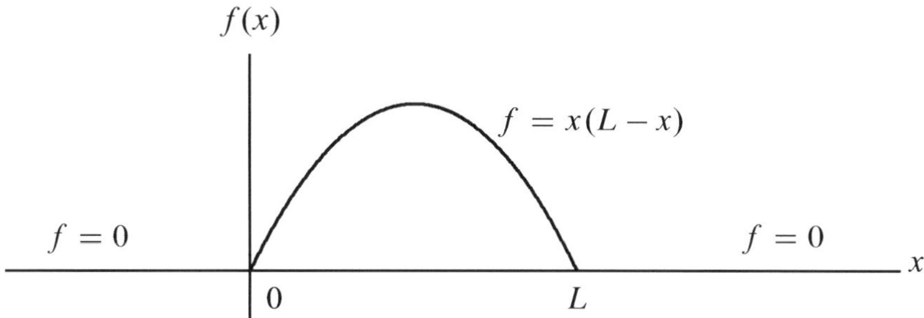

In mathematical notation, the function may be expressed as

$$f(x) = \begin{cases} 0, & -\infty < x < 0, \\ x(L-x), & 0 < x < l, \\ 0, & L < x < \infty. \end{cases}$$

For the full expansion given in equation (3.27), the coefficient $F(\lambda)$ is given by equation (3.25). For this example,

$$\begin{aligned}
F(\lambda) &= \frac{1}{\pi}\left[\int_{-\infty}^{0} 0 \ \cos\lambda x \, dx + \int_0^L x(L-x)\cos\lambda x \, dx + \int_L^{\infty} 0 \ \cos\lambda x \, dx\right] \\
&= \frac{1}{\pi}\int_0^L x(L-x)\cos\lambda x \, dx \\
&= \frac{1}{\pi}\left[x(L-x)\frac{\sin\lambda x}{\lambda}\bigg|_0^L - \int_0^L (L-2x)\frac{\sin\lambda x}{\lambda}\, dx\right] \\
&= \frac{1}{\pi}\left[0 + (L-2x)\frac{\cos\lambda x}{\lambda^2}\bigg|_0^L + 2\frac{\sin\lambda x}{\lambda^3}\bigg|_0^L\right] \\
&= \frac{1}{\pi}\left[-\frac{L(\cos\lambda L + 1)}{\lambda^2} + \frac{2\sin\lambda L}{\lambda^3}\right].
\end{aligned}$$

The coefficient $G(\lambda)$ for the sine part of the expansion may be obtained from equation (3.26). In the present case,

$$\begin{aligned}
G(\lambda) &= \frac{1}{\pi}\left[\int_{-\infty}^{0} 0 \ \sin\lambda x \, dx + \int_0^L x(L-x)\sin\lambda x \, dx + \int_0^{\infty} 0 \ \sin\lambda x \, dx\right] \\
&= \frac{1}{\pi}\int_0^L x(L-x)\sin\lambda x \, dx \\
&= \frac{1}{\pi}\left[x(L-x)\left(-\frac{\cos\lambda x}{\lambda}\right)\bigg|_0^L + \int_0^L (L-2x)\frac{\cos\lambda x}{\lambda}\, dx\right] \\
&= \frac{1}{\pi}\left[0 + (L-2x)\frac{\sin\lambda x}{\lambda^2}\bigg|_0^L - 2\frac{\cos\lambda x}{\lambda^3}\bigg|_0^L\right]
\end{aligned}$$

$$= -\frac{1}{\pi}\left[\frac{L\sin\lambda L}{\lambda^2}+\frac{2(\cos\lambda L-1)}{\lambda^3}\right].$$

The complete expansion is given by

$$f(x) = \frac{1}{\pi}\int_0^\infty\left\{\left[-\frac{L(\cos\lambda L+1)}{\lambda^2}+\frac{2\sin\lambda L}{\lambda^3}\right]\cos\lambda x\right. \tag{3.28}$$

$$\left.-\left[\frac{L\sin\lambda L}{\lambda^2}+\frac{2(\cos\lambda L-1)}{\lambda^3}\right]\sin\lambda x\right\}\,d\lambda. \tag{3.29}$$

□

Integral Expansions of Functions with a Non-Zero Limit

In order to express a constant function such as $f(x) = 1$ as a Fourier sine integral, we would encounter difficulties with integration. Consider,

$$1 = \int_0^\infty A(\lambda)\sin\lambda x\,d\lambda \tag{3.30}$$

If we follow the procedure established in equations (3.22)-(3.23), we may be tempted to try

$$A(\lambda) = \frac{2}{\pi}\int_0^\infty 1\sin\lambda x dx = \frac{2}{\pi}\left[\frac{-\cos\lambda x}{\lambda}\right]_0^\infty \tag{3.31}$$

Here, we run into the difficulty that upper limit is undefined. The condition we generally use to guarantee the existence of a Fourier transform is that

$$\int_0^\infty |f(x)|\,dx$$

is finite. This is a sufficient condition, though not necessary. If this is satisfied, then,

$$F(\lambda) = \int_0^\infty f(x)\sin\lambda x dx \tag{3.32}$$

exits. However, for $f(x) = 1$, this is clearly not the case, and we use the integral identity,

$$\int_0^\infty \frac{\sin x^*}{x^*}dx^* = \frac{\pi}{2} \tag{3.33}$$

Now, let $x^* = \lambda x$ with λ as the variable, resulting in $dx^* = xd\lambda$. Equation (3.33) becomes

$$\frac{\pi}{2} = \int_{\lambda x=0}^{\lambda x=\infty}\frac{\sin\lambda x}{\lambda}d\lambda$$

$$= \int_0^\infty\frac{\sin\lambda x}{\lambda}d\lambda,$$

and it is not difficult to see that

$$1 = \frac{2}{\pi} \int_0^\infty \frac{\sin \lambda x}{\lambda} d\lambda \tag{3.34}$$

What we want is

$$\int_0^\infty A(\lambda) \sin \lambda x d\lambda = 1 \tag{3.35}$$

By comparisn of (3.34) with (3.35), i.e.,

$$\int_0^\infty A(\lambda) \sin \lambda x d\lambda = 1 = \int_0^\infty \frac{2}{\pi} \frac{1}{\lambda} \sin \lambda x d\lambda, \tag{3.36}$$

we arrive at

$$A(\lambda) = \frac{2}{\lambda \pi}. \tag{3.37}$$

This leads us to other cases in which $\int_0^\infty \mid f(x) \mid dx$ is not finite, e.g.,

$$f(x) \to f_\infty \qquad \text{as} \quad x \to \infty \tag{3.38}$$

In order to deal with this situation, we may express $f(x)$ as

$$f(x) = f_\infty + [f(x) - f_\infty] = f_\infty + f^*(x), \tag{3.39}$$

where $f^*(x) = f(x) - f_\infty$. Now, we may construct the transform of $f^*(x)$ as

$$F^*(\lambda) = \frac{2}{\pi} \int_0^\infty f^*(x)(\lambda) \sin \lambda x dx,$$

so that

$$f^*(x) = \int_0^\infty F^*(\lambda) \sin \lambda x \, d\lambda. \tag{3.40}$$

The limit value f_∞ may also be expressed as a Fourier sine integral using equation (3.34) in the form

$$f_\infty = \frac{2 f_\infty}{\pi} \int_0^\infty \frac{1}{\lambda} \sin \lambda x d\lambda \tag{3.41}$$

Adding equations (3.40) and (3.41), and keeping in mind equation (3.39), we end up with

$$\begin{aligned} f(x) = f_\infty + f^*(x) &= \frac{2 f_\infty}{\pi} \int_0^\infty \frac{1}{\lambda} \sin \lambda x \, d\lambda + \int_0^\infty F^*(\lambda) \sin \lambda x \, d\lambda \\ &= \int_0^\infty \left[\frac{2 f_\infty}{\lambda \pi} + F^*(\lambda) \right] \sin \lambda x \, d\lambda \end{aligned} \tag{3.42}$$

Alternate Forms of Integral Representation

The integral representation (3.27) can be expressed in another standard form as follows:

$$f(x) = \int_{-\infty}^{\infty} H(\lambda)e^{i\lambda x}\, d\lambda, \tag{3.43}$$

where $H(\lambda)$ is the Fourier transform of $f(x)$, given by

$$H(\lambda) = \frac{1}{2\pi}\int_{-\infty}^{\infty} f(x)e^{-i\lambda x}\, dx, \tag{3.44}$$

[See Exercises 3.3, Problem 10]

3.6.2 Combined Series and Integral Expansions

We have discussed double Fourier series expansions for functions of two variables in section 3.5 (pages 90-93). In particular we have treated finite-range expansions in both x and y which are the independent variables for a function $f(x, y)$. The procedure can be applied to infinite range expansions in either or both variables.

Let us say that a function $f(x, y)$ needs to be expanded as a sine series in x with half-period l and a complete sine and cosine integral expansion in y. We can choose to begin with either variable. Let us begin with a series expansion in x. Thus,

$$f(x, y) = \sum_{n=1}^{\infty} A_n(y)\sin\left(\frac{n\pi x}{l}\right),$$

where

$$A_n(y) = \frac{2}{l}\int_0^l f(x, y)\sin\left(\frac{n\pi x}{l}\right)\, dx.$$

Next, we expand the coefficient $A_n(y)$ as a full Fourier integral in the form

$$A_n(y) = \int_0^{\infty} F_n(\lambda)\cos\lambda y\, d\lambda + \int_0^{\infty} G_n(\lambda)\sin\lambda y\, d\lambda,$$

where

$$F_n(\lambda) = \frac{1}{\pi}\int_{-\infty}^{\infty} A_n(y)\cos\lambda y\, dy,$$

and

$$G_n(\lambda) = \frac{1}{\pi}\int_{-\infty}^{\infty} A_n(y)\sin\lambda y\, dy.$$

The complete representation of the function $f(x, y)$ is

$$f(x, y) = \sum_{n=1}^{\infty}\left\{\int_0^{\infty} [F_n(\lambda)\cos\lambda y + G_n(\lambda)\sin\lambda y]\, d\lambda\right\}\sin\left(\frac{n\pi x}{l}\right).$$

EXERCISES 3.3

Obtain the full Fourier integral representation for

1.

$$f(x) = \begin{cases} |x|, & -\pi < x < \pi \\ 0, & -\infty < x < -\pi, \quad \pi < x < \infty \end{cases}$$

2.

$$f(x) = \begin{cases} x, & -1 < x < 1 \\ 2 - x, & 1 < x < 2, \\ -2 - x, & -2 < x < -1, \\ 0, & 2 < x < \infty, \quad -\infty < x < -2 \end{cases}$$

3.

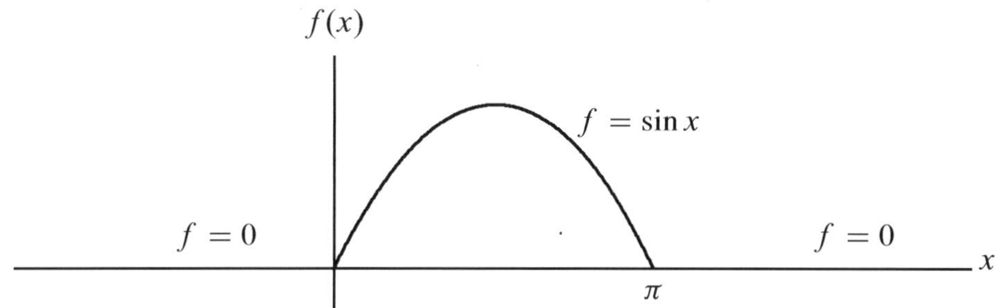

4. Expand as a Fourier sine integral taking the information over the range $0 < x < \infty$:

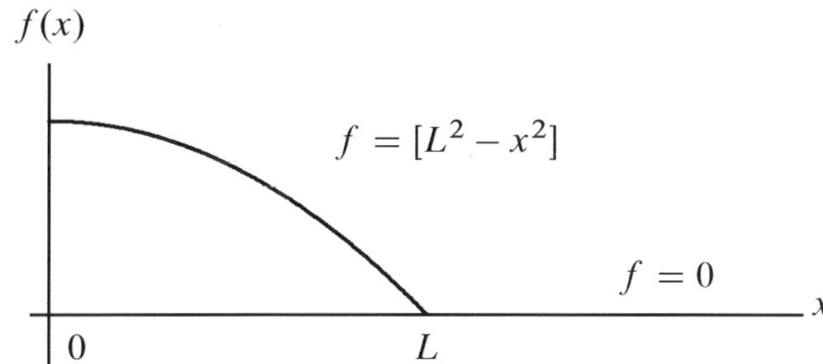

What function does the expansion represent in the range $-\infty < x < \infty$? Show by plotting approximately.

5. Expand as a Fourier integral valid over $-\infty < x < \infty$

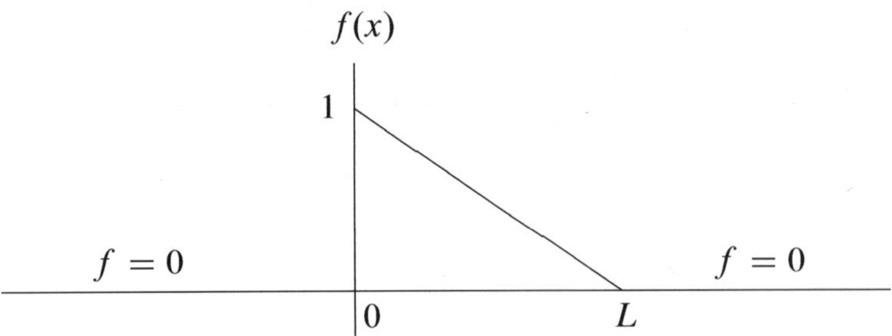

6. Expand as a double sine series to be valid over the range given:

$$f(x, y) = x^2 y, \qquad (0 < x < \pi \quad 0 < y < \pi).$$

[Take the period as 2π for each variable.]

7. Expand as a double series — sine series in x and cosine series in y.

$$f(x, y) = x(L - y), \qquad (0 < x < L \quad 0 < y < L).$$

[Take the period as $2L$ for each variable.]

8. Expand as a full series (period $2L$) in x and a full integral in y:

$$f(x, y) = |x| e^{-|y|}, \qquad (-L < x < L \quad -\infty < y < \infty).$$

9. Expand as a double sine series for $0 < x < L, \quad 0 < y < L$.

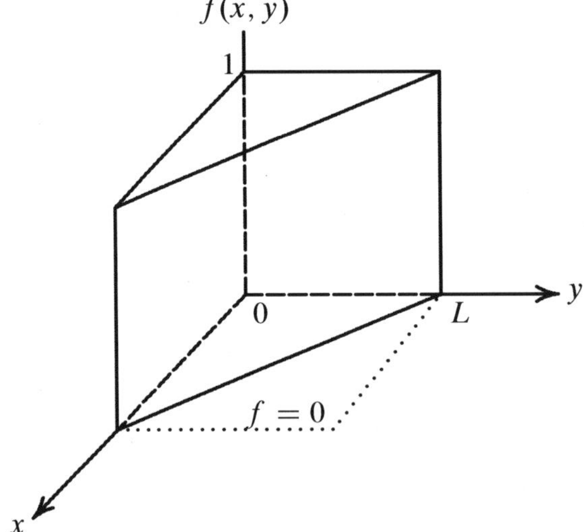

[Hint: In the double integral for the coefficients, use an upper limit on the inner integral to be a function of the outer variable. That is, the upper limit should define the line separating the zero and the nonzero regions of $f(x, y)$.]

10. Starting out with equation (3.27), together with (3.25) and (3.26), prove the relationship given by equations (3.43) and (3.44). Use the symmetry and anti-symmetry properties of $F(\lambda)$ and $G(\lambda)$ in terms of λ.

3.7 Generalization of the Fourier Expansion

In developing expansions in terms of orthogonal functions, we have considered here strictly periodic functions within the framework of sine and cosine series. However, from a broader perspective, there is a large class of orthogonal functions which are very useful in solving partial differential equations. While this application is discussed in Chapters 7-10, we focus here on the fundamentals of orthogonal functions. For this purpose it will be helpful to refer to an orthogonal set of functions as a general set $\{X_n(x)\}$ satisfying certain homogeneous end-point conditions (or boundary conditions). For example, when

$$X_n(x) = \sin\left(\frac{n\pi x}{a}\right), \quad n = 1, 2, 3, \ldots$$

we have

$$X_n(0) = X_n(a) = 0, \tag{3.45}$$

as conditions at the end points, $x = 0$ and $x = a$ in the expansion range $0 \leq x \leq a$. Similarly, the set of functions

$$X_n(x) = \cos\left(\frac{n\pi x}{a}\right), \quad n = 0, 1, 2, 3, \ldots$$

satisfies end-point homogeneous conditions

$$\frac{dX_n(0)}{dx} = \frac{dX_n(a)}{dx} = 0. \tag{3.46}$$

At the start of this chapter, we chose sine and cosine functions simply because of their periodicity, and established orthogonality for that limited class. We have now reached a point where we need to recognize that in the broader context, these functions belong to certain differential equations as their solutions. Both the functions considered here are solutions of the set of differential equations

$$\frac{d^2X_n(x)}{dx^2} + \left(\frac{n\pi}{a}\right)^2 X_n(x) = 0, \quad n = 0, 1, 2, 3, \ldots,$$

where $n = 0$ may not be relevant for the sine functions. The term multiplying $X_n(x)$ here is simply $(n\pi/a)^2$ but in general it may be more complicated, and even be a solution to some transcendental equation. To maintain generality, we shall refer to it as λ_n^2, and if we re-arrange the equation a little, we can write

$$\frac{d^2X_n(x)}{dx^2} = -\lambda_n^2 X_n(x). \tag{3.47}$$

With this arrangement, we can see that the second-derivative operation (d^2/dx^2) on $X_n(x)$ gives us back a function proportional to $X_n(x)$. We can therefore refer to $X_n(x)$ as an eigenfunction of (d^2/dx^2), and $-\lambda_n^2$ as the corresponding eigenvalue. From this vewpoint, we can see clearly that both $\sin\lambda_n x$ and $\cos\lambda_n x$ are eigenfunctions of (d^2/dx^2). The values of λ_n are generally determined by the boundary conditions such as (3.45) and (3.46). It is, of course, understood that these boundary conditions are among the simplest ones, and we will examine a wider class of boundary conditions that permit orthogonality of the eigenfunctions. For that matter, we will examine a second order linear differential equation in the most general terms. Let us first begin with an example in which we examine equation (3.47) but consider a slightly different combination of boundary conditions than (3.45) and (3.46).

Example 3.9 *Problem Statement:*

1. Obtain the eigenfunctions of the operator (d^2/dx^2), i.e., obtain expressions for a set of functions $X_n(x)$ such that equation (3.47) is satisfied.
2. Then determine the eigenvalues such that the pair of boundary conditions be satisfied
$$\left.\frac{dX_n(x)}{dx}\right|_{x=0} = 0 \qquad \text{and} \qquad X_n(x)|_{x=a} = 0 \tag{3.48}$$
3. Show that the resulting eigenfunctions for the specific set of eigenvelues are mutually orthogonal. Use this result to express an arbitrary function $f(x)$ in terms of this set of eigenfunctions, i.e.,
$$f(x) = \sum_{n=0}^{\infty} a_n X_n(x) \tag{3.49}$$

Part 1: Based the techniques for ordinary differential equations with constant coefficients in Chapter 1 (page 6), we obtain the following general solution for equation (3.47):
$$X_n(x) = A_n \sin\lambda_n x + B_n \cos\lambda_n x, \tag{3.50}$$

Part 2: Upon applying the first of the boundary conditions (3.48), we obtain
$$\left.\frac{dX_n(x)}{dx}\right|_{x=0} = A_n\lambda \cos\lambda_n 0 - B_n\lambda \sin\lambda_n 0 = 0,$$
leading to $A_n = 0$, and reducing $X_n(x)$ to
$$X_n(x) = B_n \cos\lambda_n x.$$
Now, applying the second of the boundary conditions (3.48) give
$$X_n(a) = B_n \cos\lambda_n a = 0.$$
This would lead to either $B_n = 0$ or $\cos\lambda_n a = 0$. The former is a trivial solution leading to $X_n(x) = 0$. Therefore, to obtain anything meaningful, we need to work with the latter. We realize that the cosine function is zero when the argument is from the set
$$\left\{\tfrac{1}{2}\pi, \tfrac{3}{2}\pi, \tfrac{5}{2}\pi, \ldots \left(n+\tfrac{1}{2}\right)\pi, \ldots\right\}.$$

That is, the selection of

$$\lambda = \left(n + \tfrac{1}{2}\right)\pi, \quad n = 0, 1, 2, 3, \ldots$$

will satisfy the second of the boundary conditions (3.48). There is an infinite number of these values, and we can label these individually as

$$\lambda_0 = \tfrac{1}{2}\left(\frac{\pi x}{a}\right), \lambda_1 = \tfrac{3}{2}\left(\frac{\pi x}{a}\right), \lambda_2 = \tfrac{5}{2}\left(\frac{\pi x}{a}\right), \ldots \quad \text{or} \quad \lambda_n = \left(n + \tfrac{1}{2}\right)\pi, \quad n = 0, 1, 2, 3, \ldots$$

The solution to equation (3.47) with boundary conditions (3.48) is

$$X_n(x) = B_n \cos \lambda_n x,$$

with λ_n given above. We should note at this point that for every different value of λ_n, (3.47) is a slightly different equation. It is therefore prudent that we maintain proper indices for each equation corresponding to the respactive λ_n, and include the values of n for completeness, i.e., we write (3.47) as

$$\frac{d^2 X_n(x)}{dx^2} + \lambda_n^2 X_n(x) = 0, \quad n = 0, 1, 2, 3, \ldots, \tag{3.51}$$

and the solution as

$$X_n(x) = B_n \cos \lambda_n x = B_n \cos\left[\left(n + \tfrac{1}{2}\right)\frac{\pi x}{a}\right], \quad n = 0, 1, 2, 3, \ldots.$$

At this point the coefficient B_n can be set equal to unity without loss of generality for demosntrating orthogonality or series expansion. Thus,

$$X_n(x) = \cos \lambda_n x = \cos\left[\left(n + \tfrac{1}{2}\right)\frac{\pi x}{a}\right], \quad n = 0, 1, 2, 3, \ldots. \tag{3.52}$$

Part 3: Orthogonality and Series Expansion

To demonstrate orthogonality, let us consider two arbitrary eigenfunctions from the set given by equation (3.52). Let us take

$$X_m(x) = \cos \lambda_m x = \cos\left[\left(m + \tfrac{1}{2}\right)\frac{\pi x}{a}\right] \quad \text{and} \quad X_n(x) = \cos \lambda_n x = \cos\left[\left(n + \tfrac{1}{2}\right)\frac{\pi x}{a}\right].$$

Now, let us evaluate the integral over the range $0 \le x \le a$

$$\begin{aligned}
I_{mn} &= \int_0^a \cos \lambda_m x \cos \lambda_n x \, dx \\
&= \int_0^a \tfrac{1}{2}\left[\cos(\lambda_m - \lambda_n)x + \cos(\lambda_m + \lambda_n)x\right] dx \\
&= \tfrac{1}{2}\left[\frac{\sin(\lambda_m - \lambda_n)x}{\lambda_m - \lambda_n} + \frac{\sin(\lambda_m + \lambda_n)x}{\lambda_m + \lambda_n}\right]_0^a \\
&= \tfrac{1}{2}\left[\frac{\sin(m-n)\pi - 0}{(m-n)\frac{\pi}{a}} + \frac{\sin(m+n+1)\pi - 0}{(m+n+1)\frac{\pi}{a}}\right] \\
&= \tfrac{1}{2}a\left[\frac{\sin(m-n)\pi}{(m-n)\pi}\right] \\
&= 0, \quad \text{if} \quad m \neq n.
\end{aligned}$$

This establishes orthogonality of the set under discussion. We can evaluate the integral for the case $m = n$.

$$\begin{aligned} I_{nn} &= \int_0^a \cos\lambda_n^2 x\, dx \\ &= \int_0^a \tfrac{1}{2}\left[1+\cos\left(2\lambda_n x\right)\right]\, dx \\ &= \tfrac{1}{2}\left[x + \frac{\sin\left(2\lambda_n x\right)}{2\lambda_n}\right]_0^a \\ &= \tfrac{1}{2}\left[a - 0 + \frac{\sin\left(2\lambda_n a\right) - 0}{2\lambda_n}\right] \\ &= \tfrac{1}{2}\left[a + \frac{\sin\left(2n+1\right)\pi}{(2n+1)\pi/a}\right] \\ &= \tfrac{1}{2}a \end{aligned}$$

To express an arbitrary function $f(x)$ as a Fourier series of the form (3.49) for eigenfunctions $X_n(x) = \cos\lambda_n x$, we may write

$$f(x) = \sum_{n=0}^{\infty} b_n X_n(x) = \sum_{n=0}^{\infty} b_n \cos\lambda_n x, \qquad \text{with} \quad \lambda_n = \left(n+\tfrac{1}{2}\right)\frac{\pi}{a}. \tag{3.53}$$

Let us multiply each side of equation (3.53) by $\cos\lambda_m x$ and integrate over the range $[0, a]$. Thus,

$$\begin{aligned} \int_0^a f(x)\cos\lambda_m x\, dx &= \int_0^a \sum_{n=0}^{\infty} b_x \cos\lambda_n x \cos\lambda_m x\, dx \\ &= \sum_{n=0}^{\infty} b_n \int_0^a \cos\lambda_n x \cos\lambda_m x\, dx \end{aligned} \tag{3.54}$$

With the orthogonality that we have established, the integral on the right-hand side is zero except when $m = n$ when $I_{mm} = I_{nn} = \frac{1}{2}a$. Therefore, if we open up the sum, we may write

$$\int_0^a f(x)\cos\lambda_m x\, dx = b_0{\cdot}0 + b_1{\cdot}0 + b_2{\cdot}0 + \ldots + b_m\left(\tfrac{1}{2}a\right) + b_{m+1}{\cdot}0 + \ldots + b_n{\cdot}0 + \ldots \tag{3.55}$$

It is clear that the only term under the summation that survives is the

$$b_m \int_0^a \cos\lambda_m^2 x\, dx = b_m\left(\tfrac{1}{2}a\right),$$

and we can see that

$$\int_0^a f(x)\cos\lambda_m x\, dx = b_m\left(\tfrac{1}{2}a\right),$$

from which we have the explicit expression for b_m as

$$b_m = \frac{2}{a}\int_0^a f(x)\cos\lambda_m x\, dx, \qquad m = 0, 1, 2, 3, \ldots, \tag{3.56}$$

which is slightly different from the half-range cosine expansions in which we had $\lambda_n = n$ and $a = \pi$ [see Section 3.4.2 on page 81]. In particular, with $\lambda_n = \left(n+\frac{1}{2}\right)\frac{\pi}{a}$, the leading term in the series is $b_0 \cos\left(\frac{1}{2}\pi x/a\right)$ instead of being just a constant when $X_n(x) = \cos nx$. □

This example has illustrated the idea of developing a Fourier series with the orthogonal eigenfunctions being generated from a very simple second-order ordinary differential equation involving an adjustable parameter (λ_n) determined by two homogeneous boundary conditions. We will now extend this principle to general second-order ordinary differential equations. The procedure for developing orthogonal eigenfunctions is referred to as the Sturm-Liouville Theory.

3.7.1 Sturm-Liouville Theory

In Chapter 1, we discussed the general form of second-order ordinary differential equations as

$$a_2(x)\frac{d^2X(x)}{dx^2} + a_1(x)\frac{dX(x)}{dx} + a_0(x)X(x) = 0,$$

where we are using $X(x)$ as a dependent variable instead of $y(x)$. This equation can be re-written as

$$\frac{d}{dx}\left[p(x)\frac{dX(x)}{dx}\right] + S(x)X(x) = 0.$$

In addition, if we wish to allow for an adjustable parameter, λ^2, independent of x, the general form can be written as

$$\frac{d}{dx}\left[p(x)\frac{dX(x)}{dx}\right] - s(x)X(x) + \lambda^2 q(x)X(x) = 0, \tag{3.57}$$

where we have defined $S(x) = -s(x) + \lambda^2 q(x)$. Provided $q(x)$ is not identically zero, we can rewrite this equation as

$$\frac{1}{q(x)}\frac{d}{dx}\left[p(x)\frac{dX(x)}{dx}\right] - \frac{s(x)}{q(x)}X(x) = -\lambda^2 X(x), \tag{3.58}$$

where we can see that $X(x)$ is an eigenfunction of the differential operator

$$\frac{1}{q(x)}\frac{d}{dx}\left[p(x)\frac{d}{dx}\right] - \frac{s(x)}{q(x)}$$

with eigenvalue $-\lambda^2$. Since this eigenvalue is adjustable, $X(x)$ corresponds to large spectrum of solutions that depend on the specific values of λ. In many cases, these values generally correspond the requirement that certain homogeneous boundary conditions be satisfied at the end points of the range where $X(x)$ is defined. Often, these will be discrete values [such as $\lambda = \lambda_n = \left(n+\frac{1}{2}\right)\pi/a$ as in Example 3.9]. There are as well other types of conditions (such as spatial periodicity) which will also be appropriate for establishing orthogonality and leading to discrete values of λ. It is therefore important that we refer to each λ as λ_n, and recognize that for each different value of λ_n, there is a different equation in reference to (3.57) or (3.58). It is appropriate to distinguish the different equations and their solutions with the corresponding values of λ_n. Let us then write (3.57) as

$$\frac{d}{dx}\left[p(x)\frac{dX_n(x)}{dx}\right] - s(x)X_n(x) + \lambda_n^2 q(x)X_n(x) = 0, \tag{3.59}$$

The values of λ_0, λ_1, λ_2, $\lambda_3 \ldots$ would be determined by certain pairs of boundary conditions (or periodicity conditions) for all the eigenfunctions, $X_n(x), \quad n = 0, 1, 2, 3 \ldots$. Our aim here is to establish the types of boundary conditions that would make the set of eigenfunctions orthogonal. So, let us take two arbitrary eigenfunctions, $X_j(x)$ and $X_k(x)$, with respective eigenvalues λ_j and λ_k, corresponding to $n = j$ and $n = k$. Therefore, we have

$$\frac{d}{dx}\left[p(x)\frac{dX_j(x)}{dx}\right] - s(x)X_j(x) + \lambda_j^2 q(x)X_j(x) = 0, \tag{3.60}$$

and

$$\frac{d}{dx}\left[p(x)\frac{dX_k(x)}{dx}\right] - s(x)X_k(x) + \lambda_k^2 q(x)X_k(x) = 0, \tag{3.61}$$

We are interested in the functions $X_j(x)$ and $X_k(x)$ over some range $a \leq x \leq b$, and we wish to classify the types of end-point conditions that may be imposed at $x = a$ and $x = b$ on these functions such that $X_j(x)$ is orthogonal to $X_k(x)$. Since $X_j(x)$ and $X_k(x)$ are arbitrary selections from $X_n(x), \quad n = 0, 1, 2, 3 \ldots$, the orthogonality of $X_j(x)$ with $X_k(x)$ will imply the orthogonality of the whole set of eigenfunctions with each other.

Let us multiply equation (3.60) by $X_k(x)$ and integrate over $[a, b]$. This gives

$$\int_a^b \left[p(x)X_j'(x)\right]' X_k(x)\,dx - \int_a^b s(x)X_j(x)X_k(x)\,dx + \lambda_j^2 \int_a^b q(x)X_j(x)X_k(x)\,dx = 0, \tag{3.62}$$

where the primes ($'$) denote derivatives with respect to x. Next, let us multiply equation (3.61) by $X_j(x)$ and again integrate over $[a, b]$, resulting in

$$\int_a^b \left[p(x)X_k'(x)\right]' X_j(x)\,dx - \int_a^b s(x)X_k(x)X_j(x)\,dx + \lambda_k^2 \int_a^b q(x)X_k(x)X_j(x)\,dx = 0, \tag{3.63}$$

We notice that in both (3.62) and (3.63), the middle terms are identical and the third ones are proportional. Therefore, if we subtract, say, (3.63) from (3.62), we obtain

$$\int_a^b \left[p(x)X_j'(x)\right]' X_k(x)\,dx - \int_a^b \left[p(x)X_k'(x)\right]' X_j(x)\,dx + \left(\lambda_j^2 - \lambda_k^2\right)\int_a^b q(x)X_k(x)X_j(x)\,dx = 0, \tag{3.64}$$

which may be written as

$$\left(\lambda_j^2 - \lambda_k^2\right) \int_a^b q(x) X_k(x) X_j(x)\, dx$$
$$= -\int_a^b \left[p(x) X_j'(x)\right]' X_k(x)\, dx + \int_a^b \left[p(x) X_k'(x)\right]' X_j(x)\, dx \qquad (3.65)$$

Let us now carry out integration by parts on both the integrals on the right-hand side. This results in

$$\begin{aligned}
&\left(\lambda_j^2 - \lambda_k^2\right) \int_a^b q(x) X_k(x) X_j(x)\, dx \\
&= -\left[p(x) X_j'(x)\right] X_k(x)\Big|_a^b + \int_a^b \left[p(x) X_j'(x)\right] X_k'(x)\, dx \\
&\qquad + \left[p(x) X_k'(x)\right] X_j(x)\Big|_a^b - \int_a^b \left[p(x) X_k'(x)\right] X_j'(x)\, dx \\
&= -\left[p(x) X_j'(x) X_k(x)\right]_a^b + \left[p(x) X_k'(x) X_j(x)\right]_a^b \\
&= -\left[p(b) X_j'(b) X_k(b) - p(a) X_j'(a) X_k(a)\right] \\
&\qquad + \left[p(b) X_k'(b) X_j(b) + p(a) X_k'(a) X_j(a)\right] \\
&= -p(b)\left[X_j'(b) X_k(b) - X_k'(b) X_j(b)\right] \\
&\qquad + p(a)\left[X_j'(a) X_k(a) - X_k'(a) X_j(a)\right]
\end{aligned} \qquad (3.66)$$

We notice now that the right-hand side consists of values of $X_n(x)$ and $X_n'(x)$ only at the end-points $x = a$ and $x = b$. Therefore, it is possible classfy the conditions under which this side would vanish. If this can be achieved, we can guarantee the orthogonality relationship that

$$\int_a^b q(x) X_j(x) X_k(x)\, dx = 0 \qquad (3.67)$$

unless $\lambda_j^2 = \lambda_k^2$. Here we see that need to insert a 'weighting factor' $q(x)$ in the integral to have it vanish. This is the outcome of the generalization, and a necessary parameter which was equal to unity for the Fourier sine and cosine series. Assuming that all eigenvalues are distinct, we can assume the validity of equation (3.67) based on the requirement that the right-hand side of equation (3.66) vanish. This can happen in several different ways that are classified in the discussion here.

1. *First Kind: Dirichlet Boundary Conditions*

   If the value of a function is specified at a boundary, we refer to it as a Dirichlet boundary condition. A pair of homogeneous Dirchlet conditions are

$$X_n(a) = X_n(b) = 0, \qquad \text{where } n \text{ corresponds to all possiblities of } j \text{ and } k. \qquad (3.68)$$

In this case, we can see that the right-hand side of equation (3.66) vanishes. Therefore, for the ordinary differential equation (3.59), with a pair of homogeneous boundary conditions (3.67), the eigenfunctions given by the set $X_n(x)$ are mutually orthogonal.

2. *Second Kind: Neumann Boundary Conditions*
   If, on the other hand, we have the derivative on each side set equal to zero, i.e.,

$$X_n'(a) = X_n'(b) = 0, \tag{3.69}$$

   we will again have the right-hand side of equation (3.66) going to zero, and orthogonality is established.

3. *Third Kind: Mixed Boundary Conditions*
   Other possibilities which will lead to orthogonality include mixed boundary conditions in which the specification is a linear combination of $X_n(x)$ and $X_n'(x)$ at a boundary. For example, the pair

$$X_n'(a) = h_A X_n(a), \quad \text{and} \quad X_n'(b) = h_B X_n(b), \tag{3.70}$$

   with h_A and h_B being constants, will yield the following value for the right-hand side of (3.66):

$$\begin{aligned} RHS = & -p(b)\left[h_B X_k(b) X_j(b) - h_B X_j(b) X_k(b)\right] \\ & +p(a)\left[h_A X_k(a) X_j(a) - h_A X_j(a) X_k(a)\right] = 0, \end{aligned} \tag{3.71}$$

   which, being zero, provides the guarantee for orthogonality.

Combinations of Dirichlet, Neumann or Mixed Conditions
In the above discussion, we chose each component (i.e., one at $x = a$ and another at $x = b$) of the pair to be of the same class. However, these components could be any combination of the above three classes. For example, each of the following pairs of boundary conditions will also cause right-hand side of equation (3.66) to vanish:

$$\begin{aligned} X_n'(a) = X_n(b) &= 0, \\ \text{or} \quad X_n(a) = X_n'(b) &= 0, \end{aligned} \tag{3.72}$$

and guarantee orthogonality. We may also have any one of the following pairs:

$$\begin{aligned} X_n(a) &= 0 \quad \text{and} \quad X_n'(b) - h_B X_n(b) = 0, \\ X_n'(a) &= 0 \quad \text{and} \quad X_n'(b) - h_B X_n(b) = 0, \\ X_n'(a) - h_A X_n(a) &= 0 \quad \text{and} \quad X_n(b) = 0, \\ X_n'(a) - h_A X_n(a) &= 0 \quad \text{and} \quad X_n'(b) = 0, \\ X_n'(a) - h_A X_n(a) &= 0 \quad \text{and} \quad X_n'(b) - h_B X_n(b) = 0. \end{aligned} \tag{3.73}$$

that will yield orthogonal eigenfunctions. A case of the first one of these five pairs is illustrated in the next example.

Example 3.10 Consider once again equation (3.47) so that the functions of the set $\{X_n(x)\}$ are eigenfunctions of d^2/dx^2. Let us define the range $0 \leq x \leq l$ so that the end points are $x = a = 0$ and $x = b = l$ where the following pair of boundary conditions applies:

$$X_n(0) = 0, \qquad \text{and} \qquad X_n'(l) - h_B X_n(l) = 0. \tag{3.74}$$

The general solution to equation (3.47) is the same as (3.50),

$$X_n(x) = A_n \sin \lambda_n x + B_n \cos \lambda_n x. \tag{3.75}$$

Applying the first of the boundary conditions (3.74) gives

$$X_n(0) = A_n \sin \lambda_n 0 + B_n \cos \lambda_n 0 = B_n = 0,$$

reducing $X_n(x)$ to

$$X_n(x) = A_n \sin \lambda_n x. \tag{3.76}$$

Now, applying the second of the boundary conditions (3.74) we obtain,

$$X_n'(l) - h_B X_n(l) = A_n (\lambda_n \cos \lambda_n l - h_B \sin \lambda_n l) = 0.$$

For a non-trivial solution, we need to maintain $A_n \neq 0$, leaving us with the requirement

$$\lambda_n \cos \lambda_n l - h_B \sin \lambda_n l = 0.$$

This is a transcendental equation that determines the values of λ_n which have to be obtained numerically. It may also be written in the form,

$$\lambda_n l \cos(\lambda_n l) = (h_B l) \sin(\lambda_n l),$$

or

$$\frac{\lambda_n l}{h_B l} = \tan(\lambda_n l). \tag{3.77}$$

One can graphically visualize the roots of this equation by plotting both the left-hand side and the right-hand side on the same graph, as shown in Figure 3.2 With these values of $\lambda_n, n = 0, 1, 2, \ldots$, the set of eigenfunctions given by (3.76) are orthogonal. Actually, $\lambda_0 = 0$, while satisfying the transcendental equation, is redundant since $\sin \lambda_0 x = 0$. The set of eigenfunctions $\{\sin \lambda_n x\}$ may be used to express any piecewise-continuous function $f(x)$ over the range $0 \leq x \leq l$, i.e.,

$$f(x) = \sum_{n=1}^{\infty} a_n \sin \lambda_n x, \qquad 0 \leq x \leq l. \tag{3.78}$$

Let us multiply each side by $\sin \lambda_m x$, and integrate over $[0, l]$. This results in

$$\begin{aligned} \int_0^l f(x) \sin \lambda_m x \, dx &= \int_0^l \sum_{n=1}^{\infty} a_n \sin \lambda_n x \, \sin \lambda_m x \, dx \\ &= \sum_{n=1}^{\infty} a_n \int_0^l \sin \lambda_n x \, \sin \lambda_m x \, dx. \end{aligned} \tag{3.79}$$

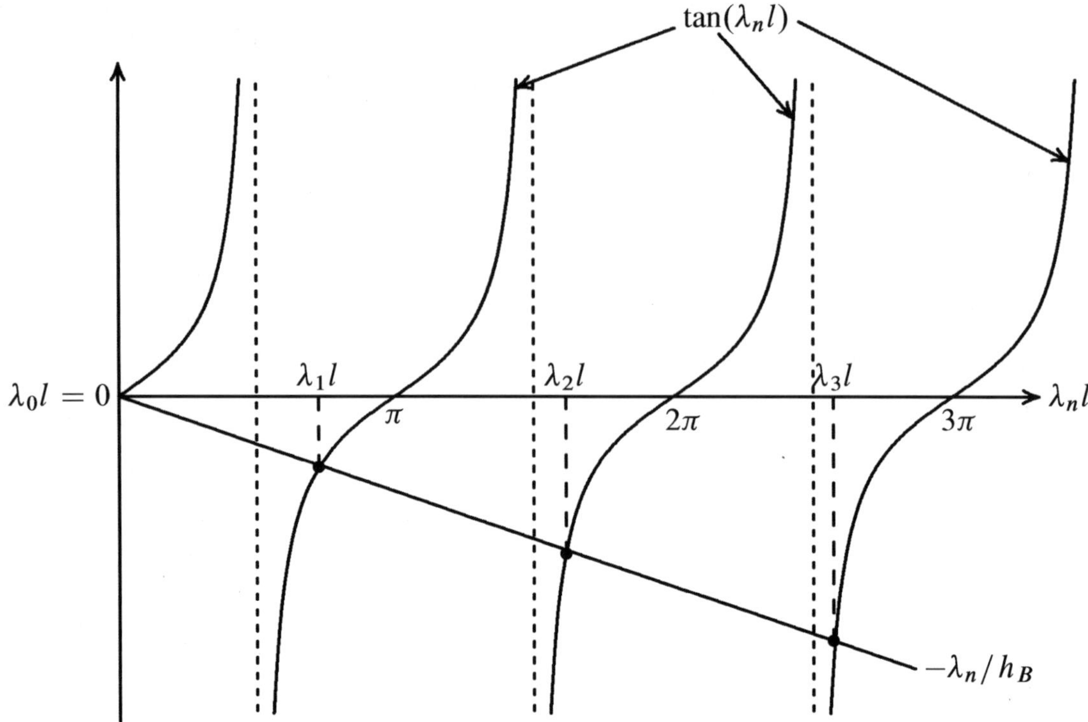

Figure 3.2: Graphical representation of the roots of equation (3.77)

Keeping in mind that

$$\int_0^l \sin\lambda_n x \ \sin\lambda_m x \ dx = 0 \qquad \text{for} \qquad m \neq n,$$

and opening up the summation in (3.79), we obtain

$$\begin{aligned}
&\int_0^l f(x)\sin\lambda_m x \ dx \\
&\quad = a_1 \cdot 0 + a_1 \cdot 0 + \cdots + a_m \int_0^l \sin\lambda_m^2 x \ dx + a_{m+1} \cdot 0 + \cdots + a_n \cdot 0 + \cdots \\
&\quad = a_m \int_0^l \sin\lambda_m^2 x \ dx \\
&\quad = a_m \int_0^l \tfrac{1}{2}\left[1 - \cos 2\lambda_m x\right] dx \\
&\quad = a_m \tfrac{1}{2}\left[x - \frac{\sin 2\lambda_m x}{2\lambda_m}\right]_0^l \\
&\quad = a_m \tfrac{1}{2}\left[l - \frac{\sin 2\lambda_m l}{2\lambda_m}\right]
\end{aligned} \tag{3.80}$$

With this result, the set of coefficients in equation (3.78) may be expressed as

$$a_m = \frac{\displaystyle\int_0^l f(x)\sin\lambda_m x\,dx}{\frac{1}{2}l\left[1 - \dfrac{\sin 2\lambda_m l}{2\lambda_m l}\right]},$$

where m, being an arbitrary index, may be replaced with n throughout this equation. With λ_n values completely determined from the transcendental equation, and if $f(x)$ is specified, we may explicitly write the Fourier series expansion for this function in terms of the series in $\sin\lambda_n x$. □

4. *Fourth Class of Conditions: Periodic Eigenfunctions*
A close examination of the right-hand side of equation (3.66) shows that there exist other possibilities to make it vanish. This happens if

$$X_n(a) = X_n(b) \qquad \text{and} \qquad X_n'(a) = X_n'(b) \tag{3.81}$$

provided $p(a) = p(b)$. Also, if $p(a) = p(b) = 0$, orthogonality is achieved without the requirement (3.81) above, and this is discussed later in Section 3.7.5 on pages 134-142. In addition, combinations such as $p(a) = 0$ along with $X_n(b) = 0$ or $X_n'(b) = 0$ are also admissible. However, for the moment, we shall discuss the former case requiring (3.81) which is typical of situations when $X_n(x)$ is required to be periodic. For example, in the case of circular geometry, with say θ as an angular coordinate, any function describing a physical parameter (such as temperature, fluid velocity, or electric field) would have 2π-periodity in θ. Such periodic continuity also requires that the angular derivatives be periodic. Keeping this in mind, for a Fourier-type expansion, we would require eigenfunctions and their derivatives that are 2π-periodic in the angular coordinate, as illustrated in Example 3.11 here.

Example 3.11 In the simple case,

$$\frac{d^2X_n}{d\theta^2} + \lambda_n^2 X_n(\theta) = 0, \quad 0 \le \theta \le 2\pi, \tag{3.82}$$

with the periodicity requirement,

$$X_n(0) = X_n(2\pi) \qquad \text{and} \qquad X_n'(0) = X_n'(2\pi) \tag{3.83}$$

with a and b defined as

$$a = 0 \qquad \text{and} \qquad b = 2\pi.$$

We have, corresponding to the general Sturm-Liouville form (3.59), the following identification of parameters:

$$x = \theta, \quad p(x) = 1, \quad s(x) = 0, \quad \text{and} \quad q(x) = 1.$$

The general solution to equation (3.82) is

$$X_n(\theta) = A_n \sin \lambda_n \theta + B_n \cos \lambda_n \theta \tag{3.84}$$

For this to be 2π-periodic we need to meet the conditions in equation (3.83). Starting with the first one of these, we have

$$A_n \sin \lambda_n 0 + B_n \cos \lambda_n 0 = A_n \sin(\lambda_n 2\pi) + B_n \cos(\lambda_n 2\pi), \tag{3.85}$$

leading to

$$B_n = B_n \cos(\lambda_n 2\pi).$$

For $B_n \neq 0$, we would require $\cos(\lambda_n 2\pi) = 1$, or $\lambda_n = n$. Now, with the second condition in (3.83),

$$A_n\lambda_n \cos \lambda_n 0 - B_n\lambda_n \sin(\lambda_n 0 = A_n\lambda_n \cos(\lambda_n 2\pi) - B_n\lambda_n \sin(\lambda_n 2\pi). \tag{3.86}$$

Taking into consideration that $\lambda_n = n$, this reduces to just a confirmation

$$A_n\lambda_n = A_n\lambda_n \cos(\lambda_n 2\pi) = A_n\lambda_n,$$

which means that to maintain generality, we should keep $A_n \neq 0$. The full eigenfunction satisfying the periodicity condition can be written as

$$X_n(\theta) = A_n \sin n\theta + B_n \cos n\theta. \tag{3.87}$$

Based on our earlier development on periodic functions (see Section 3.4.3 on page 82), we have already established that an arbitrary function $F(\theta)$ can be expanded as a Fourier series in terms of the set $\{A_n \sin n\theta + B_n \cos n\theta\}$ over the region $0 \leq \theta \leq 2\pi$, or $-\pi \leq \theta \leq \pi$, or for that matter any region covering a span of 2π. It is not difficult to show that the set $\{X_n(\theta)\}$ is orthogonal over any region with a range of 2π (see Problem 6 in Exercises 3.1). □

EXERCISES 3.4

1. For the differential equation (3.47) construct a set of eigenfunctions $\{X_n(x)\}$ in the range $0 \leq x \leq l$ with boundary conditions,

$$X_n'(0) = 0, \qquad \text{and} \qquad X_n(l) = 0.$$

Expand the function $f(x) = x$ in the range $0 \leq x \leq l$.

2. Again, for the differential equation (3.47) construct a set of eigenfunctions $\{X_n(x)\}$ in the range $0 \leq x \leq l$ with boundary conditions,

$$X_n'(0) = 0, \qquad \text{and} \qquad X_n(l) - hX_n'(l) = 0.$$

Expand the function $f(x) = x$ in the range $0 \leq x \leq l$.

In the examples we have considered so far, the weighting factor $q(x)$ has not been particularly relevant because it has remained equal to unity. With several types of differential equations, this is not the case. One important case is that of Bessel's equation [see equation (2.57) in Chapter 2]. This leads us to Fourier-Bessel series which is a very important tool for many types of partial differential equations in cylindrical geometry.

3.7.2 Fourier-Bessel Series

Referring to Section 2.4.3 in Chapter 2, the cylindrical Bessel's equation of order n is

$$x^2\frac{d^2y(x)}{dx^2} + x\frac{dy(x)}{dx} + (x^2 - n^2)y(x) = 0 \tag{3.88}$$

In cylindrical geometry with r as a radial coordinate, the differential operator that commonly appears is

$$\frac{d^2}{dr^2} + \frac{1}{r}\frac{d}{dr} - \frac{n^2}{r^2}. \tag{3.89}$$

As an example, we consider the case $n = 0$.

Example 3.12 *Problem Statement*: Construct a set of orthogonal eigenfunctions, $X_m(r)$, of the differential operator,

$$\frac{d^2}{dr^2} + \frac{1}{r}\frac{d}{dr}. \tag{3.90}$$

in the domain $0 \leq x \leq R$, with homogeneous boundary conditions

$$\left.\frac{dX_m(r)}{dr}\right|_{r=0} = 0 \qquad \text{and} \qquad X_m(r)|_{r=R} = 0 \tag{3.91}$$

Solution:
The eigenfunctions $X_m(r)$ of this operator are therefore the solutions of

$$\frac{d^2X_m(r)}{dr^2} + \frac{1}{r}\frac{dX_m(r)}{dr} = -\lambda_m^2 X_m(r), \tag{3.92}$$

where we use the index m for the different eigenvalues, $-\lambda_m^2$, and the corresponding eigenfunctions so as not to confuse with the n, the notation for the order. Proper indexing will be particularly important when we consider eigenfunctions based on the order n operator. To begin with, let us state this equation in a form as close as possible to equation (3.88) so that the solutions of (3.93) may be related to Bessel functions. Multiplying this equation by r^2 and placing all nonzero terms on the left-hand side gives

$$r^2\frac{d^2X_m(r)}{dr^2} + r\frac{dX_m(r)}{dr} + \lambda_m^2 r^2 X_m(r) = 0. \tag{3.93}$$

Setting $n = 0$ in equation (3.88) gives something very similar,

$$x^2\frac{d^2y(x)}{dx^2} + x\frac{dy(x)}{dx} + x^2y(x) = 0. \tag{3.94}$$

These last two equations would be identical if we choose the parameters

$$x = \lambda_m r, \quad \text{and} \quad y(x) = y(\lambda_m r) = X_m(r). \tag{3.95}$$

Since equation (3.94) is the cylindrical Bessel's equation of order zero, the general solution is

$$y(x) = AJ_0(x) + BY_0(x), \tag{3.96}$$

where $J_0(x)$ and $Y_0(x)$ are given by the power-series expansions in equations (2.69) and (2.70), respectively on page 68. In view of the variables defined in (3.95), we may write the solution of (3.93) as

$$X_m(r) = A_m J_0(\lambda_m r) + B_m Y_0(\lambda_m r), \tag{3.97}$$

We may now apply the boundary conditions (3.91). One of these conditions will eliminate one of the integration constants, and the other will establish the values of λ_n. With the first boundary condition,

$$X'_m(0) = A_m \lambda_m J'_0(0) + B_m \lambda_m Y'_0(0) = 0. \tag{3.98}$$

We should note that $Y_0(x)$ has a $\ln x$ type of behavior about $x = 0$. Therefore, the derivative has a $1/x$ type singularity which will cause $Y_0(\lambda_m r) \to \infty$ as $r \to 0$. To satisfy (3.98), we need to set $B_m = 0$ and effectively eliminate the singular solution $Y_0(\lambda_m r)$. The eigenfunction given by (3.97)reduces to

$$X_m(r) = A_m J_0(\lambda_m r). \tag{3.99}$$

Next, we can apply the second of the boundary conditions (3.91), i.e.,

$$X_m(R) = A_m J_0(\lambda_m R) = 0. \tag{3.100}$$

Again, as with previous examples of eigenfunctions, we need to require $A_n \neq 0$ to have a nontrivial eigenfunction, and the only choice left is $J_0(\lambda_m R) = 0$. At this point, the coefficient A_m is redundant and we may set $A_m = 1$ so that

$$X_m(r) = J_0(\lambda_m r). \tag{3.101}$$

To obtain the values of λ_m, we need to examine the roots of $J_0(x) = 0$ which is a transcendental equation. As seen in the plot of $y(x) = J_0(x)$, the roots occur at irregular intervals and are located at $x = \beta_{m0},\ \ m = 1, 2, 3, \ldots$. The values of β_{m0} are tabulated in the figure as well. With the roots known, the zeros of the eigenfunction occur at $\lambda_n R = \alpha_n$, or $\lambda_n = \alpha_n/R$ with $n = 1, 2, 3, \ldots$. The boundary conditions (3.91) are in the categories defined on page 111 as the second and first kinds, respectively, and therefore orthogonality in the form of equation (3.67) is guaranteed. However, we need to properly determine the weighting factor $q(x)$ which, in the examples we have considered so far, was equal to unity. Noting that in general,

$$\frac{d^2 X_m(r)}{dr^2} + \frac{1}{r}\frac{dX_m(r)}{dr} = \frac{1}{r}\frac{d}{dr}\left[r\frac{dX_m(r)}{dr}\right]$$

, and using this identity in (3.93), and multiplying the equation bt r we are able to cast the differential equation for the eigenfunction (3.101) in the Sturm-Liouville form (3.59). As a result, we have

$$\frac{d}{dr}\left[r\frac{dX_m(r)}{dr}\right] + \lambda_m^2 r X_m(r) = 0. \tag{3.102}$$

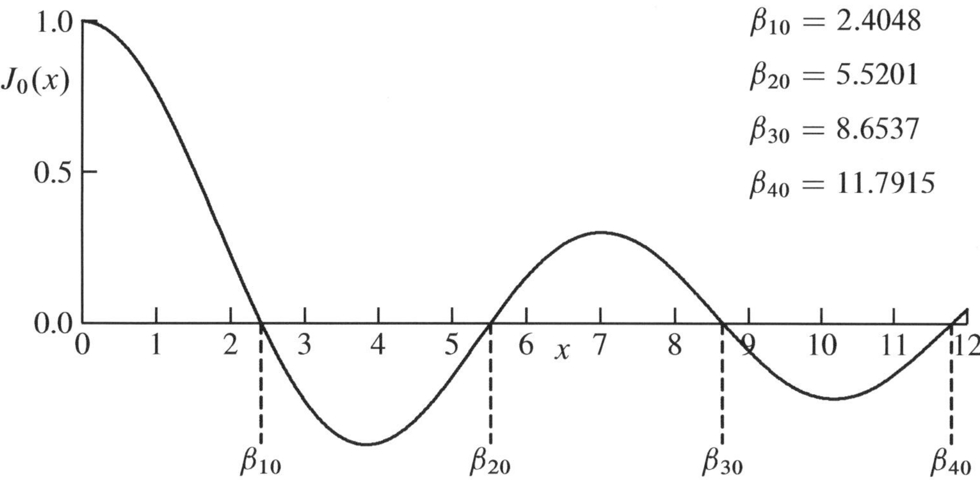

Figure 3.3: Roots of the Bessel function $J_0(x)$

We need to compare this with equation (3.59) which is stated here again for convenience with m as the index,

$$\frac{d}{dx}\left[p(x)\frac{dX_m(x)}{dx}\right] - s(x)X_m(x) + \lambda_m^2 q(x)X_m(x) = 0. \tag{3.103}$$

We can now identify the variables as follows:

$$x = r, \quad X_m(x) = X_m(r), \quad s(x) = s(r) = 0,$$

and the weighting factor

$$q(x) = q(r) = r.$$

In addition the range $0 \leq r \leq R$ defines the end-points as $a = 0$ and $b = R$ for the present example. The orthogonality relationship corresponding to equation (3.67) is therefore

$$\int_0^R rJ_0(\lambda_j r)J_0(\lambda_k r)\,dr = 0, \quad j \neq k. \tag{3.104}$$

It is now possible to express an arbitrary function $f(r)$ as a series of cylindrical Bessel functions of order zero (and higher orders as well). Such an expansion is considered in the next example. □

Example 3.13 *Problem Statement*:
Using the orthogonal set of eigenfunctions $X_m(r) = J_0(\lambda_m r)$, expand the piecewise-continuous function $f(r)$ as

$$f(r) = \sum_{m=1}^{\infty} a_m J_0(\lambda_m r), \qquad 0 \leq r \leq R, \tag{3.105}$$

where the eigenfunctions satisfy the homogeneous boundary conditions,

$$J_0'(0) = 0 \qquad \text{and} \qquad J_0(\lambda_m R) = 0.$$

Calculate the coefficients a_m or the special case, $f(r) = 1$.

Solution:
Keeping in mind that the orthogonality relation (3.104) contains the weighting factor $q(r) = r$, we multiply each side of (3.105) by $rJ_0(\lambda_k r)$ and integrate over $[0, R]$, resulting in

$$\begin{aligned}\int_0^R f(r)rJ_0(\lambda_k r)\,dr &= \int_0^R \sum_{m=1}^{\infty} a_m J_0(\lambda_m r)rJ_0(\lambda_k r)\,dr \\ &= \sum_{m=1}^{\infty} a_m \int_0^R rJ_0(\lambda_m r)J_0(\lambda_k r)\,dr \\ &= a_1\cdot 0 + a_2\cdot 0 + \cdots + a_k\int_0^R r\,[J_0(\lambda_k r)]^2\,dr + a_{k+1}\cdot 0 + \cdots \\ &= a_k\int_0^R r\,[J_0(\lambda_k r)]^2\,dr \qquad (3.106)\end{aligned}$$

It is not difficult to see that the expression for the set of coefficients a_k is

$$a_k = \frac{\int_0^R f(r)rJ_0(\lambda_k r)\,dr}{\int_0^R r\,[J_0(\lambda_k r)]^2\,dr} \qquad (3.107)$$

The integral in the denominator is available in various mathematical handbooks (see e.g., [2]), and the expression is

$$\int_0^R r\,[J_0(\lambda_k r)]^2\,dr = \tfrac{1}{2}R^2\left\{\left[J_0'(\lambda_k R)\right]^2 + [J_0(\lambda_k R)]^2\right\}. \qquad (3.108)$$

For the current set of examples, based on the boundary condition $J_0(\lambda_m R) = 0$, we can eliminate the second term on the right-hand side. In addition we have, among various Bessel function properties, the identity $J_0'(x) = -J_1(x)$. The use of these results reduces the above integral to

$$\int_0^R r\,[J_0(\lambda_k r)]^2\,dr = \tfrac{1}{2}R^2\left[J_0'(\lambda_k R)\right]^2 = \tfrac{1}{2}R^2\,[J_1(\lambda_k R)]^2. \qquad (3.109)$$

The integral in the numerator of (3.107) depends on the definition of $f(x)$. For the special case $f(x) = 1$, we can calculate this using equation (3.102). Here, we replace $X_m(r)$ with the known solution $J_0(\lambda_m r)$, and change the index from m to k so that

$$\frac{d}{dr}\left[r\frac{dJ_0(\lambda_k r)}{dr}\right] = -\lambda_k^2 rJ_0(\lambda_k r) \qquad (3.110)$$

Integrating both sides over $[0, R]$ yields

$$\begin{aligned}\lambda_k^2\int_0^R rJ_0(\lambda_k r)dr &= -\left[r\frac{dJ_0(\lambda_k r)}{dr}\right]_{r=0}^{r=R} \\ &= -r\,[-\lambda_k J_1(\lambda_k r)]_{r=0}^{r=R} \\ &= R\lambda_k J_1(\lambda_k R). \qquad (3.111)\end{aligned}$$

or

$$\int_0^R rJ_0(\lambda_k r)dr = \frac{RJ_1(\lambda_k R)}{\lambda_k}. \tag{3.112}$$

The coefficients a_k given in (3.107) may be expressed as

$$\begin{aligned} a_k &= \frac{RJ_1(\lambda_k R)/\lambda_k}{\frac{1}{2}R^2\,[J_1(\lambda_k R)]^2} \\ &= \frac{2}{R\lambda_k\,[J_1(\lambda_k R)]}, \end{aligned} \tag{3.113}$$

and the expansion for $f(r) = 1$ may be written in the explicit form as a Fourier-Bessel series,

$$1 = \sum_{k=1}^{\infty} \frac{2J_0(\lambda_k r)}{\lambda_k R\,[J_1(\lambda_k R)]},$$

where $\lambda_k = \alpha_k/R$ with α_k given by the roots of the transcendental equation $J_0(x) = 0$.

□

EXERCISES 3.5

Expand the functions below as Fourier-Bessel series of order zero,

$$f(r) = \sum_{m=0}^{\infty} a_m J_0(\lambda_m r), \qquad 0 \le r \le R,$$

where the requisite condition at $r = R$ is as specified.

1. $f(r) = r^2/R^2$, with $J_0(\lambda_m R) = 0$.
2. $f(r) = (1 - r^2/R^2)$, with $J_0'(\lambda_m R) = 0$.

Series in Order-n Bessel Functions

The Fourier-Bessel series can easily be extended to order-n Bessel functions by setting up eigenfunctions of the differential operator given by (3.89). The eigenfunctions correspond to the solutions of the differential equation,

$$\frac{d^2X(r)}{dr^2} + \frac{1}{r}\frac{dX(r)}{dr} - \frac{n^2}{r^2}X(r) = -\lambda^2 X(r). \tag{3.114}$$

We have not indexed the eigenfunctions and the eigenvalues because of the appearance of n as a parameter in addition to m that we will use to index the different eigenvalues. First of all, we need to note that (3.114) is a different equation for every value of n. Therefore, we need to use the index n for every solution $X(r)$, i.e., we need to start looking at (3.114) in the form,

$$\frac{d^2X_n(r)}{dr^2} + \frac{1}{r}\frac{dX_n(r)}{dr} - \frac{n^2}{r^2}X_n(r) = -\lambda^2 X_n(r), \tag{3.115}$$

which may also be written as

$$r^2 \frac{d^2 X_n(r)}{dr^2} + r \frac{dX_n(r)}{dr} + \left(\lambda^2 r^2 - n^2\right) X_n(r) = 0. \tag{3.116}$$

Comparing this with equation (3.88), and using the transformation, $x = \lambda r$, we may follow the procedure similar to what we did for order zero Bessel functions [equations (3.93)-(3.97) on page 117]. It is not difficult to show that the general solution to equation (3.116) is

$$X_n(r) = AJ_n(\lambda r) + BY_n(\lambda r) \tag{3.117}$$

Next, we should have a pair of boundary conditions for $X_n(r)$ that would determine the values of λ. Consider, for example,

$$X_n(r)|_{r\to 0} < \infty \qquad \text{and} \qquad X_n(r)|_{r=R} = 0. \tag{3.118}$$

We should note here that $r = 0$ is not a clear boundary but rather a point where we may have a symmetry conditions of the type given in (3.91), or perhaps something such as (3.118) above. In fact, the latter is adequate since it eliminates the solution that is singular at $r = 0$. For many physical problems, even a requirement such as

$$X_n(r) < \infty, \quad 0 \le r \le R \tag{3.119}$$

would be sufficient to eliminate one solution when it is singular somewhere in the specified range. However, based on Sturm-Liouville theory, the requirement for the set of end-point (or boundary) conditions is that a specification at these points be satisfied. Therefore, something like (3.119), while eliminating one integration constant, should somehow translate into a homogeneous condition at $r = 0$ as an outcome. This is indeed the case for Bessel functions because from the series expansions of Bessel functions, we can see that $J_0'(0) = 0$, $J_n(0) = 0$, $n \ge 1$, and $J_n'(0) = 0$, $n \ge 2$. We can be assured that upon the elimination of a singular solution, the remaining part of the eigenfunction satisfies a homogeneous condition at $r = 0$. This discussion is, of course, only relevant if $r = 0$ is included in the domain of expansion of the eigenfunctions. Since $Y_n(\lambda r) \to \infty$ as $r \to 0$, we can set $B = 0$ in equation (3.117) to require a finite solution. In addition, as with previous examples, we can set $A = 1$ without loss of generality, and eigenfunction reduces to

$$X_n(r) = J_n(\lambda r) \tag{3.120}$$

We should now notice that for every differential equation determined by a value of n, and the corresponding solution $X_n(r)$ there would be an infinite number of values of λ that would satisfy the pair of conditions at $r = 0$ and $r = R$. Therefore, within a solution group defined by one value of n, we need to label all the corresponding λ values with the label n as well as an additional label for all the different λ values within the nth group. For example, if we consider for the moment just $n = 1$ and $n = 2$, we refer to the solutions as

$$X_1(r) = J_1(\lambda r) \qquad \text{and} \qquad X_2(r) = J_2(\lambda r) \tag{3.121}$$

where we have eliminated the singular solution. The second of the pair of boundary condition (3.118) requires

$$X_1(R) = J_1(\lambda R) = 0 \qquad \text{and} \qquad X_2(R) = J_2(\lambda R) = 0 \tag{3.122}$$

which, for a non-trivial solutions, would require $A \neq 0$ and $J_1(\lambda R) = 0$, along with $J_2(\lambda R) = 0$. The values of λR that satisfy the boundary conditions are at the corresponding roots of $J_1(x) = 0$ and $J_2(x) = 0$. Let us denote the respective sets of zeros as β_{m1} and β_{m2}. It is important to note that the set β_{1m} and β_{2m} are distinctly different from each other, and we must make provision to maintain this distinction, not just for $n = 1, 2$ but for all values of n. Therefore, if we refer to the roots of $J_n(x) = 0$ as β_{mn}, then the values of λ are

$$\lambda = \lambda_{mn} = \frac{\beta_{mn}}{R}, \tag{3.123}$$

where we have indexed λ with both m and n. The values of β_{mn} are given in Table 3.1. Since we have a different equation for each value of n, and within each member of this set

m	β_{m0}	β_{m1}	β_{m2}	β_{m3}	β_{m4}	β_{m5}	β_{m6}	β_{m7}	β_{m8}
0	–	0	0	0	0	0	0	0	0
1	2.4048	3.8317	5.1356	6.3802	7.5883	8.7715	9.9361	11.0864	12.2251
2	5.5201	7.0156	8.4172	9.7610	11.0647	12.1986	13.5893	14.8213	16.0978
3	8.6537	10.1735	11.6198	13.0152	14.1725	15.7002	17.0038	18.2876	19.5545
4	11.7915	13.3237	14.7960	16.2295	17.6160	18.9801	20.3208	21.6415	22.9452
5	14.9309	16.4706	17.9598	19.4094	20.8269	22.2178	29.5861	24.9349	26.2668
6	18.0711	19.6159	21.1170	22.5827	24.0190	25.4301	26.8202	28.1912	29.5457
7	21.2116	22.7601	24.2701	25.7482	27.1991	28.6265	30.0997	31.4228	32.7958
8	24.3525	25.9097	27.4206	28.9084	30.3710	31.8117	33.2330	34.6171	36.0256
9	27.4935	29.0468	90.5692	32.0649	33.5371	34.9888	36.4220	37.8187	39.2405
10	30.6346	32.1897	33.7165	35.2187	36.6990	38.1599	39.6092	41.0308	42.4439

Table 3.1: Values of the roots (β_{mn}) of $J_n(x)$ (adapted from Abramowitz & Stegun [1])

defined by the value on n we have a subset of equations corresponding to each value of λ, we should index the solutions given by (3.120) accordingly, i.e.,

$$X_{mn}(r) = J_n(\lambda_{mn} r). \tag{3.124}$$

This represents a solution to the differential equation (3.116) which, with proper indexing, should be

$$r^2 \frac{d^2 X_{mn}(r)}{dr^2} + r \frac{dX_{mn}(r)}{dr} + \left(\lambda^2 r^2 - n^2\right) X_{mn}(r) = 0. \tag{3.125}$$

We may combine the first two terms terms and rearrange this equation to represent the Sturm-Liouville format,

$$\frac{d}{dr}\left(r \frac{dX_{mn}(r)}{dr}\right) - \frac{n^2}{r} X_{mn}(r) + \lambda_{mn}^2 r X_{mn}(r) = 0. \tag{3.126}$$

Comparing with the general second-order equation (3.103) and using $x = r$, we can identify the weighting factor as $q(x) = q(r) = r$. Following the procedure similar to Example 3.12 on pages 117-119, we can establish the orthogonality relationship similar to equation (3.104),

$$\int_0^R rJ_n(\lambda_{jn}r)J_n(\lambda_{kn}r)\,dr = 0, \quad j \neq k \tag{3.127}$$

where n remains a free index. We can now expand a function $f(r)$ as a series,

$$f(r) = \sum_{m=0}^{\infty} a_m J_n(\lambda_{mn}r). \tag{3.128}$$

After going through the procedure similar to that in Example 3.13 (see pages 119-121), the set of coefficients a_m can be found to be

$$a_m = \frac{\int_0^R f(r)rJ_n(\lambda_{mn}r)\,dr}{\int_0^R r\,[J_n(\lambda_{mn}r)]^2\,dr}. \tag{3.129}$$

The integral in the denominator can be obtained by using the orthogonality relationship (3.66) in the limit $\lambda_j \to \lambda_k$. The result is for unspecified boundary condition at $r = R$ is

$$\int_0^R r\,[J_n(\lambda_{mn}r)]^2\,dr = \tfrac{1}{2}R^2\left\{\left[J_n'(\lambda_{mn}R)\right]^2 + \left(1 - \frac{n^2}{\lambda_{mn}^2R^2}\right)[J_n(\lambda_{mn}R)]^2\right\}. \tag{3.130}$$

For the boundary condition (3.118), i.e.,

$$J_n(\lambda_{mn}R) = 0, \tag{3.131}$$

equation (3.130) reduces to

$$\int_0^R r\,[J_n(\lambda_{mn}r)]^2\,dr = \tfrac{1}{2}R^2\left[J_n'(\lambda_{mn}R)\right]^2, \tag{3.132}$$

and we may write the set of coefficients a_m in equation (3.129) as

$$a_m = \frac{\int_0^R f(r)rJ_n(\lambda_{mn}r)\,dr}{\frac{1}{2}R^2\left[J_n'(\lambda_{mn}R)\right]^2}. \tag{3.133}$$

Remarks on Boundary Conditions

With the development of the Fourier-Bessel series, we have considered only $X_{mn}(R) = 0$. However, orthogonality can be guaranteed the second and the third types (see page 112) as well. Therefore, instead of $X_{mn}(R) = 0$, if we required $X_{mn}'(R) = 0$, resulting in

$$J_n'(\lambda_{mn}R) = 0, \tag{3.134}$$

we would have

$$a_m = \frac{\int_0^R f(r) r J_n(\lambda_{mn} r)\, dr}{\frac{1}{2} R^2 \left(1 - \frac{n^2}{\lambda_{mn}^2 R^2}\right) [J_n(\lambda_{mn} R)]^2}, \tag{3.135}$$

where $\lambda_{mn} = \gamma_{mn}/R$ with γ_{mn} representing the roots of the transcendental equation $J_n'(x) = 0$. These roots are different from the set β_{mn} given in (3.123). A few of these values are tabulated below.

m	γ_{m0}	γ_{m1}	γ_{m12}	γ_{m3}	γ_{m4}	γ_{m5}	γ_{m6}	γ_{m7}	γ_{m8}
0	0.0000	1.8412	3.0542	4.2012	5.3176	6.4156	7.5013	8.5778	9.6474
1	3.8317	5.3314	6.7061	8.0152	9.2824	10.5199	11.7349	12.9324	14.1155
2	7.0156	8.5363	9.9695	11.3459	12.6819	13.9872	15.2682	16.5294	17.7740
3	10.1735	11.7060	13.1704	14.5858	15.9641	17.3128	18.6374	19.9419	21.2291
4	13.3237	14.8636	16.3475	17.7887	19.1960	20.5755	21.9317	23.2681	24.5872
5	16.4706	18.0155	19.5129	20.9725	22.4010	23.8036	25.1839	26.5450	27.8893
6	19.6159	21.1644	22.6716	24.1449	25.5898	27.0103	28.4098	29.7907	31.1553
7	22.7601	24.3113	25.8260	27.3101	28.7678	30.2028	31.6179	33.0152	34.3966
8	25.9037	27.4571	28.9777	30.4703	31.9385	33.3854	34.8134	36.2244	37.6201
9	29.0468	30.6019	32.1273	33.6269	35.1039	36.5608	37.9996	39.4223	40.8302
10	32.1897	33.7462	35.2755	36.7810	38.2653	39.7306	41.1788	42.6115	44.0300

Table 3.2: Values of the roots (γ_{mn}) of $J_n'(x)$

Special Circumstances for $\gamma_{00} = 0$: For the Fourier-Bessel series in Examples 3.12 and 3.13 on pages 117-121, if we changed the boundary condition at $r = R$ from $X_m(R) = 0$ to $X_m'(R) = 0$, then the values of λ_m are given by $J_0'(\lambda_m R) = 0$. The coefficients a_m in the series expansion (3.105) can still be expressed by (3.107) where the denominator takes the form,

$$\int_0^R r\, [J_0(\lambda_k r)]^2\, dr = \tfrac{1}{2} R^2 [J_0(\lambda_k R)]^2, \qquad k \geq 0, \tag{3.136}$$

where $\lambda_k = \gamma_{k0}/R, \quad k = 0, 1, 2, \ldots$ with γ_{k0} given in the first column of Table 3.2. It should be noted that we are including $\lambda_0 = \gamma_{00}/R = 0$ because this leads to a non-zero value of the eigenfunction, $J_0(\lambda_0 r) = J_0(0) = 1$. With higher order Bessel-function expansions, the zero root is redundant since $J_n(0) = 0$ for $n \geq 1$. The coefficient a_0 in equation (3.107) on page 120 is then equal to

$$a_0 = \frac{\int_0^R f(r) r J_0(\lambda_0 r)\, dr}{\frac{1}{2} R^2 [J_0(\lambda_k R)]^2} \tag{3.137}$$

$$= \frac{\int_0^R f(r) r\, dr}{\frac{1}{2} R^2} = \frac{2}{R^2} \int_0^R f(r) r\, dr \tag{3.138}$$

It is not difficult to see that a_0 is, in fact, the average value of the function $f(r)$ over the circular disk $r \leq R$, quite analogous to the leading term of the Fourier cosine expansion in

Section 3.4.2 on pages 81-82. However, the general expression

$$a_m = \frac{\int_0^R f(r) r J_0(\lambda_m r)\, dr}{\frac{1}{2} R^2 \left[J_0(\lambda_k R)\right]^2}, \qquad m = 0, 1, 2, 3, \ldots \tag{3.139}$$

that would be the $n = 0$ case of (3.135), and would take the place of (3.107), is valid for all indices $m \geq 0$, as indicated.

Third Kind Boundary Condition at $r = R$**:** With the boundary condition of the third kind at $r = R$ involving a linear combination of $X_{mn}(r)$ and $X'_{mn}(r)$ (see pages 112-115), we would have instead of $X_{mn}(R) = 0$ or $X'_{mn}(R) = 0$, the mixed condition

$$X'_{mn}(R) = h_R X_{mn}(R), \tag{3.140}$$

which, for $X_{mn}(r) = J_n(\lambda_{mn} r)$, becomes

$$\lambda_{mn} J'_n(\lambda_{mn} R) = h_R J_n(\lambda_{mn} R). \tag{3.141}$$

With this boundary condition, the denominator in equation (3.129), given by equation (3.130), simplifies a bit and the and the expression for the set of coefficients a_m becomes

$$a_m = \frac{\int_0^R f(r) r J_n(\lambda_{mn} r)\, dr}{\frac{1}{2} R^2 \left(\frac{h_R^2}{\lambda_{mn}^2} + 1 - \frac{n^2}{\lambda_{mn}^2 R^2} \right) \left[J_n(\lambda_{mn} R)\right]^2}. \tag{3.142}$$

EXERCISES 3.6

Expand the functions below an order n Fourier-Bessel series, i.e.,

$$f(r) = \sum_{m=0}^{\infty} a_m J_n(\lambda_{mn} r), \qquad 0 \leq r \leq R,$$

where the requisite condition at $r = R$ is as specified.

1.

$$f(r) = \begin{cases} \left(1 - \frac{r^2}{a^2}\right), & 0 \leq r < a \\ 0, & a \leq r \leq R, \end{cases}$$

with $J_n(\lambda_{mn} R) = 0$.

2. $f(r) = r^2/R^2, 0 \leq r \leq R$ with $J'_n(\lambda_{mn} R) = 0$. Treat only the cases corresponding to $n \neq 0$.

3.7.3 Spherical Bessel Function Expansions

The differential operator

$$\frac{d^2}{dr^2} + \frac{2}{r}\frac{d}{dr} - n(n+1), \quad n = 0, 1, 2, 3, \ldots. \tag{3.143}$$

arises in many types of problems in spherical geometry. If a set of functions $X_n(r)$ were eigenfunctions of this operator corresponding to each integer value of n, we could write the differential equation for $X_n(r)$ as

$$\frac{d^2 X_n(r)}{dr^2} + \frac{2}{r}\frac{dX_n(r)}{dr} - \frac{n(n+1)}{r^2} X_n(r) = -\lambda^2 X_n(r), \quad n = 0, 1, 2, 3, \ldots, \tag{3.144}$$

where $-\lambda^2$ represents the eigenvalues corresponding to the eigenfunction $X_n(r)$. This is quite similar to the spherical Bessel's equation (2.74), and if we choose $x = \lambda r$, it becomes exactly that form. As discussed in Chapter 2, Section 2.4.3 (pages 70-74), the solutions to this equation can be expressed in terms of the fractional order Bessel functions, $J_{n+\frac{1}{2}}(x)$ and $Y_{n+\frac{1}{2}}(x)$. For every integer value of n there is a set of two linearly independent solutions of equation (3.144) which may be expressed as

$$X_n(r) = A j_n(\lambda r) + B y_n(\lambda r), \tag{3.145}$$

where we can refer to equations (2.75)-(2.77) on pages 70-71 for the complete definitions of $j_n(x)$ and $y_n(x)$. The values of λ will be defined by two boundary conditions at different values of r, again for each n. Thus, for a set of values of λ defined for every n, it is appropriate to refer to them with the indices m and n, i.e., $\lambda = \lambda_{mn}$. Furthermore, for every λ_{mn}, it would make sense to refer to the solution as $X_{mn}(r)$. For expansions of piecewise continuous functions of r in terms of the solutions of equation (3.144), we need to set it up in the Sturm-Liouville form. Thus, we rearrange (3.144) as

$$\frac{d}{dr}\left(r^2 \frac{dX_{mn}(r)}{dr}\right) - n(n+1) X_{mn}(r) = -\lambda_{mn}^2 r^2 X_{mn}(r). \tag{3.146}$$

Comparing this with equation (3.103), we can identify $x = r$, and the weighting factor $q(x) = q(r) = r^2$. Thus for the class of boundary conditions discussed on pages 111-116, we have the orthogonality relationship

$$\int_{r_1}^{r_2} r^2 X_{mn}(r) X_{m'n}(r)\, dr = 0, \quad m \neq m', \tag{3.147}$$

with the boundary conditions being satisfied at $r = r_1$ and $r = r_2$. As an example, we shall consider $r_1 = 0$ and $r_2 = R$.

Example 3.14

PROBLEM STATEMENT: For the differential equation (3.146), construct a set of eigenfunctions $X_{mn}(r)$ for unspecified n (but $n \neq 0$) that satisfy $X_{mn}(0) = 0$, and $X_{mn}(R) = 0$. Then expand $f(r) = 1$ as a series of these eigenfunctions over the range $0 \leq r \leq R$.

SOLUTION: The general solution is given by equation (3.145) which we may rewrite as

$$X_{mn}(r) = A_{mn} j_n(\lambda_{mn} r) + B_{mn} y_n(\lambda_{mn} r), \tag{3.148}$$

First, for requiring that $X_{mn}(0) = 0$, we note that $y_n(\lambda_{mn} r) \to -\infty$ as $x \to 0$. Therefore, we set $B_{mn} = 0$, reducing equation (3.148) to

$$X_{mn}(r) = A_{mn} j_n(\lambda_{mn} r).$$

Now, to satisfy $X_{mn}(R) = 0$, we go ahead and put

$$A_{mn} j_n(\lambda_{mn} R) = 0.$$

This can be satisfies with $j_n(\lambda_{mn} R) = 0$, or

$$\lambda_{mn} = \frac{\alpha_{mn}}{R},$$

where α_{mn} represents the roots of $j_n(x)$ which can be computed numerically, or found in mathematical handbooks. Now let us consider the expansion

$$f(r) = \sum_{m=0}^{\infty} A_{mn} j_n(\lambda_{mn} r). \tag{3.149}$$

Now, let us multiply both sides by $j_n(\lambda_{m'n} r)$, weight it with r^2, and integrate with respect to r over the range $[0, R]$, i.e.,

$$\begin{aligned}
\int_0^R f(r) r^2 j_n(\lambda_{m'n} r)\, dr &= \int_0^R r^2 j_n(\lambda_{m'n} r) \sum_{m=0}^{\infty} A_{mn} j_n(\lambda_{mn} r)\, dr \\
&= \sum_{m=0}^{\infty} A_{mn} \int_0^R r^2 j_n(\lambda_{m'n} r) j_n(\lambda_{mn} r)\, dr
\end{aligned}$$

By orthogonality, all the terms under the summation integrate to zero, except when $m = m'$. Therefore, keeping only the surviving term on the right-hand side, we have

$$\int_0^R f(r) r^2 j_n(\lambda_{m'n} r)\, dr = A_{m'n} \int_0^R r^2 j_n(\lambda_{m'n} r) j_n(\lambda_{m'n} r)\, dr.$$

We can now replace m' with m, or any other index for that matter, and write

$$\int_0^R f(r) r^2 j_n(\lambda_{mn} r)\, dr = A_{mn} \int_0^R r^2 j_n(\lambda_{mn} r) j_n(\lambda_{mn} r)\, dr. \tag{3.150}$$

The integral on the right-hand side may be evaluated using equation (3.130) which is valid for non-integer values of n. Alternately, the orthogonality relationship (3.66) can be used in the limit $\lambda_j \to \lambda_k$ to obtain the integral. The result is

$$\int_0^R r^2 \left[j_n(\lambda_{mn} r)\right]^2 dr = \frac{1}{2} R^3 \left\{ \left[j_n'(\lambda_{mn} R)\right]^2 + \frac{j_n(\lambda_{mn} R) j_n'(\lambda_{mn} R)}{\lambda_{mn} R} + \left(1 - \frac{n(n+1)}{\lambda_{mn}^2 R^2} \left[j_n(\lambda_{mn} R)\right]^2\right) \right\}, \tag{3.151}$$

where we have replaced m' with m since it is a dummy index. For the example being considered now, we have $X_{mn}(R) = j_n(\lambda_{mn} R) = 0$. Therefore, the above integral is

$$\int_0^R r^2 \left[j_n(\lambda_{mn} r)\right]^2 dr = \frac{1}{2} R^3 \left[j_n'(\lambda_{mn} R)\right]^2. \tag{3.152}$$

Using this in equation (3.150), we may write

$$A_{mn} = \frac{\int_0^R f(r) r^2 j_n(\lambda_{mn} r)\, dr}{\frac{1}{2} R^3 \left[j_n'(\lambda_{mn} R)\right]^2} \tag{3.153}$$

For the special case, $f(r) = 1$, the integral in the numerator may be expressed in terms of Bessel and Lommel functions [2]. Alternately, we may use the power-series expansion given by equation (2.76). Thus, with $f(r) = 1$, the integral is

$$\begin{aligned} \int_0^R r^2 j_n(\lambda_{mn} r)\, dr &= R^3 \int_0^{\lambda_{mn} R} x^2 j_n(x)\, dx \\ &= R^3 \sum_{k=0}^{\infty} \frac{(-1)^k (k+n)! 2^n}{k!(2k+2n+1)!} \int_0^{\lambda_{mn} R} x^{2k+2+n}\, dx \\ &= R^3 \sum_{k=0}^{\infty} \frac{(-1)^k (k+n)! 2^n \left(\lambda_{mn} R\right)^{2k+3+n}}{k!(2k+2n+1)!(2k+n+3)}, \end{aligned} \tag{3.154}$$

resulting in the following expression for A_{mn}:

$$A_{mn} = 2 \sum_{k=0}^{\infty} \frac{(-1)^k (k+n)! 2^n \left(\lambda_{mn} R\right)^{2k+3+n}}{k!(2k+2n+1)!(2k+n+3) \left[j_n'(\lambda_{mn} R)\right]^2} \tag{3.155}$$

Using this result in (3.149), we have the following expansion for $f(r) = 1$:

$$1 = \sum_{m=1}^{\infty} \left\{ 2 \sum_{k=0}^{\infty} \frac{(-1)^k (k+n)! 2^n \left(\lambda_{mn} R\right)^{2k+3+n}}{k!(2k+2n+1)!(2k+n+3) \left[j_n'(\lambda_{mn} R)\right]^2} \right\} j_n(\lambda_{mn} r), n \neq 0. \tag{3.156}$$

□

EXERCISES 3.7

Expand the functions below as a spherical Bessel series

$$f(r) = \sum_{m=0}^{\infty} A_m j_n(\lambda_m r), \quad 0 \le r \le R$$

for the specific values of the order n and the boundary conditions as stated. It is inderstood that at $r = 0$, either $j_n(0) = 0$ or $j_n'(0) = 0$. Use the expressions for $j_n(x)$ given in equations (2.79)-(2.81) on page 72.

1. $n = 0$, $f(r) = 1$ with $j_0(\lambda_m R) = 0$.
2. $n = 0$,

$$f(r) = \begin{cases} 1, & 0 \le x < a \\ 0, & a \le x \le R \end{cases} \qquad \text{with } j_0'(\lambda_m R) = 0$$

3. $n = 1$, $f(r) = r$ with $j_1(\lambda_m R) = 0$.
4. $n = 1$, $f(r) = r$ with $j_1'(\lambda_m R) = 0$.

3.7.4 Fourier-Bessel Integral Expansions

We may recall the development from Fourier sine and cosine series to corresponding integrals by letting the domain of expansion go to infinity (see pages 94-102). Using that analysis, we are able to express a piecewise continuous function $f(x)$ as a Fourier sine and cosine integral in the range $-\infty < x < \infty$, provided the integral $\int_{-\infty}^{\infty} |f(x)| dx$ exists. We end up with a Fourier transform $H(\lambda)$ of the function $f(x)$, and by an inverse integral operation, we can fully recover the function $f(x)$ from $H(\lambda)$ [see equations (3.43) and (3.44) on page 102]. Similar transform and inverse pairs exist for integrals involving Bessel functions in the domain $0 \le r < \infty$, and such a pair can be developed from the Fourier transform. Let us consider $f(x, y)$ as a function of two variables x and y, and carry out a double Fourier expansion of this function, i.e.,

$$f(x, y) = \frac{1}{2\pi} \int_{-\infty}^{\infty} \int_{-\infty}^{\infty} H(\lambda, \mu) e^{i\lambda x} d\lambda \; e^{i\mu y} d\mu \tag{3.157}$$

which can also be written as

$$f(x, y) = \frac{1}{2\pi} \int_{-\infty}^{\infty} \int_{-\infty}^{\infty} H(\lambda, \mu) e^{i(\lambda x + \mu y)} \, d\lambda \, d\mu. \tag{3.158}$$

The double Fourier transform is given by the expression

$$\begin{aligned} H(\lambda, \mu) &= \frac{1}{\sqrt{2\pi}} \int_{-\infty}^{\infty} \left[\frac{1}{\sqrt{2\pi}} \int_{-\infty}^{\infty} f(x, y) e^{-i\lambda x} \, dx \right] e^{-i\mu y} \, dy \\ &= \frac{1}{2\pi} \int_{-\infty}^{\infty} \int_{-\infty}^{\infty} f(x, y) e^{-i(\lambda x + \mu y)} \, dx \, dy. \end{aligned} \tag{3.159}$$

Let us consider these double integrals in two-dimensional xy and $\lambda\mu$ spaces, and transform to polar coordinates,

$$x = r\cos\theta, \quad y = r\sin\theta \qquad \text{and} \qquad \lambda = \chi\cos\phi, \quad \mu = \chi\sin\phi \tag{3.160}$$

The Jacobians for these transformations are

$$\frac{\partial(x,y)}{\partial(r,\theta)} = r \qquad \text{and} \qquad \frac{\partial(\lambda,\mu)}{\partial(\chi,\phi)} = \chi, \tag{3.161}$$

and therefore the area differentials are

$$dxdy = rdrd\theta \qquad \text{and} \qquad d\lambda d\mu = \chi d\chi d\phi.$$

If we choose $f(x, y)$ to be an axisymmetric function so that in the (r, θ) coordinate system it is a function of r only, we may write

$$f(x,y) = F(r).$$

Using both sets of polar coordinates in equation (3.159), we may write it as

$$\begin{aligned} H(\lambda,\mu) &= \frac{1}{2\pi}\int_{\theta=0}^{\theta=2\pi}\int_{r=0}^{r=\infty} F(r)e^{-ir(\lambda\cos\theta+\mu\sin\theta)}\,rdrd\theta \\ &= \frac{1}{2\pi}\int_{\theta=0}^{\theta=2\pi}\int_{r=0}^{r=\infty} F(r)e^{-ir\chi(\cos\phi\cos\theta+\sin\phi\sin\theta)}\,rdrd\theta \\ &= \frac{1}{2\pi}\int_{\theta=0}^{\theta=2\pi}\int_{r=0}^{r=\infty} F(r)e^{-ir\chi[\cos(\theta-\phi)]}\,rdrd\theta \end{aligned} \tag{3.162}$$

Now, let us transform $\theta' = \theta - \phi$ so that

$$H(\lambda,\mu) = \frac{1}{2\pi}\int_{\theta'=-\phi}^{\theta'=2\pi-\phi}\int_{r=0}^{r=\infty} F(r)e^{-ir\chi\cos\theta'}\,rdrd\theta'. \tag{3.163}$$

Here, the integrand is a periodic function in θ' with a period 2π. Since we are integrating over one full period, the range does does not matter as long as we maintain a span on 2π. Therefore, we may set the range of integration for θ' to $[0, 2\pi]$ and write

$$H(\lambda,\mu) = \frac{1}{2\pi}\int_{\theta'=0}^{\theta'=2\pi}\int_{r=0}^{r=\infty} F(r)e^{-ir\chi\cos\theta'}\,rdrd\theta'$$

and after switching the order of integration in the double integral, we have

$$H(\lambda,\mu) = \int_{r=0}^{r=\infty}\left[\frac{1}{2\pi}\int_{\theta'=0}^{\theta'=2\pi} e^{-ir\chi\cos\theta'}d\theta'\right]F(r)rdr \tag{3.164}$$

Here, we should note that the integral in the brackets is actually a representation of the Bessel function of order zero, i.e.,

$$J_0(\chi r) = \left[\frac{1}{2\pi} \int_{\theta'=0}^{\theta'=2\pi} e^{-ir\chi\cos\theta'} d\theta' \right]. \tag{3.165}$$

As a result, $H(\lambda, \mu)$ is represented by

$$H(\lambda, \mu) = \int_{r=0}^{r=\infty} J_0(\chi r) F(r) r dr \tag{3.166}$$

which indicates that in $H(\lambda, \mu)$ the dependence on λ and μ appears as a function of only $\chi = \sqrt{\lambda^2 + \mu^2}$. So, let us write $H(\lambda, \mu) = h(\chi)$. Now, using the transformations in equation (3.160), and the area differential $d\lambda d\mu = \chi d\chi d\phi$, we can write equation (3.158) as

$$\begin{aligned} f(x, y) = F(r) &= \frac{1}{2\pi} \int_{\phi=0}^{\phi=2\pi} \int_{\chi=0}^{\chi=\infty} h(\chi) e^{ir\chi(\cos\phi\cos\theta + \sin\phi\sin\theta)} \chi d\chi d\phi \\ &= \frac{1}{2\pi} \int_{\phi=0}^{\phi=2\pi} \int_{\chi=0}^{\chi=\infty} h(\chi) e^{ir\chi[\cos(\phi-\theta)]} \chi d\chi d\phi \end{aligned} \tag{3.167}$$

Again, reversing the order of integration and now using the transformation $\phi' = (\phi - \theta)$, we may apply the procedure similar to equations (3.163) to (3.166), and obtain the result

$$F(r) = \int_{\chi=0}^{\chi=\infty} h(\chi) J_0(\chi r)\, \chi d\chi, \tag{3.168}$$

together with (3.166) written as

$$h(\chi) = \int_{r=0}^{r=\infty} J_0(\chi r) F(r)\, r dr. \tag{3.169}$$

Here, $h(\chi)$ in equation (3.169) is referred to as the Hankel transform of $F(r)$, and integral in equation (3.168) is the inverse Hankel transform. This result can be extended to order-n Bessel functions so that the transform-inverse pair is

$$h_n(\chi) = \int_{r=0}^{r=\infty} J_n(\chi r) F_n(r)\, r dr, \qquad \text{and} \qquad F_n(r) = \int_{\chi=0}^{\chi=\infty} h_n(\chi) J_n(\chi r)\, \chi d\chi. \tag{3.170}$$

An arbitrary, piecewise continuous function, $F(r)$, can be represented as a Bessel function integral

$$\int_0^\infty |F(r)| \sqrt{r}\, dr \qquad \text{is finite.}$$

Let us now consider the exponential function $F(r) = e^{-kr}, \quad 0 \leq r < \infty$ as an example.

Example 3.15 *Problem Statement:* Expand the function

$$F(r) = e^{-kr}$$

as a Fourier-Bessel integral so that

$$F(r) = e^{-kr} = \int_0^\infty h(\chi) J_0(\chi r) \chi d\chi. \tag{3.171}$$

Solution:
According equations (3.169) and (3.170), the Hankel transform of $F(r) = e^{-kr}$ is

$$h(\chi) = \int_0^\infty J_0(\chi r) e^{-kr} r\, dr. \tag{3.172}$$

Using the integral representation of $J_0(\chi r)$ in equation (3.165), we can write equation (3.172) as

$$h(\chi) = \int_0^\infty \left[\frac{1}{2\pi} \int_{\theta'=0}^{\theta'=2\pi} e^{-ir\chi\cos\theta'} d\theta' \right] e^{-kr} r\, dr. \tag{3.173}$$

Now, switching the order of integration and combining the exponential terms, we have

$$\begin{aligned} h(\chi) &= \frac{1}{2\pi} \int_{\theta'=0}^{\theta'=2\pi} \left[\int_0^\infty e^{-(k+i\chi\cos\theta')r} r\, dr \right] d\theta' \\ &= \frac{1}{2\pi} \int_{\theta'=0}^{\theta'=2\pi} \frac{d\theta'}{(k+i\chi\cos\theta')^2} \\ &= \frac{k}{\left[k^2 - (i\chi)^2\right]^{\frac{3}{2}}} \\ &= \frac{k}{\left(k^2 + \chi^2\right)^{\frac{3}{2}}}. \end{aligned} \tag{3.174}$$

Upon substituting this result into equation (3.171), we end up with the integral representation,

$$e^{-kr} = \int_0^\infty \frac{k J_0(\chi r) \chi d\chi}{\left(k^2 + \chi^2\right)^{\frac{3}{2}}}. \tag{3.175}$$

□

EXERCISES 3.8

Express the following functions as Fourier-Bessel integrals in terms of $J_0(\lambda r)$:

1. $f(r) = \dfrac{1}{\sqrt{k^2 + r^2}}$

2. $f(r) = e^{-a^2 r^2}$

3.7.5 Fourier-Legendre Series

In Chapter 2, Section 2.4.1 (pages 58-62), we examined Legendre's equation,

$$\left(1-x^2\right)\frac{d^2y(x)}{dx^2}-2x\frac{dy(x)}{dx}+n(n+1)y(x)=0, \tag{3.176}$$

which has the general solution of the form

$$y(x)=A_nP_n(x)+B_nQ_n(x),$$

where $P_n(x)$ and $Q_n(x)$ are the two linearly independent solutions. Of these, the set $P_n(x)$ is a group of order n polynomials given by the set of equations (2.53) on page 61 for $n=1,2,\ldots,7$. Let us now examine equation (3.176) from the context of Sturm-Liouville theory, and develop Fourier-type series expansions. Equation (3.176) can be written as

$$\frac{d}{dx}\left[\left(1-x^2\right)\frac{dy(x)}{dx}\right]=-n(n+1)y(x). \tag{3.177}$$

Here, the solutions would be eigenfunctions of the operator

$$\frac{d}{dx}\left[\left(1-x^2\right)\frac{d}{dx}\right]$$

with eigenvalues $-\lambda_n^2=-n(n+1)$. For convenience, we may change the dependent variable from $y(x)$ to $X_n(x)$, noting that we have associated an index n corresponding to the eigenvalues. Thus,

$$\frac{d}{dx}\left[\left(1-x^2\right)\frac{dX_n(x)}{dx}\right]=-n(n+1)X_n(x). \tag{3.178}$$

Upon comparing this equation with the general form (3.59), we can identify the weighting factor as $q(x)=1$ and $p(x)=(1-x^2)$. Therefore, if we can satisfy boundary conditions or periodicity conditions of the type discussed on pages 111-116, an orthogonality relationship can be established. Since we are interested in the domain $-1\le x\le 1$, and since the solutions $Q_n(x)$ are all logarithmically singular at $x=\pm 1$, we would not be using these in an expansion including these end points. Therefore, we seek an expansion of the type,

$$F(x)=\sum_{n=0}^{\infty}A_nX_n(x)=\sum_{n=0}^{\infty}A_nP_n(x), \tag{3.179}$$

and we wish to establish the orthogonality of the set $\{P_n(x)\}$ over the range of integration $-1\le x\le 1$. Unlike Bessel functions, where the eigenvalues were determined by some homogeneous boundary conditions, we have taken the eigenvalues $-n(n+1)$ as given on the basis of other consideration. Therefore, we will look for boundary conditions or periodicity conditions as an outcome. Upon examining some of the polynomial solutions, we see that none of them are zero at the end-points $x = \pm 1$. However, let us examine equation (3.66) which encompasses in the general form the possibilities of end-point conditions leading to orthogonality. This requirement is repeated here for clarity.

$$\begin{aligned}(\lambda_j^2 - \lambda_k^2) \int_a^b q(x) X_k(x) X_j(x)\, dx \\ = -p(b)\left[X_j'(b)X_k(b) - X_k'(b)X_j(b)\right] \\ +p(a)\left[X_j'(a)X_k(a) - X_k'(a)X_j(a)\right] = 0, \end{aligned} \tag{3.180}$$

where $a = -1$, $b = 1$, and $p(x) = (1 - x^2)$ has already been identified. Clearly, $p(1) = p(-1) = 0$, and with $q(x) = 1$, we can see that for the present case, we have the orthogonality relationship,

$$\int_a^b q(x) X_k(x) X_j(x)\, dx = \int_{-1}^1 P_k(x) P_j(x)\, dx = 0, \qquad \text{provided} \quad j \neq k. \tag{3.181}$$

For $j = k$, it is not difficult to show that

$$\int_{-1}^1 [P_k(x)]^2\, dx = \frac{2}{2k+1} \tag{3.182}$$

With the orthogonality established, we may expand a piecewise-continuous function $f(x)$ in terms of the set $P_n(x)$, i.e.,

$$f(x) = \sum_{n-0}^{\infty} a_n P_n(x). \tag{3.183}$$

As usual, let us multiply each side by $P_m(x)$ and integrate over $[-1, 1]$. This results is

$$\begin{aligned}\int_{-1}^1 f(x) P_m(x)\, dx &= \sum_{n=0}^{\infty} a_n \int_{-1}^1 P_n(x) P_m(x)\, dx \\ &= a_0 \cdot 0 + a_1 \cdot 0 + \cdots + a_m \int_{-1}^1 [P_m(x)]^2\, dx + \cdots + a_n \cdot 0 + \cdots \\ &= a_m \left(\frac{2}{2m+1}\right), \end{aligned} \tag{3.184}$$

leading to

$$a_m = \left(m + \tfrac{1}{2}\right) \int_{-1}^1 f(x) P_m(x)\, dx, \qquad m = 0, 1, 2, 3, \ldots \tag{3.185}$$

For a defined function $f(x)$, an explicit expansion in terms of Legendre polynomials can be obtained. An example is considered next.

Example 3.16

Problem Statement: Expand the function

$$f(x) = \begin{cases} 1, & -1 \leq x \leq 0 \\ (1-x^2), & 0 \leq x \leq 1 \end{cases} \tag{3.186}$$

as a Fourier-Legendre series over the range $-1 \leq x \leq 1$.

Solution:

Assuming a series of the type given in equation (3.183), and using the result (3.183), we may write

$$\begin{aligned} a_n &= \left(n+\tfrac{1}{2}\right)\int_{-1}^{1} f(x)P_n(x)\,dx \\ &= \left(n+\tfrac{1}{2}\right)\left[\int_{-1}^{0} 1P_n(x)\,dx + \int_{0}^{1}\left(1-x^2\right)P_n(x)\,dx\right] \\ &= \left(n+\tfrac{1}{2}\right)\left[\int_{-1}^{1} 1P_n(x)\,dx - \int_{0}^{1} x^2P_n(x)\,dx\right] \end{aligned} \tag{3.187}$$

Let us consider each integral sepaarately. For the first one, considering that $P_0(x) = 1$, and using the orthogonality of the set $P_n(x)$ over $[0, 1]$, we have

$$\begin{aligned} \int_{-1}^{1} 1P_n(x)\,dx &= \int_{-1}^{1} P_0(x)P_n(x)\,dx, \\ &= \begin{cases} \displaystyle\int_{-1}^{1} [P_0(x)]^2\,dx, & n=0 \\ 0, & n \geq 1 \end{cases} \\ &= \begin{cases} 2, & n=0 \\ 0, & n \geq 1 \end{cases} \end{aligned} \tag{3.188}$$

For the second integral, we shall carry out integration by parts, and we need to use Legendre's equation in the following form:

$$\frac{d}{dx}\left[\left(1-x^2\right)\frac{dP_n(x)}{dx}\right] = -n(n+1)P_n(x) \tag{3.189}$$

so that the indefinite integral of $P_n(x)$ is

$$\int P_n(x)\,dx = -\frac{1}{n(n+1)}\left[\left(1-x^2\right)\frac{dP_n(x)}{dx}\right] + C \qquad (n \neq 0) \tag{3.190}$$

Using this expression, and ignoring the integration constant since it is redundant in this procedure, we can carry out the second integral in equation (3.187). Thus, for $n \neq 0$

$$\int_0^1 x^2P_n(x)\,dx = \left[x^2\int P_n(x)\,dx\right]_0^1 - \int_0^1\left[2x\int P_n(x)\,dx\right]dx$$

$$
\begin{aligned}
&= -\left\{\frac{x^2}{n(n+1)}\left[\left(1-x^2\right)\frac{dP_n(x)}{dx}\right]\right\}_0^1 \\
&\qquad + \int_0^1 \left\{\frac{2x}{n(n+1)}\left[\left(1-x^2\right)\frac{dP_n(x)}{dx}\right]\right\} dx \\
&= 0 + \int_0^1 \left[\frac{\left(2x-2x^3\right)}{n(n+1)}\frac{dP_n(x)}{dx}\right] dx \\
&= \left[\frac{\left(2x-2x^3\right)}{n(n+1)}P_n(x)\right]_0^1 - \int_0^1 \left[\frac{\left(2-6x^2\right)}{n(n+1)}P_n(x)\right] dx \\
&= 0 - \frac{2}{n(n+1)}\int_0^1 P_n(x)\,dx + \frac{6}{n(n+1)}\int_0^1 x^2 P_n(x)\,dx, \quad (3.191)
\end{aligned}
$$

where the first integral may be evaluated using equation (3.190), and the second one may be attached to the right-hand side. This procedure results in

$$
\begin{aligned}
\left[1-\frac{6}{n(n+1)}\right]\int_0^1 x^2 P_n(x)\,dx &= \frac{2}{n^2(n+1)^2}\left[\left(1-x^2\right)\frac{dP_n(x)}{dx}\right]_0^1 \\
\frac{n(n+1)-6}{n(n+1)}\int_0^1 x^2 P_n(x)\,dx &= -\frac{2P_n'(0)}{n^2(n+1)^2} \\
(n+3)(n-2)\int_0^1 x^2 P_n(x)\,dx &= -\frac{2P_n'(0)}{n(n+1)} \\
\int_0^1 x^2 P_n(x)\,dx &= -\frac{2P_n'(0)}{n(n+1)(n+3)(n-2)} \\
& \qquad n \neq 0, 2 \qquad (3.192)
\end{aligned}
$$

For $n = 0$,

$$
\begin{aligned}
\int_0^1 x^2 P_0(x)\,dx &= \int_0^1 x^2 \cdot 1\,dx \\
&= \tfrac{1}{3}x^3\Big|_0^1 \\
&= \tfrac{1}{3}, \qquad (3.193)
\end{aligned}
$$

and for $n = 2$,

$$
\begin{aligned}
\int_0^1 x^2 P_2(x)\,dx &= \int_0^1 x^2\left[\tfrac{1}{2}\left(3x^2-1\right)\right] dx \\
&= \int_0^1 \left(\tfrac{3}{2}x^3 - \tfrac{1}{2}x^2\right) dx \\
&= \left[\tfrac{3}{10}x^5 - \tfrac{1}{6}x^3\right]_0^1 \\
&= \tfrac{2}{15} \qquad (3.194)
\end{aligned}
$$

Using equations (3.192)-(3.194) in the expression for a_n given by (3.187), together with (3.188), we obtain

$$
a_0 = \left(0+\tfrac{1}{2}\right)\left[2-\tfrac{1}{3}\right] = \tfrac{5}{6}
$$

$$\begin{aligned} a_2 &= \left(2+\tfrac{1}{2}\right)\left(-\tfrac{2}{15}\right) = -\tfrac{1}{3} \\ a_n &= \left(n+\tfrac{1}{2}\right)\frac{2P_n'(0)}{n(n+1)(n+3)(n-2)}, \qquad n \neq 0, 2 \end{aligned} \tag{3.195}$$

From the polynomial expressions for Legendre polynomials (Chapter 2, page 61), we can see that for even n, $P_n(x)$ is a set of even functions, and therefore, $P_n'(x) = 0$, $n = 0, 2, 4, \ldots$. Consequently, $a_n = 0$ for even n except for $n = 0, 2$. Using this information, we may write the expression for the Fourier-Legendre expansion of the function $f(x)$ described by equation (3.186) as

$$f(x) = \tfrac{5}{6} - \tfrac{1}{3}P_2(x) + \sum_{n=1,3,5,\ldots}^{\infty} \frac{(2n+1)P_n'(0)P_n(x)}{n(n+1)(n+3)(n-2)}. \tag{3.196}$$

Looking at the expressions for $P_n(x)$ once again (page 61), it is not difficult to show that for odd values of n,

$$\begin{aligned} P_n'(0) &= (-1)^{\frac{1}{2}(n-1)} \frac{1\cdot 3\cdot 5\cdots(n-2)n}{2\cdot 4\cdot 6\cdots(n-3)(n-1)} \\ &= (-1)^{\frac{1}{2}(n-1)} \frac{n!}{\left(\left[\frac{1}{2}(n-1)\right]!2^{\frac{1}{2}(n-1)}\right)^2} \\ &= (-1)^{\frac{1}{2}(n-1)} \frac{n!}{\left(\left[\frac{1}{2}(n-1)\right]!\right)^2 2^{(n-1)}}, \qquad n = 1, 3, 5, \ldots \end{aligned} \tag{3.197}$$

or equivalently,

$$P_{2n+1}'(0) = (-1)^n \frac{(2n+1)!}{n!^2 2^{2n}}, \qquad n = 0, 1, 2, 3, \ldots \tag{3.198}$$

□

Half-Range Legendre Series

For situations with a function defined in the range $0 \leq x \leq 1$ or $-1 \leq x \leq 0$, it is possible to expand the function in either range as a Legendre series using only odd or only even polynomials, i.e.,

$$g(x) = \sum_{n=1,3,5,\ldots}^{\infty} a_n P_n(x), \tag{3.199}$$

or

$$f(x) = \sum_{n=0,2,4,\ldots}^{\infty} b_n P_n(x). \tag{3.200}$$

Let us limit our discussion at this time to $0 \leq x \leq 1$ since it is quite straightforward to extended the results to the negative range $-1 \leq x \leq 0$. In either case (odd or even), the expansion series would be a complete set, i.e., any integrable piecewise continuous function in the half range can be expressed as an even series or an odd series. Let us begin with the odd series (3.199), and recall the property that for any odd function that is continuous at $x = 0$, the function value at that point is zero, i.e., $X_n(0) = P_n(0) = 0$ for $n = 1, 3, 5, \ldots$. To test the orthogonality of this odd set, let us examine Legendre's equation (3.178), and compare it with the general second-order form (3.59). As we did earlier on page 134, the parameter $p(x)$ in equation (3.59) can be identified as $p(x) = \left(1 - x^2\right)$. Now, let us look at the general orthogonality relationship (3.180) with end points $a = 0$ and $b = 1$. With $p(1) = 0$, the first term on the right-hand side of (3.180) vanishes. The second term also vanishes on account of $X_n(a) = P_n(0) = 0$ for $n = 1, 3, 5, \ldots$, and we have orthogonality of the odd set over [0, 1]. Similarly, for even functions with continuous derivatives at $x = 0$, we have the corresponding derivatives vanishing, and the second term on the right-hand side of (3.180) is therefore zero. Thus, with $P_n'(0) = 0$ for $n = 0, 2, 4, \ldots$, the even set is orthogonal over [0, 1]. As discussed on page 115 following equation (3.81), we have here a combination of conditions at $x = 0$ and $x = 1$ that lead to orthogonality, i.e.,

$$X_n(a) = X_n(0) = 0 \qquad \text{or} \qquad X_n'(a) = X_n'(0) = 0,$$

and

$$p(b) = p(1) = 0.$$

Now that we have established the orthogonality of the sets $P_n(x)$, $n = 1, 3, 5, \ldots$ and $P_n(x)$, $n = 0, 2, 4, \ldots$, the respective coefficients a_n and b_n of the expansions (3.199) and (3.200) may be obtained. Let us start with (3.199). We multiply both sides by $P_m(x)$ and integrate over [0, 1], i.e.,

$$\begin{aligned} \int_0^1 g(x) P_m(x)\, dx &= \sum_{n=1,3,5,\ldots}^{\infty} a_n \int_0^1 P_n(x) P_m(x)\, dx \\ &= a_m \int_0^1 [P_m(x)]^2\, dx \end{aligned} \tag{3.201}$$

Since the integrand $[P_m(x)]^2$ is the *square* of a function, it is an even function if extended to the negative range. Therefore, it is not difficult to see that

$$\begin{aligned} \int_0^1 [P_m(x)]^2\, dx &= \tfrac{1}{2} \int_{-1}^1 [P_m(x)]^2\, dx \\ &= \tfrac{1}{2} \left(\frac{2}{2m+1} \right). \\ &= \frac{1}{2m+1}, \qquad m = 0, 1, 2, 3, \ldots \end{aligned} \tag{3.202}$$

It should be noted here that while we are dealing with an odd series, the result (3.202) is valid for both even and odd Legendre polynomials. Using this expression in equation (3.201), we end up with

$$a_m = (2m+1) \int_0^1 g(x) P_m(x)\, dx, \qquad m = 1, 3, 5, \ldots \tag{3.203}$$

for the series (3.199). Similarly, for the even Legendre series (3.200), the coefficients can be shown to be

$$b_m = (2m+1)\int_0^1 f(x)P_m(x)\,dx, \qquad m = 0, 2, 4, \ldots \tag{3.204}$$

As an example, let us consider $g(x) = 1$ for the odd series.

Example 3.17 *Problem Statement:* Expand the function $g(x) = 1$ as an odd Legendre series, i.e.,

$$1 = \sum_{n=1,3,5,\ldots}^{\infty} a_n P_n(x), \tag{3.205}$$

Solution:
Using equation (3.203), we have the following expression for a_n:

$$a_n = (2n+1)\int_0^1 P_n(x)\,dx. \tag{3.206}$$

Now using equation (3.190), it is not difficult to show that

$$\begin{aligned} a_n &= (2n+1)\left\{-\frac{1}{n(n+1)}\left[\left(1-x^2\right)\frac{dP_n(x)}{dx}\right]\right\}_{x=0}^{x=1} \\ &= \frac{2n+1}{n(n+1)}P_n'(0), \end{aligned} \tag{3.207}$$

and the series expansion for $g(x) = 1$ becomes

$$\begin{aligned} 1 &= \sum_{n=1,3,5,\ldots}^{\infty} \frac{2n+1}{n(n+1)}P_n'(0)P_n(x) \\ &= \sum_{n=1}^{\infty} \frac{4n+3}{(2n+1)(2n+2)}P_{2n+1}'(0)P_{2n+1}(x) \\ &= \sum_{n=1}^{\infty} \frac{4n+3}{(2n+1)(2n+2)}(-1)^n\frac{(2n+1)!}{n!^2 2^{2n}}P_{2n+1}(x), \\ &= \sum_{n=1}^{\infty} \frac{(-1)^n(4n+3)(2n)!}{2(n+1)n!^2 2^{2n}}P_{2n+1}(x), \\ &= \sum_{n=1}^{\infty} \frac{(-1)^n(4n+3)(2n)!}{n!(n+1)!2^{2n-1}}P_{2n+1}(x), \qquad 0 \le x \le 1. \end{aligned} \tag{3.208}$$

where we have used the expression for $P_{2n+1}'(0)$ given by (3.200). □

Next, as an even series expansion, let us consider the following example.

Example 3.18 *Problem Statement:* Expand the function $f(x) = x$ as an even Legendre series, i.e.,

$$x = \sum_{n=0,2,4,\ldots}^{\infty} b_n P_n(x), \tag{3.209}$$

Solution:
Using equation (3.204), we have the following expression for b_n:

$$b_n = (2n+1)\int_0^1 xP_n(x)\,dx. \tag{3.210}$$

To obtain the expression for the integral here, we make use of equation (3.190) once again. Thus,

$$\begin{aligned}
\int_0^1 xP_n(x)\,dx &= \left\{-x\frac{1}{n(n+1)}\left[\left(1-x^2\right)\frac{dP_n(x)}{dx}\right]\right\}_{x=0}^{x=1} \\
&\quad + \int_0^1 \frac{1}{n(n+1)}\left[\left(1-x^2\right)\frac{dP_n(x)}{dx}\right]dx \\
&= 0 + \frac{1}{n(n+1)}\left[\left(1-x^2\right)P_n(x)\Big|_0^1 + \int_0^1 2xP_n(x)\,dx\right] \\
&= -\frac{P_n(0)}{n(n+1)} + \frac{1}{n(n+1)}\int_0^1 2xP_n(x)\,dx
\end{aligned} \tag{3.211}$$

If we combine the integrals on the left and the right-hand sides, we obtain

$$\begin{aligned}
\left[1-\frac{2}{n(n+1)}\right]\int_0^1 xP_n(x)\,dx &= -\frac{P_n(0)}{n(n+1)} \\
\frac{n^2+n-2}{n(n+1)}\int_0^1 xP_n(x)\,dx &= -\frac{P_n(0)}{n(n+1)} \\
\frac{(n+2)(n-1)}{n(n+1)}\int_0^1 xP_n(x)\,dx &= -\frac{P_n(0)}{n(n+1)} \\
\int_0^1 xP_n(x)\,dx &= -\frac{P_n(0)}{(n+2)(n-1)}.
\end{aligned} \tag{3.212}$$

From the expressions for $P_n(x)$ given in Chapter 2, page 61, it is not difficult to show that for even values of n,

$$\begin{aligned}
P_n(0) &= (-1)^{\frac{1}{2}n}\frac{1\cdot 3\cdot 5\cdots(n-3)(n-1)}{2\cdot 4\cdot 6\cdots(n-2)n} \\
&= (-1)^{\frac{1}{2}n}\frac{(n-1)!}{\left(\frac{1}{2}n\right)!\left(\frac{1}{2}n-1\right)!2^{n-1}}, \qquad n = 2,4,4,\ldots
\end{aligned} \tag{3.213}$$

or equivalently,

$$P_{2n}(0) = (-1)^n\frac{(2n-1)!}{n!\,(n-1)!2^{2n-1}}, \qquad n = 1,2,3,\ldots, \tag{3.214}$$

and for $n = 0$, $P_0(0) = 1$. The series expansion for $f(x) = x$ can be written as

$$x = -\sum_{n=0,2,4\ldots}^{\infty}\frac{2n+1}{(n+2)(n-1)}P_n(0)P_n(x)$$

$$
\begin{aligned}
&= \tfrac{1}{2} - \sum_{n=1}^{\infty} \frac{4n+1}{(2n+2)(2n-1)} P_{2n}(0) P_{2n}(x) \\
&= \tfrac{1}{2} - \sum_{n=1}^{\infty} \frac{4n+1}{(2n+2)(2n-1)} (-1)^n \frac{(2n-1)!}{n!\,(n-1)!2^{2n-1}} P_{2n}(x) \\
&= \tfrac{1}{2} - \sum_{n=1}^{\infty} \frac{(-1)^n (4n+1)(2n-2)!}{(n+1)!\,(n-1)!2^{2n}} P_{2n}(x), \qquad 0 \le x \le 1. \quad (3.215)
\end{aligned}
$$

where we have used the expression for $P_{2n}(0)$ given by (3.214). □

Generating Function

For every type of orthogonal polynomials, say $\{X_n(x),\ n = 0, 1, 2, 3, \ldots\}$, it is possible to define a generating function $f(x,t)$ such that

$$f(x,t) = \sum_{n=0}^{\infty} t^n X_n(x).$$

In the case of Legendre polynomials, we have

$$f(x,t) = \frac{1}{\sqrt{1-2xt+t^2}}.$$

By expanding as a power series in t, it is possible to show that the coefficient of t^n is actually $P_n(x)$, i.e.,

$$\frac{1}{\sqrt{1-2xt+t^2}} = \sum_{n=0}^{\infty} t^n P_n(x). \tag{3.216}$$

This result is quite useful in expressing parameters involving the distance between two points in space as a Legendre series (see Example 6.2 on pages 238-241)

EXERCISES 3.9

1. Expand $f(x) = x^2$ as a full-range $(-1 \le x \le 1)$ Legendre polynomial series.
2. Now Expand $f(x) = x^2$ as an odd half-range $(0 \le x \le 1)$ Legendre polynomial series.
3. Expand the function,

$$f(x) = \begin{cases} 1, & -1 \le x \le 0 \\ (1-x), & 0 \le x \le 1 \end{cases}$$

as a Legendre polynomial series over the range $-1 \le x \le 1$.

4. Expand $f(x) = x + x^2$ as an even half-range $(0 \le x \le 1)$ Legendre polynomial series.

Chapter 4

Application to Partial Differential Equations

For physical systems governed by more than one independent variable, the differential equations describing the system are generally *partial* differential equations. To qualify in this category, there must be derivatives with respect to two or more independent variables. The simplest case would involve two independent variables. For most systems of physical interest, these independent variables are usually spatial coordinates and/or time. The general form of a linear second order partial differential equation involving two independent variables is

$$A'\frac{\partial^2 u}{\partial x^2} + a\frac{\partial u}{\partial x} + B'\frac{\partial^2 u}{\partial x \partial y} + C'\frac{\partial^2 u}{\partial y^2} + c\frac{\partial u}{\partial y} + D'u = 0 \tag{4.1}$$

where the coefficients A', a, B', C', c and D' may not in general be constants. However, for simplicity we shall assume these coefficients to be constants. Here, the dependent variable, u, is a function of two independent variables, (x, y), i.e.,

$$u = u(x, y).$$

4.1 Classification

The character of the equation is determined by these coefficients and their relation with each other, particularly the coefficients of the highest derivatives. In fact, it is not difficult to show that the above equation can be reduced to

$$A\frac{\partial^2 u'}{\partial x'^2} + B\frac{\partial^2 u'}{\partial x' \partial y'} + C\frac{\partial^2 u'}{\partial y'^2} = 0, \tag{4.2}$$

provided that A' and C' are not zero. Here $x'(x, y)$ and $y'(x, y)$ are new independent variables and the dependent variable, $u(x, y)$, may need to be modified with the introduction of another function, $g(x, y)$, so that

$$u'(x', y') = u(x, y)g(x, y)$$

Regardless of the presence of the lower order derivatives, the character of the second order differential equation is determined by the coefficients of its second order derivatives. Therefore, we shall examine equation (4.2).

We first linearly transform the independent variables as follows:

$$\xi = \alpha x' + \beta y', \qquad \eta = \gamma x' + \delta y'.$$

As a result, we have

$$\left(A\alpha^2 + B\alpha\beta + C\beta^2\right)\frac{\partial^2 u'}{\partial \xi^2} + \left[2A\alpha\gamma + B(\alpha\gamma + \beta\delta) + 2C\beta\delta\right]\frac{\partial^2 u'}{\partial \xi \partial \eta} + \left(A\gamma^2 + B\gamma\delta + C\delta^2\right)\frac{\partial^2 u'}{\partial \eta^2} = 0$$

We now attempt to make the coefficients of $\partial^2 u'/\partial \xi^2$ and $\partial^2 u'/\partial \eta^2$ equal to zero. This will require

$$A\alpha^2 + B\alpha\beta + C\beta^2 = 0,$$

and

$$A\gamma^2 + B\gamma\delta + C\delta^2 = 0.$$

It is not difficult to see that

$$\left(\frac{\beta}{\alpha}, \frac{\delta}{\gamma}\right) = \frac{-B \pm \sqrt{B^2 - 4AC}}{2C}.$$

Depending on the SIGN of the discriminant, $(B^2 - 4AC)$, the values of β/α and δ/γ, are real or complex. The differential equation is classified accordingly, as given in the table below

DISCRIMINANT VALUE	CHARACTER OF ROOTS	CHARACTER OF EQUATION
$(B^2 - 4AC) > 0$	real roots	hyperbolic
$(B^2 - 4AC) < 0$	complex roots	elliptic
$(B^2 - 4AC) = 0$	repeated roots	parabolic

Prototypes of Partial Differential Equations

EQUATION	MATHEMATICAL FORM	CLASSIFICATION
Laplace's Equation	$\frac{\partial^2 u}{\partial x^2} + \frac{\partial^2 u}{\partial y^2} = 0$	elliptic
Wave Equation	$\frac{1}{c^2}\frac{\partial^2 u}{\partial t^2} = \frac{\partial^2 u}{\partial x^2}$	hyperbolic
Diffusion Equation	$\frac{1}{\alpha}\frac{\partial T}{\partial t} = \frac{\partial^2 T}{\partial x^2}$	parabolic

In three or more variables,

$$\nabla^2 u = \frac{\partial^2 u}{\partial x^2} + \frac{\partial^2 u}{\partial y^2} + \frac{\partial^2 u}{\partial z^2} = 0 \qquad \text{Elliptic} \tag{4.3}$$

$$\frac{1}{c^2}\frac{\partial^2 u}{\partial t^2} = \nabla^2 u \qquad \text{Hyperbolic} \tag{4.4}$$

$$\frac{1}{\alpha}\frac{\partial T}{\partial t} = \nabla^2 T \qquad \text{Parabolic} \tag{4.5}$$

4.2 Initial and Boundary Conditions

Let us start by considering Laplace's equation in two dimensions. This is an elliptic equation.

Laplace's Equation

$$\frac{\partial^2 u}{\partial x^2} + \frac{\partial^2 u}{\partial y^2} = 0 \tag{4.6}$$

If this was over a two-dimensional rectangular domain, say, $0 < x < a$, and $0 < y < b$, there should be one condition on each segment of the boundary. For example, we may have

$$u(0,y) = u_1(y), \quad \left.\frac{\partial u(x,y)}{\partial x}\right|_{x=a} = q_1(y), u(x,0) = u_2(x), \quad \text{and} \quad u(x,b) = u_3(x), \tag{4.7}$$

as shown in Figure 4.1 For an arbitrarily shaped boundary, the boundary condition may be

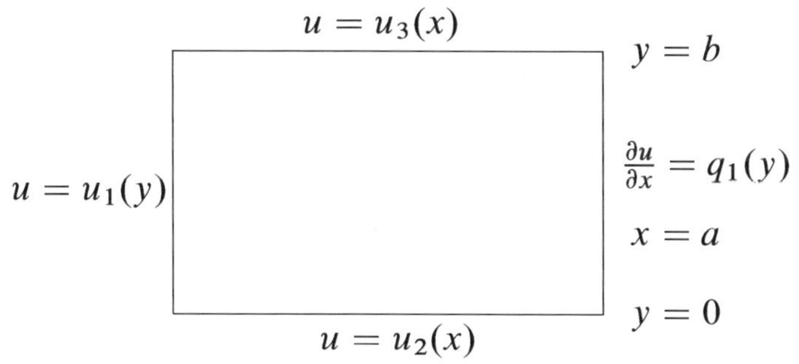

$u = u_3(x)$ $y = b$

$u = u_1(y)$ $\frac{\partial u}{\partial x} = q_1(y)$ $x = a$

$y = 0$

$u = u_2(x)$

Figure 4.1: An example of boundary conditions on a rectangle for an elliptic equation.

stated as a set of conditions on various segments of the boundary so that each point has only one condition, as shown in the figure 4.2.

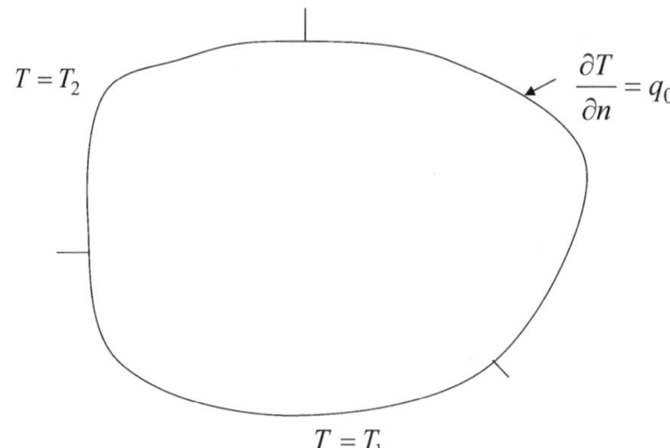

Figure 4.2: An example of boundary conditions for an elliptic equation on an arbitrarily shaped two-dimensional region.

Heat Equation

This is an example of a parabolic equation. For the one-dimensional time-dependent equation,

$$\frac{1}{\alpha}\frac{\partial T}{\partial t} = \frac{\partial^2 T}{\partial x^2} \tag{4.8}$$

the initial condition may be stated as

$$T(x, 0) = T_0(x) \tag{4.9}$$

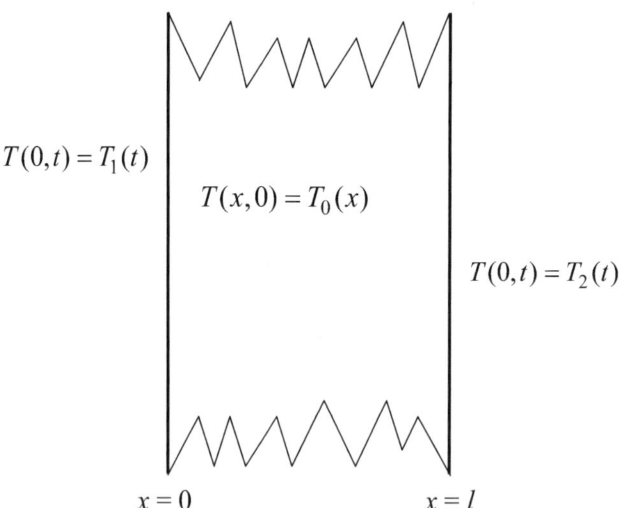

Figure 4.3: Initial and boundary conditions for one-dimensional parabolic equation

Since the equation has only a first derivative in time, there is only one initial condition required. There are, however, two boundary conditions, one for each segment of the boundary, as shown in figure 4.3. Mathematically, the boundary conditions are

$$T(0, t) = T_1(t) \qquad \text{and} \qquad T(l, t) = T_2(t).$$

In two dimensions, there would be a single condition on every point on the boundary, while there is one one initial condition. That is, for the differential equation,

$$\frac{1}{\alpha}\frac{\partial T}{\partial t} = \frac{\partial^2 T}{\partial x^2} + \frac{\partial^2 T}{\partial y^2} \tag{4.10}$$

the initial condition is

$$T(x, y, 0) = T_0(x, y),$$

and examples of boundary conditions are

$$\begin{aligned} T(0, y, t) &= T_1(y, t), \\ T(x, b, t) &= T_2(x, t), \\ T(a, y, t) &= T_3(y, t), \\ \left.\frac{\partial T(x, y, t)}{\partial y}\right|_{y=0} &= q_0(x, t), \end{aligned}$$

where we emphasize that there is a single condition on every point on the boundary. While we have stated the boundary condition as time-dependent, we shall restrict our analysis to time-independent cases. Also, the boundary conditions can involve a linear combination of $T(x, y, t)$ and its normal derivative as a single condition (known as a mixed condition).

Wave Equation

For the wave equation in two dimensions,

$$\frac{1}{c^2}\frac{\partial^2 u}{\partial t^2} = \frac{\partial^2 u}{\partial x^2} + \frac{\partial^2 u}{\partial y^2}, \tag{4.11}$$

boundary conditions on a rectangle are similar in character to the two-dimensional heat equation, i.e.,

$$u = 0 \quad \text{at} \quad x = 0, \quad x = a, \quad y = 0, \quad y = b.$$

Again, there is a single condition at every point on the boundary. However, the equation has a second derivative in time, and therefore, there should be two initial conditions. These may be stated as

$$\begin{aligned} u(x, y, 0) &= u_0(x, y) \\ \frac{\partial u(x, y, 0)}{\partial t} &= v_0(x, y). \end{aligned}$$

4.3 Method of Characteristics

4.3.1 D'Alembert's Solution for the One-Dimensional Wave Equation

Consider an infinitely long string,

$$\frac{1}{c^2}\frac{\partial^2 u}{\partial t^2} = \frac{\partial^2 u}{\partial x^2} \tag{4.12}$$

with initial conditions

$$u(x,0) = \phi(x) \tag{4.13}$$

and

$$\frac{\partial u}{\partial t}(x,0) = \theta(x). \tag{4.14}$$

There are two boundary conditions

$$u \to 0 \quad \text{as} \quad x \to \pm\infty. \tag{4.15}$$

Let us transform the independent variables as follows:

$$\xi = x - ct, \tag{4.16}$$
$$\eta = x + ct, \tag{4.17}$$

so that

$$\begin{aligned}
\frac{\partial u}{\partial x} &= \frac{\partial u}{\partial \xi}\frac{\partial \xi}{\partial x} + \frac{\partial u}{\partial \eta}\frac{\partial \eta}{\partial x} && (4.18)\\
&= \left(\frac{\partial u}{\partial \xi} + \frac{\partial u}{\partial \eta}\right) && (4.19)\\
&= \left(\frac{\partial}{\partial \xi} + \frac{\partial}{\partial \eta}\right) u && (4.20)\\
\frac{\partial^2 u}{\partial x^2} &= \left(\frac{\partial}{\partial \xi} + \frac{\partial}{\partial \eta}\right)^2 u && (4.21)\\
\frac{\partial u}{\partial t} &= \frac{\partial u}{\partial \xi}\frac{\partial \xi}{\partial t} + \frac{\partial u}{\partial \eta}\frac{\partial \eta}{\partial t} && (4.22)\\
&= \frac{\partial u}{\partial \xi}(-c) + \frac{\partial u}{\partial \eta}(c) && (4.23)\\
&= c\left(-\frac{\partial}{\partial \xi} + \frac{\partial}{\partial \eta}\right) u && (4.24)
\end{aligned}$$

$$\begin{aligned}
\frac{\partial^2 u}{\partial t^2} &= c^2\left(-\frac{\partial}{\partial \xi} + \frac{\partial}{\partial \eta}\right)^2 u && (4.25)\\
&= c^2\left(\frac{\partial^2}{\partial \xi^2} - 2\frac{\partial^2}{\partial \xi \partial \eta} + \frac{\partial^2}{\partial \eta^2}\right) u && (4.26)
\end{aligned}$$

Now substituting these expressions for $(\partial^2 u/\partial t^2)$ and $(\partial^2 u/\partial x^2)$ into the wave equation.

$$\frac{1}{c^2}\frac{\partial^2 u}{\partial t^2} - \frac{\partial^2 u}{\partial x^2} = 0, \tag{4.27}$$

we obtain

$$\left(\frac{\partial^2 u}{\partial \xi^2} - 2\frac{\partial^2 u}{\partial \xi \partial \eta} + \frac{\partial^2 u}{\partial \eta^2}\right) - \left(\frac{\partial^2 u}{\partial \xi^2} + 2\frac{\partial^2 u}{\partial \xi \partial \eta} + \frac{\partial^2 u}{\partial \eta^2}\right) = 0, \tag{4.28}$$

which reduces to

$$-4\frac{\partial^2 u}{\partial \xi \partial \eta} = 0, \tag{4.29}$$

or

$$\frac{\partial^2 u}{\partial \xi \partial \eta} = 0 \tag{4.30}$$

Integration with respect to η gives

$$\frac{\partial u}{\partial \xi} = F(\xi) = f'(\xi), \tag{4.31}$$

and then integration with respect to ξ leads to

$$u = f(\xi) + g(\eta), \tag{4.32}$$

which is the same as

$$u(x,t) = f(x - ct) + g(x + ct) \tag{4.33}$$

Satisfy the initial conditions

$$u(x,0) = f(x) + g(x) = \phi(x) \tag{4.34}$$

$$\frac{\partial u(x,0)}{\partial t} = -cf'(x) + cg'(x) = \theta(x) \tag{4.35}$$

$$-f'(x) + g'(x) = \frac{1}{c}\theta(x) \tag{4.36}$$

Integration with respect to x from $x = x_0$ to x yields

$$-f(x) + f(x_0) + g(x) - g(x_0) = \frac{1}{c}\int_{x_0}^{x} \theta(x)\,dx \tag{4.37}$$

$$-f(x) + g(x) = \frac{1}{c}\int_{x_0}^{x} \theta(x)dx - f(x_0) + g(x_0) \tag{4.38}$$

$$f(x) + g(x) = \phi(x) \tag{4.39}$$

$$2g(x) = \phi(x) + \frac{1}{c}\int_{x_0}^{x} \theta(x)\,dx - f(x_0) + g(x_0) \tag{4.40}$$

$$g(x) = \frac{1}{2}\left\{\phi(x) + \frac{1}{c}\int_{x_0}^{x} \theta(x)\,dx - f(x_0) + g(x_0)\right\} \tag{4.41}$$

$$f(x) = \frac{1}{2}\left\{\phi(x) - \frac{1}{c}\int_{x_0}^{x} \theta(x)dx + f(x_0) - g(x_0)\right\} \tag{4.42}$$

$$u(x,t) = f(x-ct) + g(x+ct) = \frac{1}{2}\left\{\phi(x-ct) - \frac{1}{c}\int_{x_0}^{x-ct} \theta(x)dx + f(x_0) - g(x_0)\right\} \tag{4.43}$$

$$+ \quad \frac{1}{2}\left\{\phi(x+ct) + \frac{1}{c}\int_{x_0}^{x+ct} \theta(x)dx - f(x_0) + g(x_0)\right\} \tag{4.44}$$

$$= \quad \frac{1}{2}\left\{\phi(x-ct) + \phi(x+ct) + \frac{1}{c}\int_{x_0}^{x+ct} \theta(x)dx + \frac{1}{c}\int_{x-ct}^{x_0} \theta(x)dx\right\} \tag{4.45}$$

$$= \quad \frac{1}{2}\left\{\phi(x-ct) + \phi(x+ct) + \frac{1}{c}\int_{x-ct}^{x+ct} \theta(x)dx\right\} \tag{4.46}$$

Example 4.1 As an example consider the initial conditions,

$$u(x,0) = \phi(x) = \frac{1}{1+x^2} \qquad \text{and} \qquad \left.\frac{\partial u(x,t)}{\partial t}\right|_{t=0} = 0.$$

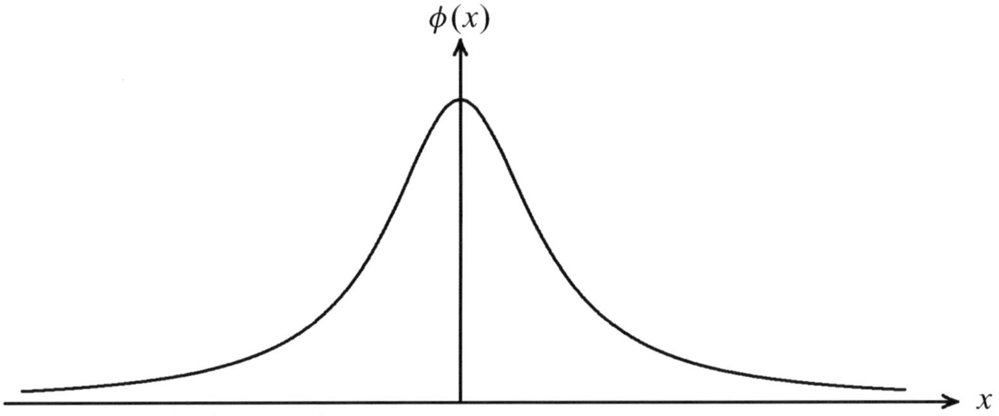

The expression for the displacement is

$$u(x,t) = \tfrac{1}{2}\left[\phi(x-ct) + \phi(x+ct)\right] = \tfrac{1}{2}\left[\frac{1}{(x-ct)^2} + \frac{1}{(x+ct)^2}\right]. \tag{4.47}$$

To understand the behavior of this solution, let us examine the separate parts

$$u(x,t) = \tfrac{1}{2}\phi(x-ct) \qquad \text{and} \qquad \tfrac{1}{2}\phi(x+ct),$$

at a time $t = t_1 > 0$. The function, $\phi(x)$ peaks at $x = 0$. Therefore, to follow the peak, we need to see where the respective arguments of $\phi(x - ct_1)$ and $\phi(x + ct_1)$ are zero. We notice that

$$x - t_1 \quad \text{is zero if} \quad x = ct_1,$$

and

$$x + ct_1 \quad \text{is zero if} \quad x = -ct_1.$$

$$u(x,t_1) = \tfrac{1}{2}\left[\phi(x-ct_1) + \phi(x+ct_1)\right] \tag{4.48}$$

Let us visualize the functions $\frac{1}{2}\phi(x - ct_1)$ and $\frac{1}{2}\phi(x + ct_1)$, as well as $u(x,t_1)$. The picture shows the wave splitting into two smaller wavelets, each propagating away from the center. This is illustrated in the next figure where the displacement $u(x,t)$ is shown for $ct = 0, 0.8, 1.5, 2.5$.

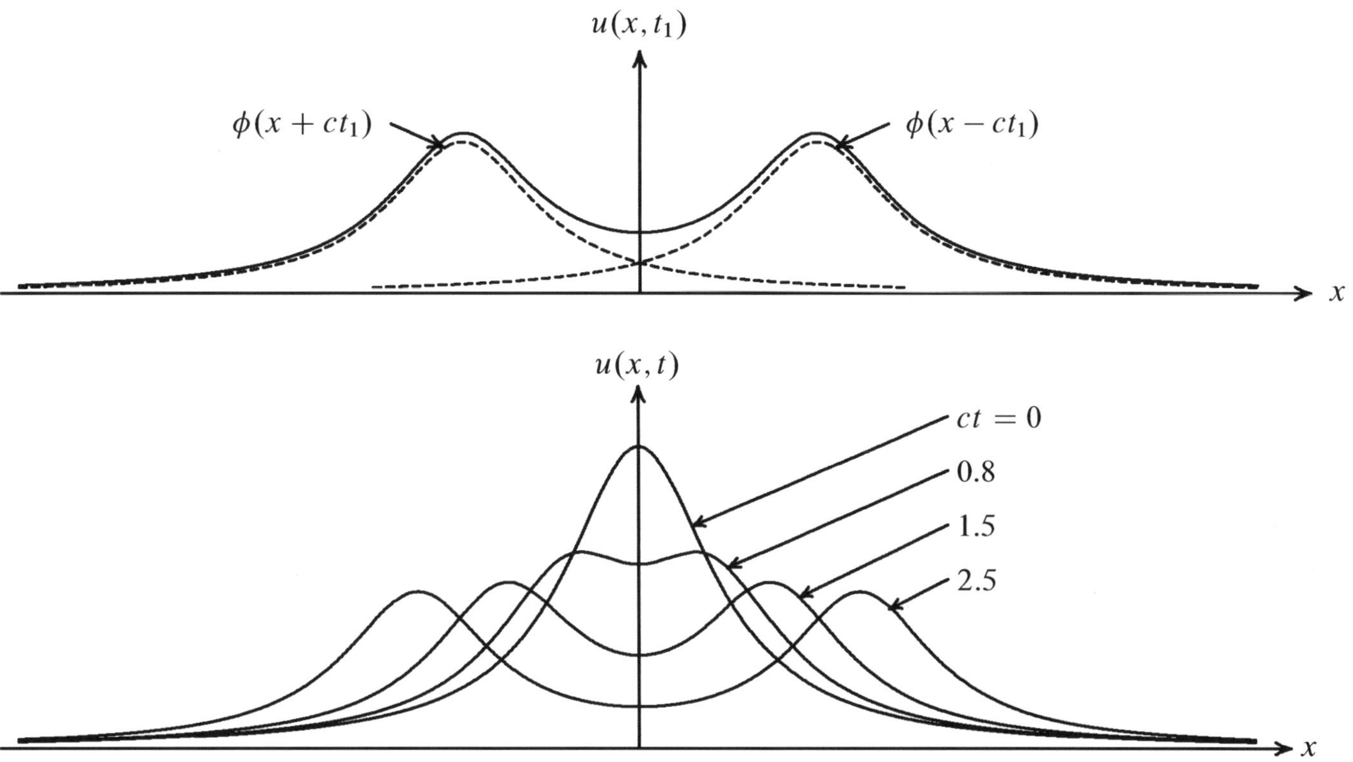

4.3.2 The Characteristics

With the general linear transformation,

$$\xi = \alpha x + \beta t$$

$$\eta = \gamma x + \delta t$$

we eliminate

$$\frac{\partial^2 u}{\partial \xi^2} \quad \text{and} \quad \frac{\partial^2 u}{\partial \eta^2}$$

by choosing $\alpha = 1, \quad \beta = c, \quad \gamma = 1, \quad \delta = c$. The lines defined by constant ξ and constant η are the characteristics of the wave equation. For the previous example involving the standard form of the one-dimensional wave equation (4.12), the characteristics are the lines defined by $x - ct =$ constant and $x + ct =$ constant. In the next exercise, the solution procedure will lead to a different set of characteristics pertaining to the equation therein.

EXERCISES 4.1 Obtain the 'general' solution for the differential equation:

$$3\frac{\partial^2 u}{\partial t^2} - 4\frac{\partial^2 u}{\partial t \partial x} + \frac{\partial^2 u}{\partial x^2} = 0.$$

Proceed as follows: Assume

$$\xi = \alpha x + \beta t,$$

$$\eta = \gamma x + \delta t,$$

and write the differential equation in terms of ξ and η. Choose $\alpha = \gamma = 1$ and obtain the values of β and δ so that there are only the cross derivative terms left. Then obtain the general solution in terms of ξ and η.

Now apply the initial conditions

$$u(x, 0) = \phi(x)$$

and

$$\left.\frac{\partial u}{\partial t}\right|_{t=0} = 0.$$

Consider the special case when $\phi(x)$ has the form shown below, and plot $u(x, t)$ for $t = 1$ and $t = 2$.

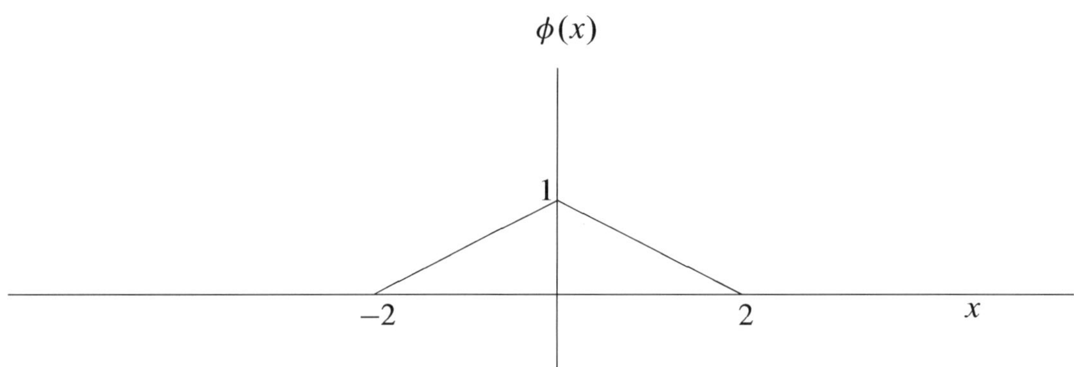

4.4 The Method of Separation of Variables

The method of separation of variables is applicable to a large class of linear partial differential equations. The technique can be better appreciated with an example.

Example 4.2 Let us consider the two-dimensional Laplace's equation,

$$\frac{\partial^2 T}{\partial x^2} + \frac{\partial^2 T}{\partial y^2} = 0, \tag{4.49}$$

for the rectangular region shown. Here we assume that a solution may be expressed as a product of two single-variable functions, $X(x)$ and $Y(y)$ so that

$$T(x, y) = X(x)Y(y).$$

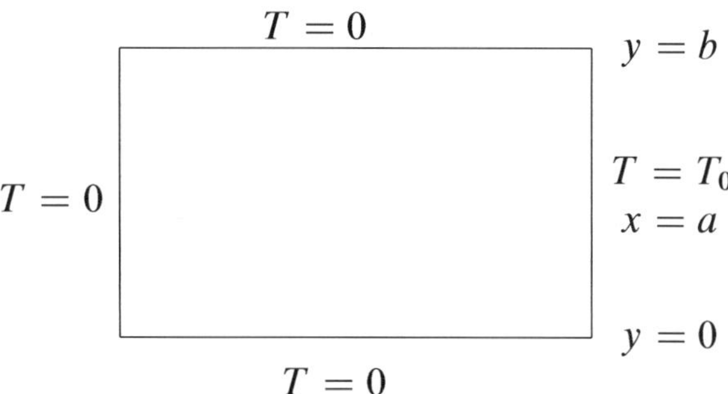

Figure 4.4: Boundary conditions for example 4.2

This may be only one of many possible solutions. The derivatives of $T(x, y)$ may be written as

$$\begin{aligned} \frac{\partial T}{\partial x} &= X'(x)Y(y), \\ \frac{\partial^2 T}{\partial x^2} &= X''(x)Y(y), \end{aligned} \tag{4.50}$$

$$\begin{aligned} \frac{\partial T}{\partial y} &= X(x)Y'(y), \\ \frac{\partial^2 T}{\partial y^2} &= X(x)Y''(y), \end{aligned} \tag{4.51}$$

Upon the substitution of these expressions (4.50) and (4.51) for the partial derivatives into the original differential equation (4.49) gives

$$X''(x)Y(y) + X(x)Y''(y) = 0. \tag{4.52}$$

Let us divide this equation by by $X(x)Y(y)$. As a result, we have

$$\frac{X''(x)Y(y)}{X(x)Y(y)} + \frac{X(x)Y''(y)}{X(x)Y(y)} = 0, \tag{4.53}$$

or

$$\frac{X''(x)}{X(x)} + \frac{Y''(y)}{Y(y)} = 0. \tag{4.54}$$

This may be written as

$$\frac{X''(x)}{X(x)} = -\frac{Y''(y)}{Y(y)}. \tag{4.55}$$

This equation states that a function x on the left-hand side equals to a function of y on the right-hand side. The only way we can have a function of x being equal to a function of y (which are independent variables) is that the function must be a constant function. Therefore, we let

$$\frac{X''}{X} = -\frac{Y''}{Y} = \lambda^2, \tag{4.56}$$

where λ^2, at this point could be any constant, real or complex. The actual determination of this constant will be made on the basis of several other considerations which we shall see later in this example. However, first we must solve the ordinary differential equations represented by equation (4.56). These are

$$X''(x) - \lambda^2 X(x) = 0, and \tag{4.57}$$

$$Y''(y) + \lambda^2 Y(y) = 0. \tag{4.58}$$

These are simple ordinary differential equations with constant coefficients. Following the procedure in Chapter 1, we assume an exponential solution for $X(x)$ of the form

$$X(x) = e^{mx} \tag{4.59}$$

and substitute into Equation (4.57). This leads to

$$\begin{aligned} m^2 e^{mx} - \lambda^2 e^{mx} &= 0 \\ \left(m^2 - \lambda^2\right) e^{mx} &= 0 \\ \left(m^2 - \lambda^2\right) &= 0 \\ m^2 &= \lambda^2 \\ m &= \pm\lambda \end{aligned} \tag{4.60}$$

With $m = \lambda$ and $m = -\lambda$, we may construct the general solution of the form

$$X(x) = A^* e^{\lambda x} + B^* e^{-\lambda x} = A \sinh \lambda x + B \cosh \lambda x. \tag{4.61}$$

Following the same procedure for Equation (4.58), and taking $Y(y) = e^{my}$ as a solution form, we find $m = \pm i\lambda$. Thus the general solution for $Y(y)$ is

$$Y(y) = C \sin \lambda y + D \cos \lambda y. \tag{4.62}$$

The product $T(x, y) = X(x)Y(y)$ can now be written as

$$X(x)Y(y) = [A \sinh \lambda x + B \cosh \lambda x]\,[C \sin \lambda y + D \cos \lambda y]. \tag{4.63}$$

As stated earlier, the value of λ has not been specified. In fact, any value of λ will yield a product $X(x)Y(y)$ that will satisfy the differential equation (4.49). For most situations, the values of λ are restricted to certain class of numbers, and we shall come to that later. At the moment, let us take λ to be an arbitrary set of numbers,

$$\lambda = \lambda_1, \lambda_2, \lambda_3, \lambda_4, \ldots.$$

For each one of these values of λ, we may write the solution as:

$$\begin{aligned} T_1 &= X_1(x)Y_1(y) = (A_1 \sinh \lambda_1 x + B_1 \cosh \lambda_1 x)(C_1 \sin \lambda_1 y + D_1 \cos \lambda_1 y) \\ T_2 &= X_2(x)Y_2(y) = (A_2 \sinh \lambda_2 x + B_2 \cosh \lambda_2 x)(C_2 \sin \lambda_2 y + D_2 \cos \lambda_2 y) \\ &\vdots \\ T_n &= X_n Y_n = (A_n \sinh \lambda_n x + B_n \cosh \lambda_n x)(C_n \sin \lambda_n y + D_n \cos \lambda_n y), \\ &\vdots \end{aligned}$$

Since the differential equation is linear, all these solutions may be added and the sum will satisfy the differential equation. Therefore, the expression

$$\begin{aligned} T(x,y) &= T_1(x,y) + T_2(x,y) + T_3(x,y) + \dots && (4.64)\\ &= X_1(x)Y_1(y) + X_2(x)Y_2(y) + X_3(x)Y_3(y) + \dots\dots && (4.65)\\ &= \sum_{n=1}^{\infty} X_n(x)Y_n(y) && (4.66)\\ &= \sum_{n=1}^{\infty} (A_n \sinh\lambda_n x + B_n \cosh\lambda_n x)(C_n \sin\lambda_n y + D_n \cos\lambda_n y). && (4.67) \end{aligned}$$

will satisfy the differential equation (4.49). This is true for any set of values of $\lambda_1, \lambda_2, \dots$. However, the boundary conditions will limit them to certain specific values. If we refer back to Figure 4.4, we see that three of the boundary conditions are homogeneous, i.e., they have zero on the right-hand side. Some of these can be satisfied rather easily by suitable selection of some of the integration constants, A_n, B_n, C_n, and D_n. Let us start with

$$T(x,0) = 0.$$

This means that we require

$$T(x,0) = 0 = \sum_{n=1}^{\infty} X_n(x)Y_n(0). \tag{4.68}$$

This can be satisfied by setting

$$Y_n(0) = 0, \quad n = 1, 2, 3, \dots$$

If this condition is met, then

$$\begin{aligned} T(x,0) &= \sum_{n=1}^{\infty} X_n(x)Y_n(0)\\ &= \sum_{n=1}^{\infty} X_n(x)\cdot 0\\ &= X_1(x)\cdot 0 + X_2(x)\cdot 0 + X_3(x)\cdot 0 + \dots\\ &= 0. && (4.69) \end{aligned}$$

The distinctive feature here is that the condition $T(x,0) = 0$ is met by just $Y_n(0) = 0, \quad n = 1, 2, 3, \dots$ because every term in the series become zero, and the entire sum vanishes.

Let us now go ahead and satisfy $Y_n(0) = 0$. By referring to equation (4.63), we may write

$$Y_n(y) = C_n \sin\lambda_n y + D_n \cos\lambda_n y. \tag{4.70}$$

Letting $y = 0$ gives

$$\begin{aligned} Y_n(0) &= 0 = C_n \sin \lambda_n 0 + D_n \cos(\lambda_n 0) \\ &= C_n \cdot 0 + D_n \cdot 1 \\ &= D_n. \end{aligned}$$

Therefore,

$$D_n = 0.$$

and

$$Y_n(y) = C_n \sin \lambda_n y. \tag{4.71}$$

As in the case of the boundary condition just satisfied, the condition at the opposite boundary $T(x, b) = 0$ is also a homogeneous one and can be satisfied by setting $Y_n(b) = 0$. Therefore, letting $y = b$ in equation (4.71) leads to

$$Y_n(b) = C_n \sin(\lambda_n b) = 0. \tag{4.72}$$

The only way that we have the left-hand side equal to zero in a nontrivial manner is if

$$\sin(\lambda_n b) = 0, \tag{4.73}$$

as opposed to $C_n = 0$. Equation (4.73) requires that the argument of the sine function be limited to integer multiples of π, i.e.,

$$\lambda_n b = n\pi. \tag{4.74}$$

or

$$\lambda_n = \frac{n\pi}{b}. \tag{4.75}$$

With λ_n having been defined, we may now write $Y_n(y)$ in the form

$$Y_n(y) = C_n \sin\left(\frac{n\pi y}{b}\right), \tag{4.76}$$

and $X_n(x)$ as

$$X_n(x) = A_n \sinh\left(\frac{n\pi x}{b}\right) + B_n \cosh\left(\frac{n\pi x}{b}\right). \tag{4.77}$$

The solution now takes the form

$$T(x, y) = \sum_{n=1}^{\infty} \left[A_n \sinh\left(\frac{n\pi x}{b}\right) + B_n \cosh\left(\frac{n\pi x}{b}\right)\right] C_n \sin\left(\frac{n\pi y}{b}\right), \tag{4.78}$$

where, by defining $a_n = A_n C_n$ and $b_n = B_n C_n$ we end up with

$$T(x, y) = \sum_{n=1}^{\infty} \left[a_n \sinh\left(\frac{n\pi x}{b}\right) + b_n \cosh\left(\frac{n\pi x}{b}\right)\right] \sin\left(\frac{n\pi y}{b}\right), \tag{4.79}$$

The boundary condition, $T(0, y) = 0$, can be satisfied by setting $X_n(0) = 0, \quad n = 1, 2, 3, \ldots$. Again, this is possible term-by-term because the boundary condition is a homogeneous one, and setting $X_n(0) = 0$ for all n makes

$$T(0, y) = \sum_{n=1}^{\infty} X_n(0) Y_n(y) = 0.$$

Thus,

$$X_n(0) = a_n \sinh 0 + b_n \cosh 0 = 0$$

results in

$$a_n \cdot 0 + b_n \cdot 1 = 0 \tag{4.80}$$

or

$$b_n = 0. \tag{4.81}$$

The expression for $X_n(x)$ reduces to

$$X_n(x) = a_n \sinh\left(\frac{n\pi x}{b}\right), \tag{4.82}$$

and $T(x, y)$ may be written as

$$T(x, y) = \sum_{n=1}^{\infty} a_n \sinh\left(\frac{n\pi x}{b}\right) \sin\left(\frac{n\pi y}{b}\right). \tag{4.83}$$

We still need to satisfy one more boundary condition which is

$$T(x, a) = T_0,$$

i.e.,

$$T(a, y) = T_0 = \sum_{n=1}^{\infty} \left[a_n \sinh\left(\frac{n\pi a}{b}\right)\right] \sin\left(\frac{n\pi y}{b}\right) \tag{4.84}$$

This is a Fourier sine series expansion in y, with a half period b. Using the orthogonality of Fourier series (see Chapter 3), the coefficient $[a_n \sinh(n\pi a/b)]$ may be written as

$$\begin{aligned}
\left[a_n \sinh\left(\frac{n\pi a}{b}\right)\right] &= \frac{2}{b}\int_0^b T_0 \sin\left(\frac{n\pi y}{b}\right) dy \\
&= \left(\frac{2}{b}\right) T_0 \left[-\frac{\cos\left(\frac{n\pi y}{b}\right)}{\frac{n\pi}{b}}\right] \\
&= \left(\frac{2}{b}\right) T_0 \left(\frac{1-(-1)^n}{n\pi/b}\right) \\
&= \frac{2T_0}{n\pi}\left[1-(-1)^n\right].
\end{aligned}$$

$$a_n = \frac{2T_0}{n\pi}\left[\frac{1-(-1)^n}{\sinh\left(\frac{n\pi a}{b}\right)}\right] \tag{4.85}$$

The expression for $T(x,y)$ can now be written in the following explicit form:

$$T(x,y) = 2T_0 \sum_{n=1}^{\infty} \frac{1-(-1)^n}{n\pi} \left[\frac{\sinh\left(\frac{n\pi x}{b}\right)}{\sinh\left(\frac{n\pi a}{b}\right)} \right] \sin\left(\frac{n\pi y}{b}\right). \tag{4.86}$$

□

In the preceding example we need to understand the behavior of the the set of functions

$$Y_n(y) = \sin\left(\frac{n\pi y}{b}\right), \qquad n = 1, 2, 3, \ldots$$

In terms of the argument $\lambda_n y$, this is a simple sine function that crosses the horizontal axis at $\lambda_n y = n\pi$, as shown here.

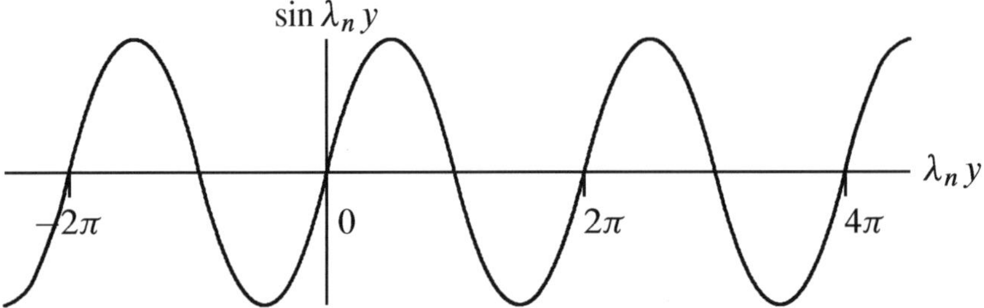

Here the parameter λ_n is adjusted so that a segment of the sine function between (0,0) and another point where it crosses the horizontal axis [such as (π,0), (2π,0), (3π,0), ...] is made to fit into the space $0 < y < b$. This is shown in Figure 4.5 on page 159. Each of the functions of the set $Y_n(y) = \sin \lambda_n y$ satisfies

$$Y_n(y) = 0 \quad \text{at} \quad y = 0, b. \tag{4.87}$$

Therefore, if we can identify a set of functions, all of which satisfy the imposed pair of homogeneous boundary conditions, then we may construct the solution as an expansion in that set of functions. These functions must also be eigenfunctions of certain differential operators relevant to the partial differential equation being considered. An eigenfunction of a differential operator is a function which, upon carrying out the differential operation, gives back a function proportional to the original function. For example, in the present case we have

$$\frac{d^2}{dy^2} Y_n(y) = -\lambda_n^2 Y_n(y).$$

Here, the solutions $Y_n(y)$ satisfying the above ordinary differential equation are eigenfunctions of the differential operator d^2/dy^2. In particular, $\sin \lambda_n y$ and $\cos \lambda_n y$ are both eigenfunctions of d^2/dy^2.

This takes us to another method of solution known as the method of eigenfunction expansions.

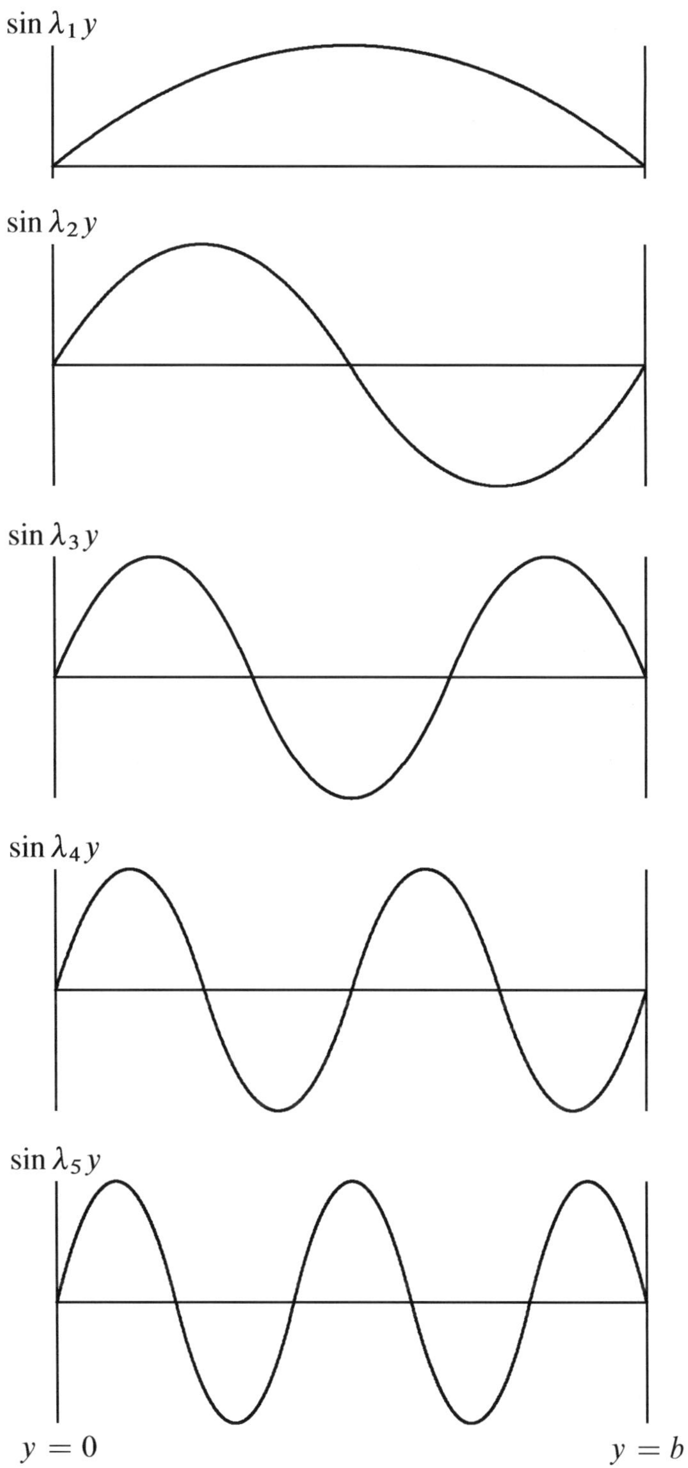

Figure 4.5: Plots of the $\sin \lambda_n y$ fitted into the space $0 \leq y \leq b$.

4.5 The Method of Eigenfunction Expansions

While this technique is an alternative to the method of separation of variables, its fundamental principles are the same as the latter. However, as we shall see, the systemization of the procedure provides for better understanding and quicker construction of the solution. Let us re-examine Example 4.2 on pages 152-158. Here, we first identify a pair of opposite homogeneous boundary conditions for that problem. Referring to Figure 4.4, we see that $T = 0$ at $y = 0$ and $y = b$. This identification indicates to us that we need eigenfunctions in the y variable that are equal to zero at $y = 0$ and $y = b$. We also need to identify the appropriate operator for which to construct eigenfunctions. This is done by isolating the y-operations in the original differential equation. Therefore, in

$$\frac{\partial^2 T}{\partial x^2} + \frac{\partial^2 T}{\partial y^2} = 0, \tag{4.88}$$

we isolate the operations in y as the operator $\partial^2/\partial y^2$. From this we construct the eigenvalue-eigenfunction equation,

$$\frac{d^2}{dy^2} Y_n(y) = -\lambda_n^2 Y_n(y), \tag{4.89}$$

and ensure that the eigenvalues are negative with respect to the highest derivative in the differential operator. Negative eigenvalues generally yield oscillating eigenfunctions and lead to the possibility of being able to satisfy a pair of homogeneous boundary conditions. Furthermore, the oscillatory aspect of the eigenfunctions is also the element that leads to orthogonality.

The solutions for the above equation are the same as given by equation (4.70), i.e.,

$$Y_n(y) = C_n \sin \lambda_n y + D_n \cos \lambda_n y. \tag{4.90}$$

Application of the boundary conditions $Y_n(0) = 0$ and $Y_n(b) = 0$ reduces this to

$$Y_n(y) = \sin \lambda_n y, \quad \text{with} \quad \lambda_n = \frac{n\pi}{b}.$$

Using this eigenfunction, we carry out an expansion of $T(x, y)$ in the form

$$T(x, y) = \sum_{n=1}^{\infty} T_n(x) \sin\left(\frac{n\pi y}{b}\right) \tag{4.91}$$

With this expansion in y the boundary conditions at $y = 0, b$ are unconditionally satisfied. Substitution of this expansion into the differential equation (4.88) gives

$$\begin{aligned}
\sum \left\{ T_n''(x) \sin\left(\frac{n\pi y}{b}\right) + T_n(x)\left[-\left(\frac{n\pi}{b}\right)^2\right] \sin\left(\frac{n\pi y}{b}\right)\right\} &= 0 \\
\sum_{n=0}^{\infty} \left[T_n''(x) - \left(\frac{n\pi}{b}\right)^2 T_n(x)\right] \sin\left(\frac{n\pi y}{b}\right) &= 0 \\
T_n''(x) - \left(\frac{n\pi}{b}\right)^2 T_n(x) &= 0 \qquad (4.92)
\end{aligned}$$

The solution to this ordinary differential equation is

$$T_n(x) = a_n \sinh\left(\frac{n\pi x}{b}\right) + b_n \cosh\left(\frac{n\pi x}{b}\right), \tag{4.93}$$

and we may write the expression for $T(x, y)$ as

$$T(x, y) = \sum_{n=1}^{\infty}\left[a_n \sinh\left(\frac{n\pi x}{b}\right) + b_n \cos\left(\frac{n\pi x}{b}\right)\right]\sin\left(\frac{n\pi y}{b}\right) \tag{4.94}$$

The rest of it is the same as in the last example from equation (4.79) onward.

We shall now consider a different combination of boundary conditions in the next example to illustrate the method of eigenfunction expansions.

Example 4.3 Once again, let us consider Laplace's equation (4.88) in a rectangular region as shown in Figure 4.6. In comparison with Example 4.2, there is a slight

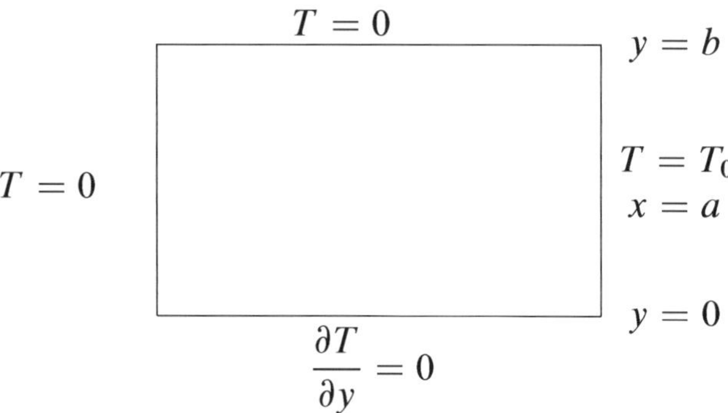

Figure 4.6: Boundary conditions for example 4.3.

variation of the boundary condition at $y = 0$. This time, the derivative condition

$$\left.\frac{\partial T(x, y)}{\partial y}\right|_{y=0}$$

is specified. This condition, along with $T(x, b) = 0$, forms a pair of opposite homogeneous conditions. We should therefore seek an eigenfunction expansion in y for the operator d^2/dy^2 in the form

$$T(x, y) = \sum_{n=0}^{\infty} T_n(x) Y_n(y) \tag{4.95}$$

where $Y_n(y)$ satisfies the same eigenvalue-eigenfunction equation (4.89) as in the reworked Example 4.2 on page 160. We may therefore use a similar expression for $Y_n(y)$ as equation (4.90), i.e.,

$$Y_n(y) = A_n \sin \lambda_n y + B_n \cos \lambda_n y \tag{4.96}$$

The boundary condition at $y = 0$ requires

$$\left.\frac{\partial T(x, y)}{\partial y}\right|_{y=0} = \sum_{n=0}^{\infty} T_n(x) Y_n'(0) = 0 \tag{4.97}$$

which, being a homogeneous condition, may be satisfied term by term with

$$Y_n'(0) = 0.$$

Therefore,

$$Y_n'(0) = A_n \lambda_n \cos 0 - B_n \lambda_n \sin 0. = A_n \lambda_n = 0 \tag{4.98}$$

This reduces $Y_n(y)$ to

$$Y_n(y) = B_n \cos \lambda_n y, \tag{4.99}$$

and we may write

$$T(x, y) = \sum_{n=0}^{\infty} T_n(x) \cos \lambda_n y.$$

The boundary condition at $y = b$ requires that

$$T(x, b) = \sum_{n=0}^{\infty} T_n(x) \cos \lambda_n b = 0, \tag{4.100}$$

which may be satisfied by

$$\cos(\lambda_n b) = 0 \tag{4.101}$$

leading to

$$\lambda_n b = (n + \tfrac{1}{2})\pi, \tag{4.102}$$

i.e.,

$$\lambda_n = (n + \tfrac{1}{2})\pi/b. \tag{4.103}$$

The expression for $T(x, y)$ may now be written as

$$T(x, y) = \sum_{n=0}^{\infty} T_n(x) \cos(\lambda_n y) = \sum_{n=0}^{\infty} T_n(x) \cos\left[(n + \tfrac{1}{2})\pi y/b\right] \tag{4.104}$$

Substitution into

$$\frac{\partial^2 T}{\partial x^2} + \frac{\partial^2 T}{\partial y^2} = 0 \tag{4.105}$$

leads to

$$\left[T_n''(x) - \lambda_n^2 T_n(x)\right] \cos \lambda_n y = 0 \tag{4.106}$$

from where it is not difficult to see that

$$T_n''(x) - \lambda_n^2 T_n(x) = 0. \tag{4.107}$$

The solution to this ordinary differential equation is

$$T_n(x) = A_n \sinh(\lambda_n x) + B_n \cosh(\lambda_n x) \tag{4.108}$$

At $x = 0$,

$$T_n(0) = 0$$

will satisfy

$$T(0, y) = 0. \tag{4.109}$$

Implementing this requirement gives

$$T_n(0) = 0 = A_n \sinh 0 + B_n \cosh 0 \qquad \text{or} \qquad B_n = 0.$$

Therefore,

$$T_n(x) = A_n \sinh(\lambda_n x), \tag{4.110}$$

and

$$\begin{aligned} T(x, y) &= \sum_{n=0}^{\infty} A_n \sinh(\lambda_n x) \cos(\lambda_n y) \\ &= \sum_{n=0}^{\infty} A_n \sinh\left[(n + \tfrac{1}{2})\pi x/b\right] \cos\left[(n + \tfrac{1}{2})\pi y/b\right] \end{aligned} \tag{4.111}$$

The last boundary condition is $T = T_0$ at $x = a$ which would be satisfied if

$$T(a, y) = T_0 = \sum_{n=0}^{\infty} \left\{A_n \sinh\left[(n + \tfrac{1}{2})\pi a/b\right]\right\} \cos\left[(n + \tfrac{1}{2})\pi y/b\right] \tag{4.112}$$

This is a Fourier cosine series for which the coefficients can be obtained by following the procedure in Chapter 3. This results in

$$\begin{aligned} \left\{A_n \sinh\left[(n + \tfrac{1}{2})\pi a/b\right]\right\} &= \frac{2}{b} \int_0^b T_0 \cos\left[\left(n + \tfrac{1}{2}\right) \pi y/b\right] dy \\ &= \frac{2}{b} T_0 \left. \frac{\sin\left[\left(n + \frac{1}{2}\right) \pi y/b\right]}{\left(n + \frac{1}{2}\right) \pi/b} \right|_0^b \\ &= \frac{2T_0}{b} \frac{(-1)^n}{\left(n + \frac{1}{2}\right) \pi/b} = \frac{2T_0(-1)^n}{\left(n + \frac{1}{2}\right) \pi} \end{aligned} \tag{4.113}$$

leading to

$$A_n = 2T_0 \cdot \frac{(-1)^n}{\left(n + \frac{1}{2}\right) \pi} \cdot \frac{1}{\sinh\left[\left(n + \frac{1}{2}\right) \pi a/b\right]}. \tag{4.114}$$

With this result, the expression for $T(x, y)$ may be explicitly expressed as

$$T(x, y) = \frac{2T_0}{\pi} \sum_{n=0}^{\infty} \frac{(-1)^n}{\left(n + \frac{1}{2}\right)} \frac{\sinh\left[\left(n + \frac{1}{2}\right) \pi x/b\right]}{\sinh\left[\left(n + \frac{1}{2}\right) \pi a/b\right]} \cos\left[\left(n + \tfrac{1}{2}\right) \pi y/b\right] \tag{4.115}$$

□

So far, we have studied examples in which the boundary conditions are simple specification of the independent variable, or its normal derivative. In situations when the boundary condition is a linear combination of the two, an explicit analytical solution may still be found. Let us consider the following example.

Example 4.4 *Problem Statement:*
Solve the one-dimensional diffusion equation in the slab $0 \leq x \leq 0$,

$$\frac{1}{\alpha}\frac{\partial T}{\partial t} = \frac{\partial^2 T}{\partial x^2}, \tag{4.116}$$

with the initial condition

$$T(x, 0) = T_0. \tag{4.117}$$

The boundary conditions are

$$T(0, t) = 0 \tag{4.118}$$

and

$$k\frac{\partial T}{\partial x} + hT = 0 \qquad \text{at} \quad x = l. \tag{4.119}$$

This is a mixed boundary condition that was referred to as the 'third kind' of condition in Chapter 3 under the Sturm-Liouville Theory in Section 3.7.1. In addition, Example 3.9 on pages 3.9-3.10 deals with the Fourier series expansion emanating from eigenfunctions that would result from this set of boundary conditions at $x = 0$ and $x = l$. Looking at Example 4.6 on pages 4.6-4.5.1, the domain of the problem, the differential equation, the boundary condition at $x = 0$, and the initial condition are all the same as the current example. Therefore, it is not difficult to show that the solution would have the form

$$T(x, t) = \sum_{n=0}^{\infty} A_n e^{-\alpha\lambda_n^2 t} \sin \lambda_n x \tag{4.120}$$

satisfying the differential equation (4.116) and the boundary condition (4.118). The boundary condition at $x = l$ requires

$$k\sum_{n=0}^{\infty} A_n e^{-\alpha\lambda_n^2 t}\lambda_n \cos \lambda_n l + h\sum_{n=0}^{\infty} a_n e^{-\alpha\lambda_n^2 t} \sin \lambda_n l = 0 \tag{4.121}$$

$$\sum_{n=0}^{\infty} a_n e^{-\alpha\lambda_n^2 t}\left[k\lambda_n \cos \lambda_n l + h \sin \lambda_n l\right] = 0 \tag{4.122}$$

$$k\lambda_n \cos \lambda_n l + h \sin \lambda_n l = 0 \tag{4.123}$$

$$-\frac{l}{hl}(\lambda_n l) = \tan(\lambda_n l) \tag{4.124}$$

This equation determines what values of λ_n satisfy the right boundary condition. These values of λ_n have to be obtained numerically. For these values of λ_n, the orthogonality of the set of functions $\sin \lambda_n x$ is guaranteed (see Chapter 3, page 112.

Here, $\lambda_0 = 0$ is one of the roots but does not contribute to the solution since $\sin 0 = 0$. Thus the expression for $T(x,t)$ given by (4.120) may be modified slightly in the form

$$T(x,t) = \sum_{n=1}^{\infty} A_n e^{-\alpha\lambda_n^2 t} \sin(\lambda_n x). \tag{4.125}$$

The initial condition $T(x,0) = T_0$ gives

$$T_0 = \sum_{n=1}^{\infty} A_n \sin(\lambda_n x) \tag{4.126}$$

The set of coefficients $\{A_n\}$ may be obtained by using the orthogonality property of this Fourier series. Let us multiply each side by $\sin \lambda_m x$ and integrate over the range $0 \le x \le l$.

$$\int_0^l T_0 \sin(\lambda_m x) dx \;=\; \sum_{n=1}^{\infty} A_n \int_0^l \sin(\lambda_n x)\sin(\lambda_m x) dx \tag{4.127}$$

Since the set of eigenfunctions $\{\sin \lambda_n x\}$ are orthogonal, we have for $m \neq n$,

$$\int_0^l \sin(\lambda_m x)\sin(\lambda_n x) dx = 0.$$

For $m = n$,

$$\begin{aligned}\int_0^l \sin^2(\lambda_m x) dx &= \int_0^l \tfrac{1}{2}\left[1-\cos(2\lambda_m x)\right]\, dx \\ &= \tfrac{1}{2}\left[x - \frac{\sin(2\lambda_m x)}{4\lambda_m}\right]_0^l \\ &= \tfrac{1}{2}\left[l - \frac{\sin(2\lambda_m l)}{4\lambda_m}\right]\end{aligned} \tag{4.128}$$

The left-hand side of equation (4.127) may be integrated, giving

$$LHS = \int_0^l T_0 \sin(\lambda_m x) dx = T_0 \left(\frac{-\cos(\lambda_m x)}{\lambda_m}\bigg|_0^l \right) = \frac{T_0[1-\cos(\lambda_m l)]}{\lambda_m}. \tag{4.129}$$

On the right-hand side of (4.127), every term in the summation is zero except when $m = n$. In that case, using the result in equation (4.128), we have

$$RHS \;= A_m \left[\frac{1}{2} l - \frac{\sin(2\lambda_m l)}{4\lambda_m}\right]. \tag{4.130}$$

Equating $LHS = RHS$, it is not difficult to see that

$$A_m = \frac{T_0[1-\cos(\lambda_m l)]}{\lambda_m \left[\frac{1}{2} l - \frac{\sin(2\lambda_m m l)}{4\lambda_m}\right]} = \frac{2T_0[1-\cos(\lambda_m l)]}{\lambda_m l - \frac{1}{2}\sin(2\lambda_m l)} \tag{4.131}$$

leading to

$$T(x,t) = 2T_0 \sum_{n=1}^{\infty} \frac{1-\cos(\lambda_n l)}{\lambda_n l - \frac{1}{2}\sin(2\lambda_n l)} e^{-\alpha\lambda_n^2 t} \sin \lambda_n x \tag{4.132}$$

where the values of λ_n are given by equation (4.125). □

4.5.1 Superposition Principle

The examples considered up to now are those in which there are three homogeneous boundary conditions and one nonhomogeneous one. This facilitates the solution because it is easy to identify a pair of opposite homogeneous boundary conditions which is necessary for suitable eigenfunction expansions. In the next example we consider the situation with two adjacent non-homogeneous conditions.

Example 4.5 The two-dimensional Laplace's equation (4.88) is considered again. The boundary conditions are as shown in Figure 4.7. Here we decompose the problem

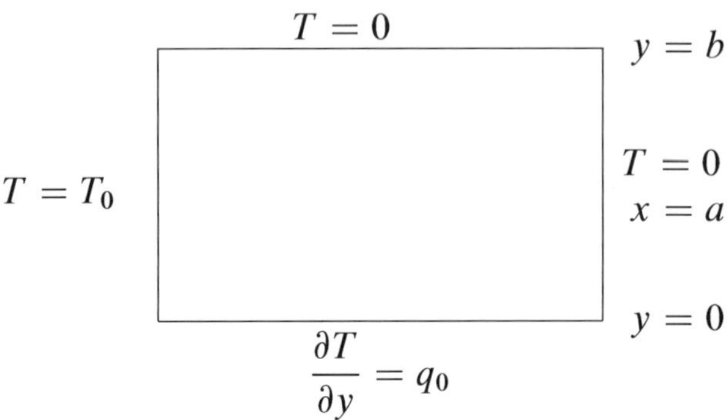

Figure 4.7: Adjacent non-homogeneous boundary conditions

into two parts,

$$T(x, y) = T_1(x, y) + T_2(x, y) \tag{4.133}$$

where each component satisfies Laplace's equation,

$$\frac{\partial^2 T_1}{\partial x^2} + \frac{\partial^2 T_1}{\partial y^2} = 0 \qquad \text{and} \qquad \frac{\partial^2 T_2}{\partial x^2} + \frac{\partial^2 T_2}{\partial y^2} = 0.$$

The boundary conditions take the form shown here. We begin with $T_1(x, y)$ which takes on a solution in the following form

$$T_1(x, y) = \sum_{n=0}^{\infty} (A_n \sinh \lambda_n x + B_n \cosh \lambda_n x) \cos \lambda_n y \tag{4.134}$$

where

$$\lambda_n = (n + \tfrac{1}{2})\pi / b.$$

Next, we consider the homogeneous boundary condition at $x = a$, i.e.,

$$T_1(a, y) = 0. \tag{4.135}$$

Letting $x = a$ in the expression (4.134) for $T_1(x, y)$ and setting it equal to zero gives

$$\sum_{n=0}^{\infty} [A_n \sinh(\lambda_n a) + B_n \cosh(\lambda_n a)] \cos \lambda_n y = 0$$

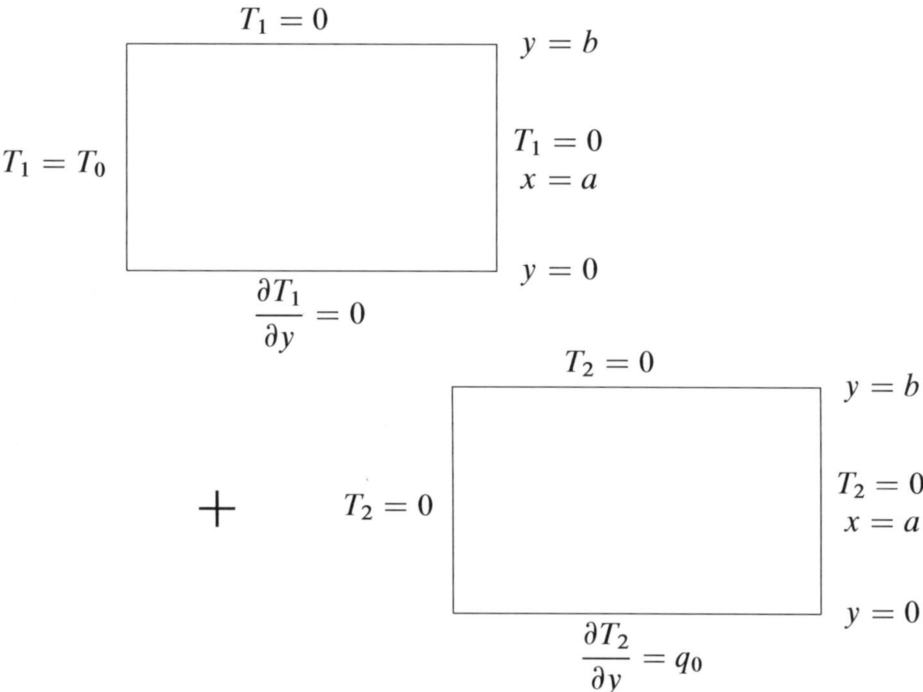

Figure 4.8: Decomposition of example 4.5

The boundary condition at $x = a$, being homogeneous, can of course be satisfied term-by-term, i.e.,

$$A_n \sinh(\lambda_n a) + B_n \cosh(\lambda_n a) = 0$$

leading to

$$B_n = -A_n \frac{\sinh(\lambda_n a)}{\cosh(\lambda_n a)}$$

Therefore, the expression for $T_1(x, y)$ may be written as

$$T_1(x, y) = \sum_{n=0}^{\infty} A_n \left[\sinh(\lambda_n x) - \frac{\sinh \lambda_n a}{\cosh \lambda_n a} \cosh \lambda_n x \right] \cos \lambda_n y \tag{4.136}$$

$$= \sum_{n=0}^{\infty} A_n \left[\frac{\cosh \lambda_n a \sinh \lambda_n x - \sinh \lambda_n a \cosh \lambda_n x}{\cosh \lambda_n a} \right] \cos \lambda_n y$$

$$= \sum_{n=0}^{\infty} A_n \frac{-\sinh[\lambda_n (a - x)]}{\cosh(\lambda_n a)} \cos(\lambda_n y) \tag{4.137}$$

we could have assumed from the beginning

$$T_1(x, y) = \sum_{n=0}^{\infty} A_n^* \sinh[\lambda_n (a - x)] \cos(\lambda_n y) \tag{4.138}$$

This solution satisfies the three homogenous boundary conditions for T_1. The last boundary condition for T_1 will have to be satisfied is a series, i.e.,

$$T_1(0, y) = T_0 = \sum_{n=0}^{\infty} A_n^* \sinh(\lambda_n a) \cos \lambda_n y \tag{4.139}$$

$$\begin{aligned}
(A_n^* \sinh \lambda_n a) &= \frac{2}{b}\int_0^b T_0 \cos \lambda_n y \, dy \\
&= \frac{2}{b} T_0 \frac{\sin \lambda_n y}{\lambda_n}\bigg|_0^b \\
&= \frac{2}{b}\frac{\sin[(n+\frac{1}{2})\pi y/b]}{(n+\frac{1}{2})\pi/b}\bigg|_0^b \\
&= \frac{2T_0(-1)^n}{(n+\frac{1}{2})\pi}
\end{aligned}$$

$$T_1(x, y) = \frac{2T_0}{\pi}\sum_{n=0}^{\infty}\frac{(-1)^n}{(n+\frac{1}{2})}\frac{\sinh[\lambda_n(a-x)]\cos\lambda_n y}{\sinh \lambda_n a} \tag{4.140}$$

or

$$T_1(x, y) = \frac{2T_0}{\pi}\sum_{n=0}^{\infty}\frac{(-1)^n}{(n+\frac{1}{2})}\frac{\sinh[(n+\frac{1}{2})\pi(a-x)/b]}{\sinh[(n+\frac{1}{2})\pi a/b]}\cos\left[(n+\tfrac{1}{2})\frac{\pi y}{b}\right] \tag{4.141}$$

For $T_2(x, y)$ we write the solution as

$$T_2(x, y) = \sum_{m=1}^{\infty}\left[a_m \sinh\left(\frac{m\pi y}{a}\right) + b_m \cosh\left(\frac{m\pi y}{a}\right)\right]\sin\left(\frac{m\pi x}{a}\right)$$

which, after considering that the boundary condition at $y = b$ is homogeneous, may be written as

$$T_2(x, y) = \sum_{m=1}^{\infty} B_m \sinh\left[\frac{m\pi}{a}(b-y)\right]\sin\left(\frac{m\pi x}{a}\right) \tag{4.142}$$

This step is analogous to the procedure leading to equation (4.138) for $T_1(x, y)$ The last condition at $y = 0$ is non-homogenous.

$$\frac{\partial T_2}{\partial y}\bigg|_{y=0} = q_0 = -\sum_{m=1}^{\infty}\left[B_m\left(\frac{m\pi}{a}\right)\cosh\left(\frac{m\pi}{a}b\right)\right]\sin\left(\frac{m\pi x}{a}\right) \tag{4.143}$$

$$\begin{aligned}
-\left[B_m\left(\frac{m\pi}{a}\right)\cosh\left(\frac{n\pi b}{a}\right)\right] &= \frac{2}{a}\int_0^a q_0 \sin\left(\frac{m\pi x}{a}\right) dx \\
&= \left(\frac{2q_0}{a}\right)\frac{-\cos\left(\frac{m\pi x}{a}\right)}{(m\pi/a)}\bigg|_0^a \\
&= \frac{2q_0}{m\pi}[1-(-1)^m]
\end{aligned} \tag{4.144}$$

$$B_m = -\frac{2q_0}{n\pi} \frac{[1-(-1)^n]}{\left(\frac{n\pi}{a}\right)\cosh\left(\frac{m\pi b}{a}\right)} \tag{4.145}$$

$$T_2(x,y) = -\frac{2q_0}{\pi^2} \sum_{m=1}^{\infty} \frac{[1-(-1)^m]}{m^2/a} \frac{\sinh\left[\frac{m\pi}{a}(b-y)\right]}{\cosh\left[\frac{m\pi}{a}b\right]} \sin\left(\frac{m\pi x}{a}\right) \tag{4.146}$$

With $T_1(x, y)$ and $T_2(x, y)$ both determined, the solution in the form of equation (4.133) is complete. □

We next consider a parabolic equation. This time, we take an example of the one-dimensional diffusion equation.

Example 4.6

$$\frac{1}{\alpha}\frac{\partial T}{\partial t} = \frac{\partial^2 T}{\partial x^2}$$

The boundary conditions are given as

$$T = 0 \quad \text{at} \quad x = 0, l.$$

and the initial condition is

$$T(x,0) = T_0 \quad \text{at} \quad t = 0.$$

With homogeneous boundary conditions at $x = 0$ and $x = l$, we can carry out an eigenfunction expansion in x. Once again, we need eigenfunctions of the operator $\partial^2/\partial x^2$. In addition, we can satisfy the homogeneous boundary conditions at the same time. Therefore, we let

$$X_n(x) = \sin\left(\frac{n\pi x}{l}\right)$$

and write

$$T(x,t) = \sum_{n=1}^{\infty} T_n(t) \sin\left(\frac{n\pi x}{l}\right) \tag{4.147}$$

Substitution into the differential equation gives

$$\begin{aligned}
\sum_{n=1}^{\infty}\left[\frac{1}{\alpha}T_n'(t)\sin\left(\frac{n\pi x}{l}\right)\right] - T_n(t)\left[-\left(\frac{n\pi}{l}\right)^2\right]\sin\left(\frac{n\pi x}{l}\right) &= 0, \\
\sum_{n=1}^{\infty}\left[\frac{1}{\alpha}T_n' + \left(\frac{n\pi}{l}\right)^2 T_n\right]\sin\left(\frac{n\pi x}{l}\right) &= 0, \quad (4.148) \\
\frac{1}{\alpha}T_n' + \left(\frac{n\pi}{l}\right)^2 T_n &= 0. \quad (4.149)
\end{aligned}$$

At this point, we let

$$T_n(t) = e^{mt} \tag{4.150}$$

and substitute into Equation (4.149), giving us

$$\frac{1}{\alpha}m + \left(\frac{n\pi}{l}\right)^2 = 0. \tag{4.151}$$

$$m = -\alpha(n\pi/l)^2 \tag{4.152}$$

$$T_n(t) = A_n e^{-\alpha(n\pi/l)^2 t} \tag{4.153}$$

$$T(x,t) = \sum_{n=1}^{\infty} A_n e^{-\alpha(n\pi x/l)^2 t} \sin\left(\frac{n\pi x}{l}\right) \tag{4.154}$$

Initial condition $T(x,0) = T_0$ gives

$$T(x,0) = T_0 = \sum_{n=1}^{\infty} A_n \sin\left(\frac{n\pi x}{l}\right) \tag{4.155}$$

This is just a Fourier series for which the orthogonality principle (see Chapter 3) can be used to obtain the set of coefficients $\{A_n\}$. Doing so leads to

$$A_n = \frac{2}{l}\int_0^l T_0 \sin\left(\frac{n\pi x}{l}\right) dx = \frac{2T_0}{l}\left[\frac{-\cos(n\pi x/l)}{n\pi/l}\right]_0^l = \frac{2T_0}{n\pi}\left[1-(-1)^n\right] \tag{4.156}$$

The complete solution can now be written as

$$T(x,t) = 2T_0 \sum_{n=0}^{\infty} \frac{[1-(-1)^n]}{n\pi} e^{-\alpha(n\pi x/l)^2 t} \sin\left(\frac{n\pi x}{l}\right).$$

□

The above procedure worked because both the boundary conditions were homogeneous so that a suitable set of eigenfunctions, $X_n(x) = \sin(n\pi x/l)$, could be used. In case one or both of the boundary conditions are non-homogeneous, we can decompose the problem into steady state and time-dependent parts. The steady state can be used to take care of the non-homogeneous conditions if these conditions are no dependent on t. For such a case, we consider the following example:

Example 4.7

PROBLEM STATEMENT:

A fluid of kinematic viscosity ν is placed between two infinitely large horizontal parallel plates, a distance l apart. The fluid is initially at rest. At time $t = 0$, the upper plate is impulsively moved, parallel to itself, at a horizontal velocity U_0 while the lower plate remains fixed. With laminar flow, the fluid motion takes place unidirectionally with only a horizontal component, $u(x,t)$. This velocity field is described by the diffusion equation,

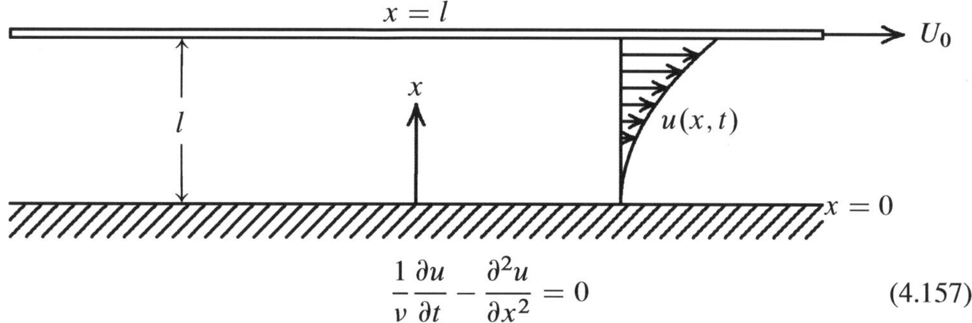

$$\frac{1}{\nu}\frac{\partial u}{\partial t} - \frac{\partial^2 u}{\partial x^2} = 0 \tag{4.157}$$

with boundary conditions

$$u(l,t) = U_0 \quad \text{and} \quad u(0,t) = 0. \tag{4.158}$$

Since the fluid is at rest initially, the initial condition is taken to be

$$u(x,0) = 0. \tag{4.159}$$

We start out with the decomposition

$$u(x,t) = u_1(x) + u_2(x,t). \tag{4.160}$$

Substituting into the differential equation gives

$$\left(0 - \frac{d^2 u_1}{dx^2}\right) + \left(\frac{1}{\nu}\frac{\partial u_2}{\partial t} - \frac{\partial^2 u_2}{\partial x^2}\right) = 0 \tag{4.161}$$

This can be satisfied if we set

$$\frac{d^2 u_1}{dx^2} = 0 \quad \text{with} \quad u_1(0) = 0 \quad \text{and} \quad u_1(l) = U_0, \tag{4.162}$$

along with

$$\frac{1}{\nu}\frac{\partial u_2}{\partial t} - \frac{\partial^2 u_2}{\partial x^2} = 0 \quad \text{with} \quad u_2(0,t) = 0 \quad \text{and} \quad u_2(l,t) = 0. \tag{4.163}$$

The initial condition for $u_2(x,t)$ is

$$u_2(x,0) = -u_1(x) \tag{4.164}$$

The differential equation for $u_1(x)$ may be integrated to give

$$u_1(x) = Ax + B. \tag{4.165}$$

Applying the boundary condition at $x = 0$ gives

$$A(0) + B = 0, \quad \text{i.e.,} \quad B = 0. \tag{4.166}$$

At $x = l$, the boundary condition for $u_1(x)$ leads to

$$Al + B = U_0,$$

or

$$A = \frac{U_0}{l} \tag{4.167}$$

With A and B determined, the solution for $U_1(x)$ ends up as

$$u_1(x) = U_0\left(\frac{x}{l}\right). \tag{4.168}$$

The differential equation for $u_2(x,t)$ is the same as the one in the previous example 4.6, and the same boundary conditions, replacing $u_2(x,t)$ for $T(x,t)$. Only the initial condition is different. The solution can therefore be written as

$$u_2(x,t) = \sum_{n=1}^{\infty} A_n e^{-\nu(n\pi/l)^2 t} \sin\left(\frac{n\pi x}{l}\right). \tag{4.169}$$

Applying the initial condition gives

$$u_2(x,0) = -U_0(\frac{x}{l}) = \sum_{n=1}^{\infty} A_n \sin\left(\frac{n\pi x}{l}\right) \tag{4.170}$$

$$\begin{aligned}
A_n &= \frac{2}{l}\int_0^l -U_0\left(\frac{x}{l}\right)\sin(n\pi x/l)dx && (4.171)\\
&= -\frac{2T_0}{l}\left\{\left(\frac{x}{l}\right)\frac{-\cos(n\pi x/l)}{(n\pi/l)}\int_0^l + \frac{1}{l}\int_0^l \frac{\cos(n\pi x/l)}{n\pi/l}dx\right\} && (4.172)\\
&= -\frac{2T_0}{l}\left\{\frac{-(-1)^n}{n\pi/l} + 0\right\} && (4.173)\\
&= \frac{2T_0}{n\pi}(-1)^n && (4.174)
\end{aligned}$$

The solution for $T_2(x,t)$ can now be written as

$$u_2(x,t) = \frac{2U_0}{\pi}\sum_{n=1}^{\infty}\frac{(-1)^n}{n}e^{-\nu(n\pi/l)^2 t}\sin\left(\frac{n\pi x}{l}\right) \tag{4.175}$$

and the complete solution for $T(x,t)$ is

$$u(x,t) = U_0\left\{\frac{x}{l} + \frac{2}{\pi}\sum_{n=1}^{\infty}\frac{(-1)^n}{n}e^{-\nu(n\pi/l)^2 t}\sin\left(\frac{n\pi x}{l}\right)\right\} \tag{4.176}$$

□

Example 4.8

Wave Equation: Let us consider an example involving the one-dimensional wave equation. Here we examine a taut string confined to the region $0 \le x \le l$. The displacement $u(x,t)$ of the string is given by the solution of

$$\frac{1}{c^2}\frac{\partial^2 u}{\partial t^2} - \frac{\partial^2 u}{\partial x^2} = 0 \tag{4.177}$$

The string is fixed at the points $x = 0$ and $x = l$ so that the appropriate boundary conditions are

$$u(0,t) = 0 \quad \text{and} \quad u(l,t) = 0 \tag{4.178}$$

We have an initial displacement

$$u(x,0) = f(x) \tag{4.179}$$

and zero initial velocity so that

$$\frac{\partial u(x,0)}{\partial t} = 0. \tag{4.180}$$

With the two homogeneous boundary conditions at $x = 0$ and $x = l$, we should look for eigenfunctions of the operator $(\partial^2/\partial x^2)$, satisfying those two conditions, i.e.,

$$X_n(x) = \sin\left(\frac{n\pi x}{l}\right).$$

Therefore, we assume the expansion

$$u(x,t) = \sum_{n=1}^{\infty} U_n(t) \sin\left(\frac{n\pi x}{l}\right) \tag{4.181}$$

and substitute into the differential equation (4.177). As a result, we have

$$\sum_{n=1}^{\infty} \left[\frac{1}{c^2} U_n'' + \left(\frac{n\pi}{l}\right)^2 U_n(t)\right] \sin\left(\frac{n\pi x}{l}\right) = 0, \tag{4.182}$$

which leads to

$$\frac{1}{c^2} U_n'' + \left(\frac{n\pi}{l}\right)^2 U_n = 0, \tag{4.183}$$

or

$$U_n'' + \left(\frac{n\pi c}{l}\right)^2 U_n = 0 \tag{4.184}$$

The solution of this equation is

$$U_n(t) = A_n \sin\left(\frac{n\pi ct}{l}\right) + B_n \cos\left(\frac{n\pi ct}{l}\right). \tag{4.185}$$

The zero velocity condition

$$\frac{\partial u(x,0)}{\partial t} = 0$$

can be satisfied by

$$U_n'(0) = 0,$$

i.e.,

$$U_n'(0) = \left(\frac{n\pi c}{l}\right)\left[A_n \cos\left(\frac{n\pi c0}{l}\right) - B_n \sin\left(\frac{n\pi c0}{l}\right)\right] = \left(\frac{n\pi c}{l}\right) A_n = 0, \tag{4.186}$$

leading to

$$A_n = 0. \tag{4.187}$$

The solution for $U_n(t)$ given by (4.185) reduces to

$$U_n(t) = B_n \cos\left(\frac{n\pi ct}{l}\right) \tag{4.188}$$

and the expression for $u(x,t)$ becomes

$$u(x,t) = \sum_{n=1}^{\infty} B_n \cos\left(\frac{n\pi ct}{l}\right) \sin\left(\frac{n\pi x}{l}\right) \tag{4.189}$$

The initial displacement condition (4.179) requires

$$u(x,0) = f(x) = \sum_{n=1}^{\infty} B_n \sin\left(\frac{n\pi x}{l}\right). \tag{4.190}$$

We assume $f(x)$ is given piecewise as

$$= \begin{cases} \frac{2x}{l} & 0 < x < \frac{1}{2}l \\ \frac{2}{l}(l-x) & \frac{1}{2}l < x < l, \end{cases} \tag{4.191}$$

corresponding to a string plucked symmetrically about its center. Thus, the set of coefficients, B_n, of the Fourier series (4.190) is given by the integral

$$\begin{aligned}
B_n &= \frac{2}{l}\int_0^l f(x)\sin\left(\frac{n\pi x}{l}\right) dx \\
&= \frac{2}{l}\left\{\int_0^{\frac{l}{2}} \left(\frac{2x}{l}\right)\sin\left(\frac{n\pi x}{l}\right) dx + \int_{\frac{1}{2}l}^{l} \frac{2}{l}(l-x)\sin\left(\frac{n\pi x}{l}\right) dx\right\} \\
&= \frac{2}{l}\left\{\left(\frac{2x}{l}\right)\left(-\frac{l}{n\pi}\right)\cos\left(\frac{n\pi x}{l}\right)\Big|_0^{\frac{l}{2}} - \int_0^{\frac{l}{2}} \frac{2}{l}\left(-\frac{l}{n\pi}\right)\cos\left(\frac{n\pi x}{l}\right) dx\right. \\
&\quad \left. + \frac{2}{l}(l-x)\left(-\frac{l}{n\pi}\right)\cos\left(\frac{n\pi x}{l}\right)\Big|_{\frac{1}{2}l}^{l} - \int_{\frac{1}{2}l}^{l}\left(-\frac{2}{l}\right)\left(-\frac{l}{n\pi}\right)\cos\left(\frac{n\pi x}{l}\right) dx\right\} \\
&= \frac{2}{l}\left\{\left(-\frac{l}{n\pi}\right)\cos\left(\tfrac{1}{2}n\pi\right) + \frac{2}{l}\left(\frac{l}{n\pi}\right)^2 \sin\left(\frac{n\pi x}{l}\right)\Big|_0^{\frac{l}{2}}\right. \\
&\quad \left. + -\left(-\frac{l}{n\pi}\right)\cos\left(\tfrac{1}{2}n\pi\right) - \frac{2}{l}\left(\frac{l}{n\pi}\right)^2 \sin\left(\frac{n\pi x}{l}\right)\Big|_{\frac{l}{2}}^{l}\right\} \\
&= \frac{8}{(n\pi)^2}\sin\left(\tfrac{1}{2}n\pi\right)
\end{aligned} \tag{4.192}$$

Using some identities for sine and cosine functions, we may write the complete solution as

$$\begin{aligned}
u(x,t) &= \sum_{n=1}^{\infty} B_n \cos\left(\frac{n\pi ct}{l}\right)\sin\left(\frac{n\pi x}{l}\right) && (4.193) \\
&= \sum_{n=1}^{\infty} \tfrac{1}{2}B_n \left\{\sin\left[\frac{n\pi}{l}(x+ct)\right] + \sin\left[\frac{n\pi}{l}(x-ct)\right]\right\} && (4.194)
\end{aligned}$$

This form clearly identifies the two characteristics of the wave equation, $(x + ct)$ and $(x - ct)$. □

4.5.2 Problems Involving Infinite Domain

In principle, the case of an infinite domain in any of the variables is not all that different from a finite-domain case. The mathematical limit of a dimension going to infinity, and the relevant implications such as a Fourier series becoming an integral will be applicable. Let us consider an example.

Example 4.9 Solve

$$\frac{\partial^2 T}{\partial x^2} + \frac{\partial^2 T}{\partial y^2} = 0 \tag{4.195}$$

with the boundary conditions as shown Here, we may view the boundary condition at

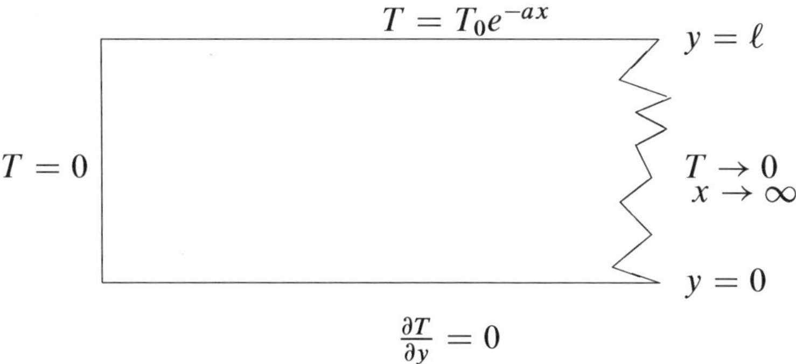

Figure 4.9: Two-dimensional domain, infinite in the x-direction

$x = 0$ and the one as $x \to \infty$ as the pair of opposite homogeneous conditions. We therefore need an eigenfunction expansion in x. In this case we will not be able to limit the values of λ to discrete quantities (such as $\lambda_n = n\pi/l$) but rather have to take the whole spectrum in the range $0 \leq \lambda < \infty$. The eigenfunction expansion takes the form

$$T(x, y) = \int_0^\infty T_\lambda(y) \sin(\lambda x)\, d\lambda \tag{4.196}$$

Here, $\sin \lambda x$ is chosen so that it is an eigenfunction of $\partial^2/\partial x^2$, and at the same time it facilitates satisfying the condition $T = 0$ at $x = 0$. Substitution into the differential equation (4.195) gives

$$\begin{aligned} \frac{\partial^2 T}{\partial x^2} + \frac{\partial^2 T}{\partial y^2} = \int_0^\infty T_\lambda(y)(-\lambda^2) \sin(\lambda x)\, d\lambda + \int_0^\infty T_\lambda''(y) \sin(\lambda x)\, d\lambda &= 0 \\ \int_0^\infty \left[T_\lambda''(y) - \lambda^2 T_\lambda(y)\right] \sin(\lambda x)\, d\lambda &= 0. \end{aligned}$$

In order for this to be valid for all values of x, we need

$$T_\lambda''(y) - \lambda^2 T_\lambda(y) = 0. \tag{4.197}$$

This ordinary differential equation has the general solution

$$T_\lambda(y) = A(\lambda)\sinh\lambda y + B(\lambda)\cosh\lambda y.$$

The substitution of this expression into equation (4.196) gives us

$$T(x,y) = \int_0^\infty [A(\lambda)\sinh\lambda y + B(\lambda)\cosh\lambda y]\sin(\lambda x)\,d\lambda \tag{4.198}$$

The boundary condition $\partial T(x,y)/\partial y|_{y=0} = 0$, being a homogeneous condition, may be satisfied under the integral sign (analogous to satisfying a homogeneous condition term-by-term for a series solution). Thus, requiring $T'_\lambda(0) = 0$ will cause $\partial T(x,0)/\partial y$ to be zero as well. To carry this out, we require

$$T'_\lambda(0) = A(\lambda)\lambda\cosh(\lambda 0) + B(\lambda)\lambda\sinh(\lambda 0) = \lambda A(\lambda) = 0,$$

reducing $T_\lambda(y)$ to

$$T_\lambda(y) = B(\lambda)\cosh\lambda y,$$

and the expression (4.198) for $T(x,y)$ becomes

$$T(x,y) = \int_0^\infty B(\lambda)\cosh\lambda y\ \sin(\lambda x)\,d\lambda \tag{4.199}$$

The boundary condition at $y = b$ requires

$$T(x,b) = T_0 e^{-ax} = \int_0^\infty B(\lambda)\cosh\lambda b\ \sin(\lambda x)\,d\lambda. \tag{4.200}$$

Using the formula for the Fourier sine integral expansion (see page 96 in Chapter 3), we have

$$B(\lambda)\cosh\lambda b = \frac{2}{\pi}\int_{x=0}^\infty T_0 e^{-ax}\sin\lambda x\,dx = \frac{2}{\pi}\frac{T_0\lambda}{a^2+\lambda^2}.$$

or

$$B(\lambda) = \frac{2}{\pi}\frac{T_0\lambda}{(a^2+\lambda^2)\cosh\lambda b},$$

and therefore, the integral expression (4.199) for $T(x,y)$ may be written as

$$T(x,y) = \frac{2T_0}{\pi}\int_0^\infty \frac{\lambda\cosh\lambda y}{(a^2+\lambda^2)\cosh\lambda b}\sin(\lambda x)\,d\lambda \tag{4.201}$$

□

EXERCISES 4.2

Solve Laplace's equation,

$$\frac{\partial^2 T}{\partial x^2} + \frac{\partial^2 T}{\partial y^2} = 0,$$

for the rectangular regions with the boundary conditions as shown.

Problem 1

$T = T_0 x(a - x)/a^2$ $\quad y = b$

$T = 0$ $\quad T = 0$, $x = a$

$y = 0$

$T = 0$

Problem 2

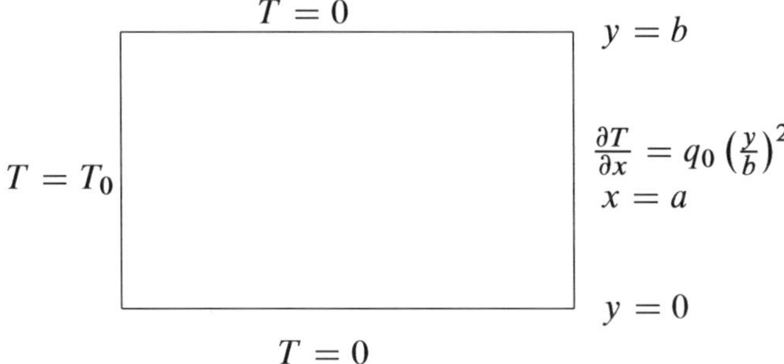

Problem 3

$T = T_0 x/a$ $\quad y = b$

$T = 0$ $\quad T = T_0$, $x = a$

$y = 0$

$\frac{\partial T}{\partial y} = 0$

Problem 4

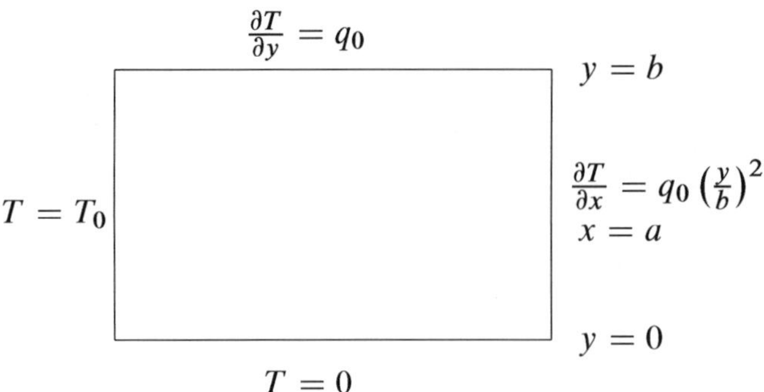

Problem 5

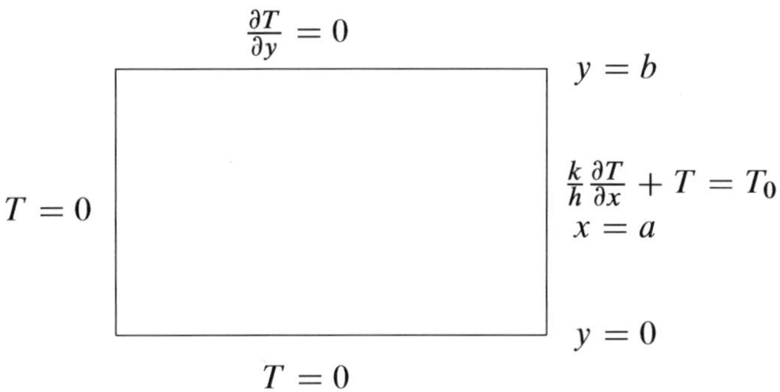

Problem 6

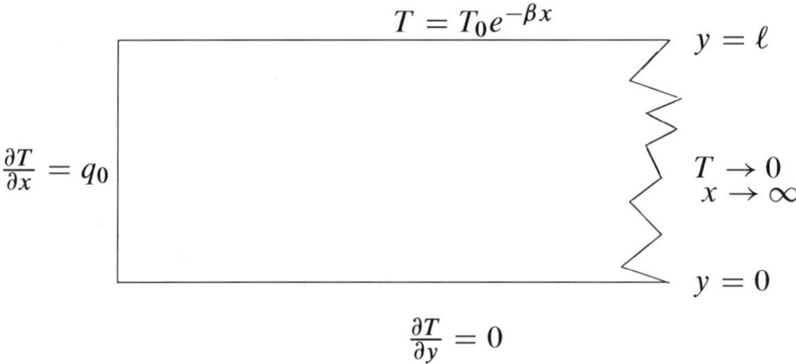

Problem 7
Solve

$$\frac{1}{\alpha}\frac{\partial T}{\partial t} = \frac{\partial^2 T}{\partial x^2}, \quad 0 \le x \le l,$$

with initial condition

$$T(x,0) = T_0 \left(\frac{x}{l}\right)^2,$$

and boundary conditions

$$T(0,t) = T_0 \quad \text{and} \quad \left.\frac{\partial T}{\partial x}\right|_{x=l} = 0.$$

Problem 8
The torsion of a long rod of rectangular cross-section is described by Poissons equation,

$$\frac{\partial^2\Psi}{\partial x^2}+\frac{\partial^2\Psi}{\partial y^2}=F_0,$$

where $\Psi(x, y)$ is the torsion function, $F_0 = 2G\theta_0$ with G as the shear modulus of the rod material, and θ_0 as the angle of twist. At the free surfaces of the rod, $\Psi = 0$, as shown.

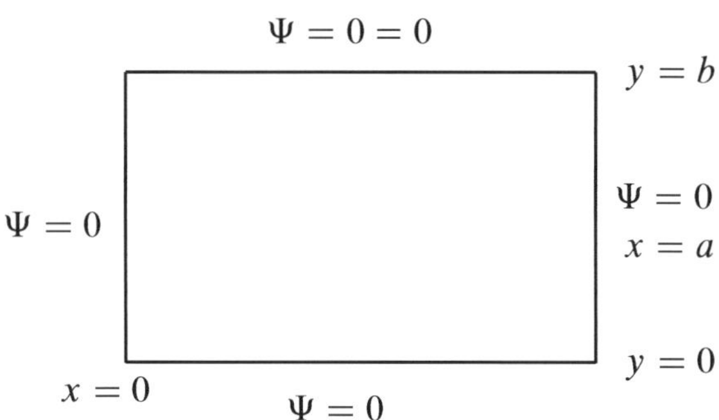

This nonhomogeneous PDE may be solved by the method of eigenfunction expansions. You may choose any variable (x or y) to expand in since both have pairs of opposite homogeneous boundary conditions. Lets pick x. Set up eigenfunctions $X_m(x)$ of the operator d^2/dx^2 and obtain expressions for $X_m(x)$ that satisfy both homogeneous the boundary conditions at $x = 0$ and $x = a$. Expand both $\Psi(x, y)$ and F_0 as Fourier series x. Substitute both expansions in the PDE above. Match term-by-term and obtain a nonhomogeneous ODE for $Y_m(y)$. Solve that ODE and satisfy the homogeneous boundary conditions at $y = 0$ and $y = b$. Then complete the solution by writing the expression for $\Psi(x, y)$.

4.6 Problems in Three Variables

Typically, for a problem involving three variables, we would need eigenfunction expansions in two variables. To understand the principles involved, we begin with an example with the least possible complexity.

Example 4.10 Let is consider the two-dimensional time-dependent heat equation,

$$\frac{1}{\alpha}\frac{\partial T}{\partial t}=\frac{\partial^2 T}{\partial x^2}+\frac{\partial^2 T}{\partial y^2} \tag{4.202}$$

with initial and boundary conditions are as shown here in Figure 4.10. Here, we have two pairs of opposite homogeneous boundary conditions, one set at $x = 0$ and $x = a$, and another at $y = 0$ and $y = b$. Based on the principles we established earlier for two-variable systems, we need to look for eigenfunctions of the operators, $\partial^2/\partial x^2$ and $\partial^2/\partial y^2$, that each respectively satisfies the corresponding pair of opposite homogeneous boundary conditions. It is not difficult to identify

$$X_m(x)=\sin\left(\frac{m\pi x}{a}\right)\quad\text{and}\quad Y_n(x)=\sin\left(\frac{n\pi y}{b}\right)$$

as the suitable eigenfunctions for our purpose. Thus, we may write

$$T(x,y,t)=\sum_{n=1}^{\infty}\sum_{m=1}^{\infty}T_{mn}(t)\sin\left(\frac{m\pi x}{a}\right)\sin\left(\frac{n\pi y}{b}\right) \tag{4.203}$$

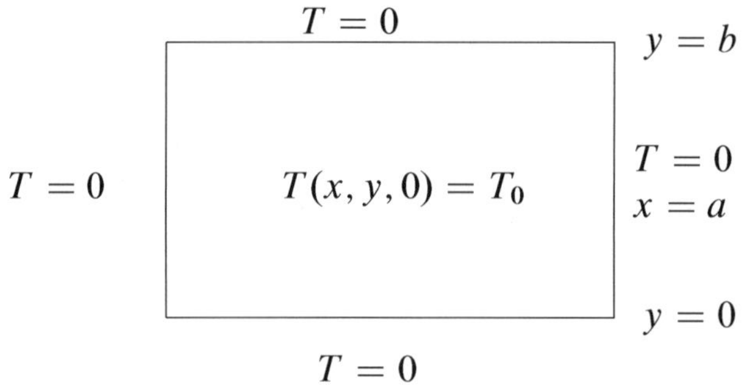

Figure 4.10: Initial and boundary conditions for example 4.10.

Substitution of this expression into the differential equation (4.202) gives us

$$\frac{1}{\alpha}\sum_{n=1}^{\infty}\sum_{m=1}^{\infty} T'_{mn}(t)\sin\left(\frac{m\pi x}{a}\right)\sin\left(\frac{n\pi y}{b}\right) \tag{4.204}$$

$$= \sum_{n=1}^{\infty}\sum_{m=1}^{\infty}\left\{T_{mn}(t)\left[-\left(\frac{m\pi}{a}\right)^2\sin\left(\frac{m\pi x}{a}\right)\sin\left(\frac{n\pi y}{b}\right)\right]\right.$$
$$\left.+T_{mn}(t)\sin\left(\frac{m\pi x}{a}\right)\left[-\left(\frac{n\pi}{b}\right)^2\sin\left(\frac{n\pi y}{b}\right)\right]\right\}$$
$$= \sum_{n=1}^{\infty}\sum_{m=1}^{\infty} T_{mn}(t)\left[-\left(\frac{m\pi}{a}\right)^2 - \left(\frac{n\pi}{b}\right)^2\right]\sin\left(\frac{m\pi x}{a}\right)\sin\left(\frac{n\pi y}{b}\right) \tag{4.205}$$

which, after some rearrangement, becomes

$$\sum_{n=1}^{\infty}\sum_{m=1}^{\infty}\left\{\frac{1}{\alpha}T'_{mn}(t) + \left[\left(\frac{m\pi}{a}\right)^2 + \left(\frac{n\pi}{b}\right)^2\right]T_{mn}(t)\right\}\sin\left(\frac{m\pi x}{a}\right)\sin\left(\frac{n\pi y}{b}\right) = 0. \tag{4.206}$$

For this result to be valid for all values of x and y in the ranges $0 \le x \le a$ and $0 \le y \le b$, respectively, we can easily see that $T_{mn}(t)$ satisfies the first-order ordinary differential equation

$$\frac{1}{\alpha}T'_{mn}(t) + \left[\left(\frac{m\pi}{a}\right)^2 + \left(\frac{n\pi}{b}\right)^2\right]T_{mn}(t) = 0,$$

for which the general solution is

$$T_{mn}(t) = A_{mn}e^{-\alpha\left[\left(\frac{m\pi}{a}\right)^2+\left(\frac{n\pi}{b}\right)^2\right]t}. \tag{4.207}$$

Upon inserting this expression for $T_{mn}(t)$ into equation (4.203), we may write

$$T(x, y, t) = \sum_{n=1}^{\infty}\sum_{m=1}^{\infty} A_{mn}e^{-\alpha[(\frac{m\pi}{a})^2+(\frac{n\pi}{b})^2]t}\sin\left(\frac{m\pi x}{a}\right)\sin\left(\frac{n\pi y}{b}\right) \tag{4.208}$$

The application of the initial condition, $T(x, y, 0) = T_0$ gives

$$T(x,y,t) = T_0 = \sum_{n=1}^{\infty}\sum_{m=1}^{\infty} A_{mn} \sin\left(\frac{m\pi x}{a}\right) \sin\left(\frac{n\pi y}{b}\right) \tag{4.209}$$

which is a double Fourier sine series. To facilitate expansion of one variable at a time, we may bracket out the variables as follows

$$T_0 = \sum_{n=1}^{\infty}\left[\sum_{m=1}^{\infty} A_{mn} \sin\left(\frac{m\pi x}{a}\right)\right] \sin\left(\frac{n\pi y}{b}\right) \tag{4.210}$$

The coefficient of the $\sin(n\pi y/b)$ series is given by

$$\begin{aligned}
\left[\sum_{m=1}^{\infty} A_{mn} \sin\left(\frac{m\pi x}{a}\right)\right] &= \frac{2}{b}\int_0^b T_0 \sin\left(\frac{n\pi y}{b}\right) dy \\
&= \frac{2T_0}{b}\left[\frac{-\cos\left(\frac{n\pi y}{b}\right)}{\frac{n\pi}{b}}\right]_0^b \\
&= \frac{2T_0\,[1-(-1)^n]}{n\pi}
\end{aligned} \tag{4.211}$$

We may now treat the left-hand side of (4.211) as a Fourier sine series in x. The set of coefficients A_{mn} may be expressed as

$$\begin{aligned}
A_{mn} &= \frac{2}{a}\int_0^a \frac{2T_0\,[1-(-1)^n]}{n\pi} \sin\left(\frac{m\pi x}{a}\right) dx \\
&= \frac{2T_0\,[1-(-1)^n]}{n\pi}\,\frac{2[1-(-1)^m]}{m\pi} \\
&= \frac{4T_0\,[1-(-1)^n]\,[1-(-1)^m]}{mn\pi^2}.
\end{aligned} \tag{4.212}$$

This may now be substituted into equation (4.207) to give an explicit form for $T(x, y, t)$,

$$\begin{aligned}
T(x,y,t) = \sum_{n=1}^{\infty}\sum_{m=1}^{\infty} & \frac{4T_0\,[1-(-1)^n]\,[1-(-1)^m]}{mn\pi^2} \\
& \times e^{-\alpha[(\frac{m\pi}{a})^2+(\frac{n\pi}{b})^2]t} \sin\left(\frac{m\pi x}{a}\right) \sin\left(\frac{n\pi y}{b}\right)
\end{aligned} \tag{4.213}$$

□

The approach of the double eigenfunction expansion worked because both pairs of boundary conditions were homogeneous. For cases in which the boundary conditions are non-homogeneous, we need to decompose the problem into a steady-state part and a time-dependent one. The steady-state part takes care of the non-homogeneous conditions while the time-dependent part sees only homogeneous boundary conditions. Let us consider another example.

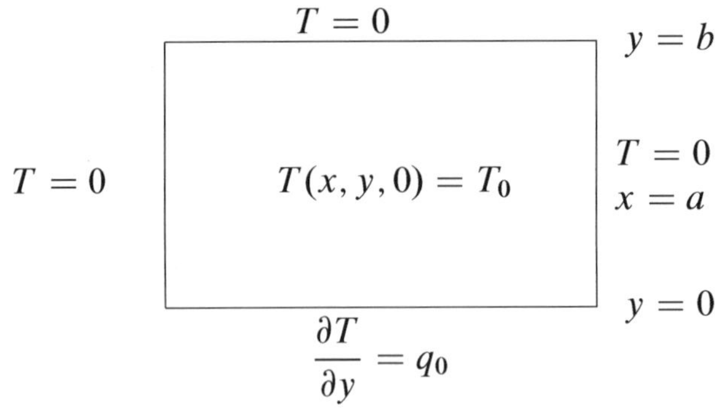

Figure 4.11: Initial and boundary conditions for example 4.11.

Example 4.11 Again, we look at the time-dependent heat equation,

$$\frac{1}{\alpha}\frac{\partial T}{\partial t} = \frac{\partial^2 T}{\partial x^2} + \frac{\partial^2 T}{\partial y^2} \tag{4.214}$$

with one non-homogeneous boundary condition at $y = 0$ as shown in Figure 4.11. In order to be able to apply the method of double eigenfunction expansion, we need to have both pairs of boundary conditions to be homegeneous. Therefore, just as we did for the one-dimensional case in Example 4.7 on page 170, we decompose $T(x, y, t)$ into a steady-state or time-independent part, $T_1(x, y)$, and a time-dependent part, $T_2(x, y, t)$, i.e.,

$$T(x, y, t) = T_1(x, y) + T_2(x, y, t). \tag{4.215}$$

Here, we will require that $T_1(x, y)$ will satisfy the non-homogeneous boundary condition(s) and $T_2(x, y, t)$ will then have all homogeneous boundary conditions, thus facilitation a double eigenfunction expansion. The boundary and initial conditions distribution is illustrated in Figure 4.12. For the steady-state part, the the time-derivative term in the differential equation vanishes, leaving us with Laplace's equation,

$$\frac{\partial^2 T_1}{\partial x^2} + \frac{\partial^2 T_1}{\partial y^2} = 0, \tag{4.216}$$

with boundary conditions as shown. This part takes care of all the non-homogeneous conditions (only one in this case at $y = 0$). In the time-dependent part, the equation for $T_2(x, y, t)$ is the same as for $T(x, y, t)$ but the boundary and initial conditions are different. Thus, $T_2(x, y, t)$ satisfies

$$\frac{1}{\alpha}\frac{\partial^2 T_1}{\partial t^2} = \frac{\partial^2 T_2}{\partial x^2} + \frac{\partial^2 T_2}{\partial y^2} \tag{4.217}$$

with the initial condition

$$T_2(x, y, 0) = T_0 - T_1(x, y), \tag{4.218}$$

and boundary conditions as shown. Since the one non-homogeneous boundary condition at $y = 0$ has been taken care of by $T_1(x, y)$, this part has all homogeneous boundary conditions, and an initial conditions modified from that for $T(x, y, t)$.

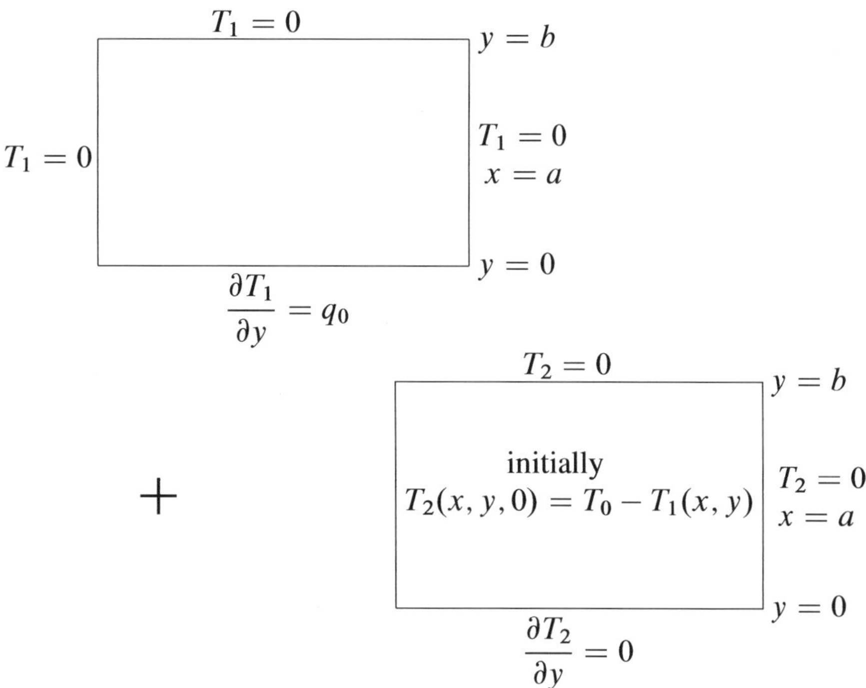

Figure 4.12: Distribution of initial and boundary conditions after decomposition into steady-state and time-dependent parts

For $T_1(x, y)$, there is a pair of opposite homogeneous boundary conditions at $x = 0$ and $x = a$, and a non-homogeneous condition at $y = 0$ exactly as in the second part of Example 4.5. This calls for an expansion in x in terms of eigenfunctions of $\partial^2/\partial x^2$ that satisfy homogeneous conditions at $x = 0$ and $x = a$. In this case, we can expand $T_1(x, y)$ in terms of $X_n(x) = \sin(m\pi x/a)$, i.e.,

$$T_1(x, y) = \sum_{m=1}^{\infty} B_m \sinh\left(\frac{m\pi(b-y)}{a}\right) \sin\left(\frac{m\pi x}{a}\right), \tag{4.219}$$

as in equation (4.138) on page 167. This expression for $T_1(x, y)$ satisfies equation (4.216) and the three homoheneous boundary conditions at $x = 0$, $x = a$ and $y = b$. After satisfying the last boundary condition at $y = 0$, we have

$$B_m = -\frac{2q_0 a}{m^2\pi^2} \frac{[1-(-1)^n]}{\cosh\left(\frac{m\pi b}{a}\right)} \tag{4.220}$$

as in equation (4.145) on page 169. As a result, the explicit expression for $T_1(x, y)$ can be written as [see (4.146) on page 169]

$$T_1(x, y) = -2q_0 a \sum_{m=1}^{\infty} \frac{[1-(-1)^m]}{m^2\pi^2} \frac{\sinh\left[\frac{m\pi}{a}(b-y)\right]}{\cosh\left(\frac{m\pi}{a}b\right)} \sin\left(\frac{m\pi x}{a}\right) \tag{4.221}$$

Having obtained the solution for $T_1(x, y)$, the initial condition for $T_2(x, y, t)$ as given in equation (4.218) is completely determined. With two pairs of homogeneous boundary conditions for $T_2(x, y, t)$, we can carry out a double eigenfunction expansion as in Example 4.10, i.e.,

$$T_2(x, y, t) = \sum_{n=0}^{\infty} \sum_{m=1}^{\infty} T_{mn}(t) \sin \lambda_m x \cos \mu_n y, \tag{4.222}$$

Here, $X_m(x) = \sin \lambda_m x$ and $Y_n(y) \cos \mu_n y$ are are eigenfunctions of $\partial^2/\partial x^2$ and $\partial^2/\partial y^2$, respectively. We have already used the condition $T_2 = 0$ at $x = 0$ by selecting the sine function for $X_m(x)$, and similarly, $\partial T_2/\partial y = 0$ at $y = 0$ is satisfied by the selection of a cosine function for $Y_n(y)$. The boundary condition $T_2 = 0$ at $x = a$ requires that $\lambda_m = m\pi/a$, and with $T_2 = 0$ at $y = b$ we need $\mu_n = \left(n + \frac{1}{2}\right) \pi/b$. By substituting equation (4.222) into the differential equation (4.218) leads to

$$\frac{1}{\alpha} \sum_{n=0}^{\infty} \sum_{m=1}^{\infty} T'_{mn}(t) \sin \lambda_m x \cos \mu_n y$$
$$= \sum_{n=0}^{\infty} \sum_{m=1}^{\infty} T_{mn}(t) \left(-\lambda_m^2\right) \sin \lambda_m x \cos \mu_n y + \sum_{n=0}^{\infty} \sum_{m=1}^{\infty} T_{mn}(t) \sin \lambda_m x \left(-\mu_n^2\right) \cos \mu_n y. \tag{4.223}$$

Since the eigenfunctions $X_m(x) = \sin \lambda_m x$ and $Y_n(y) \cos \mu_n y$ are preserved (which by definition of being eigenfunction they should be) under this operation, it is possible to all the terms under a common double sum, i.e.,

$$\sum_{m=1}^{\infty} \sum_{n=0}^{\infty} \left[\frac{1}{\alpha} T'_{mn}(t) + \left(\lambda_m^2 + \mu_n^2\right) T_{mn}(t)\right] \sin \lambda_m x \cos \mu_n y = 0.$$

It is not difficult to show that the ordinary differential equation for $T_{mn}(t)$ is

$$\frac{1}{\alpha} T'_{mn}(t) + \left(\lambda_m^2 + \mu_n^2\right) T_{mn}(t) = 0.$$

whose solution is

$$T_{mn}(t) = A_{mn} e^{-\alpha\left(\lambda_m^2 + \mu_n^2\right)t}, \tag{4.224}$$

and by using this expression in (4.222), we end up with

$$T_2(x, y, t) = \sum_{n=0}^{\infty} \sum_{m=1}^{\infty} A_{mn} e^{-\alpha\left(\lambda_m^2 + \mu_n^2\right)t} \sin \lambda_m x \cos \mu_n y, \tag{4.225}$$

Now, we can determine the set of constants $\{A_{mn}\}$ by applying the initial condition (4.218). By placing the expression for $T_2(x, y, 0)$ on the left-hand side of (4.218) and using (4.219) on the right-hand side, we may write

$$\sum_{n=0}^{\infty} \sum_{m=1}^{\infty} A_{mn} \sin \lambda_m x \cos \mu_n y = T_0 - \sum_{m=1}^{\infty} B_m \sinh\left(\frac{m\pi(b-y)}{a}\right) \sin\left(\frac{m\pi x}{a}\right). \tag{4.226}$$

Noting that $\lambda_m = m\pi/a$, and switching the order of the double sum, this equation becomes

$$\sum_{m=1}^{\infty}\sum_{n=0}^{\infty} A_{mn}\cos\mu_n y \sin\left(\frac{m\pi x}{a}\right) = T_0 - \sum_{m=1}^{\infty} B_m \sinh\left(\frac{m\pi(b-y)}{a}\right)\sin\left(\frac{m\pi x}{a}\right). \tag{4.227}$$

We could either apply the orthogonality principle [for the set $\{\sin(m\pi/a)\}$] formally to equation (4.227), or express T_0 as a sine series,

$$T_0 = \sum_{m=1}^{\infty} a_m \sin\left(\frac{m\pi x}{a}\right),$$

and drop the sum in m. Taking the latter route,

$$a_m = \frac{2}{a}\int_{x=0}^{a} T_0 \sin\left(\frac{m\pi x}{a}\right)\, dx = \frac{2T_0[1-(-1)^m]}{m\pi}.$$

Equation (4.227) can now be written as

$$\sum_{m=1}^{\infty}\sum_{n=0}^{\infty}[A_{mn}\cos\mu_n y]\sin\left(\frac{m\pi x}{a}\right)$$
$$= \sum_{m=1}^{\infty} a_m \sin\left(\frac{m\pi x}{a}\right) - \sum_{m=1}^{\infty} B_m \sinh\left(\frac{m\pi(b-y)}{a}\right)\sin\left(\frac{m\pi x}{a}\right) \tag{4.228}$$

Upon comparing the coefficients of $\sin(m\pi/a)$ term by term results in effectively dropping the summation in m, i.e.,

$$\sum_{n=0}^{\infty}[A_{mn}\cos\mu_n y] = a_m - B_m \sinh[\lambda_m(b-y)], \tag{4.229}$$

where we revert to $\lambda_m = (m\pi/a)$ for convenience. This is a Fourier cosine series in y on which we may apply orthogonality and determine the coefficients A_{mn}. Thus,

$$\begin{aligned}
A_{mn} &= \frac{\int_0^b \{a_m - B_m\sinh[\lambda_m(b-y)]\}\cos\mu_n y\, dy}{\int_0^b \cos^2\mu_n y\, dy} \\
&= \frac{a_m\int_0^b \cos\mu_n y\, dy - B_m\int_0^b \sinh[\lambda_m(b-y)]\cos\mu_n y}{\frac{1}{2}b} \\
&= \frac{2}{b}\left[a_m\frac{\sin\mu_n b}{\mu_n} - \frac{B_m\lambda_m\mu_n\cosh\lambda_m b}{(\lambda_m^2+\mu_n^2)}\right] \\
&= \left[\frac{2T_0[1-(-1)^m](-1)^n}{m\pi\left(n+\frac{1}{2}\right)} + \frac{4q_0 a}{\pi^2 b}\frac{[1-(-1)^m]}{m^2\cosh\left(\frac{m\pi b}{a}\right)}\frac{\lambda_m\cosh\lambda_m b}{(\lambda_m^2+\mu_n^2)}\right] \\
&= \frac{2[1-(-1)^m]}{m\pi^2}\left[\frac{T_0(-1)^n}{\left(n+\frac{1}{2}\right)} + \frac{2q_0 b}{\frac{m^2b^2}{a^2}+\left(n+\frac{1}{2}\right)^2}\right]
\end{aligned} \tag{4.230}$$

The complete solution is given by combining the expressions for $T_1(x, y)$ and $T_2(x, y, t)$ given by equations (4.221) and (4.225), i.e.,

$$\begin{aligned} T(x,y,t) &= T_1(x,y) + T_2(x,y,t) \\ &= -2q_0 a \sum_{m=1}^{\infty} \frac{[1-(-1)^m]}{m^2\pi^2} \frac{\sinh\left[\frac{m\pi}{a}(b-y)\right]}{\cosh\left(\frac{m\pi}{a}b\right)} \sin\left(\frac{m\pi x}{a}\right) \\ &+ \sum_{n=0}^{\infty}\sum_{m=1}^{\infty} \frac{2[1-(-1)^m]}{m\pi^2}\left[\frac{T_0(-1)^n}{\left(n+\frac{1}{2}\right)} + \frac{2q_0 b}{\frac{m^2b^2}{a^2} + \left(n+\frac{1}{2}\right)^2}\right] \\ &\times\; e^{-\alpha\left[\left(\frac{m\pi}{a}\right)^2 + \left(n+\frac{1}{2}\right)^2\left(\frac{\pi}{b}\right)^2\right]t} \sin\left(\frac{m\pi x}{a}\right)\cos\left[\left(n+\frac{1}{2}\right)\frac{\pi y}{b}\right] \end{aligned} \tag{4.231}$$

□

EXERCISES 4.3

1. A membrane is mounted on a square-shaped frame that covers the region $-\frac{1}{2}a \leq x \leq \frac{1}{2}a, -\frac{1}{2}a \leq y \leq \frac{1}{2}a$. The membrane is initially displaced in the shape,

$$u(x,y,0) = u_0\left(1-\frac{4x^2}{a^2}\right)\left(1-\frac{4y^2}{a^2}\right),$$

and let go (zero initial velocity). Assuming that the membrane oscillates according to the wave equation,

$$\frac{1}{c^2}\frac{\partial^2 u}{\partial t^2} = \frac{\partial^2 u}{\partial x^2} + \frac{\partial^2 u}{\partial y^2},$$

obtain the displacement $u(x, y, t)$. over the square region. Take the boundary conditions as $u = 0$ on all the edges.

2. Solve

$$\frac{1}{\alpha}\frac{\partial T}{\partial t} = \frac{\partial^2 T}{\partial x^2} + \frac{\partial^2 T}{\partial y^2},$$

with boundary and initial conditions as shown.

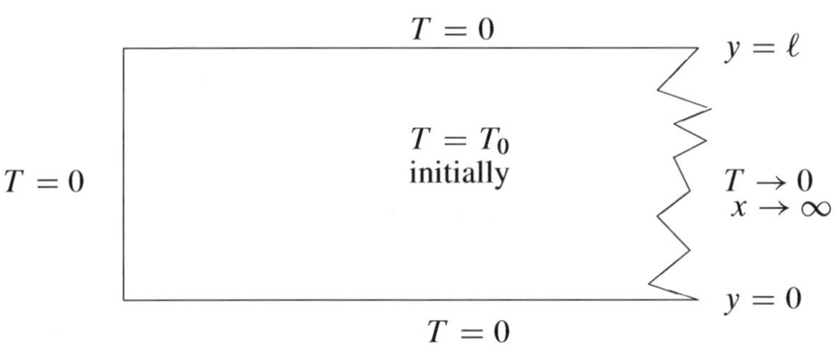

3. Solve

$$\frac{1}{\alpha}\frac{\partial T}{\partial t} = \frac{\partial^2 T}{\partial x^2} + \frac{\partial^2 T}{\partial y^2},$$

with boundary and initial conditions as shown.

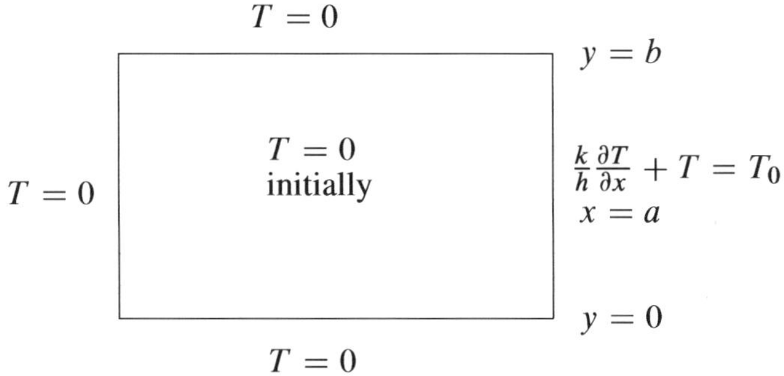

4. Solve

$$\frac{\partial^2 \psi}{\partial x^2} + \frac{\partial^2 \psi}{\partial y^2} + 2\frac{\partial \psi}{\partial y} + \psi = 0,$$

with boundary conditions as shown.

$\frac{\partial \psi}{\partial y} = q_0$

$y = b$

$\psi = e^{-y}$

$\psi = 0$

$x = a$

$y = 0$

$\psi = 0$

4.7 Geometrical Limitations

In dealing with partial differential equations, we have seen that we were successful with the eigenfunction expansions in the rectangular coordinate system as long as we were dealing with rectangular geometry. In fact, the method of separation of variables relies on the identification of each segment of the boundary of the domain of interest by a constant single value of a coordinate such as $x = 0$, or $y = b$. This is so because we depend on satisfying the homogeneous boundary conditions by requiring each eigenfunction in terms of the variable defining the boundaries to satisfy that set of homogeneous conditions or other conditions (such as periodicity) that may be satisfied term-by-term with each eigenfunction. At the same time, any linear combination of these eigenfunctions still satisfies the same set of conditions. To illustrate this principle, let us consider Example 4.2 on pages 152-158. With that example, it is clear from equation (4.69) that a homogeneous condition satisfied by each eigenfunction will indeed be satisfied by the whole sum as well. Therefore a set of eigenfunctions given by (4.76), or equivalently, $Y_n(y) = \sin(n\pi y/b)$ which satisfy, $Y_n(0) = Y_n(b) = 0$, a solution with the structure of the form (4.83), i.e.,

$$T(x, y) = \sum_{n=1}^{\infty} a_n \sinh\left(\frac{n\pi x}{b}\right) \sin\left(\frac{n\pi y}{b}\right). \tag{4.232}$$

will be zero at $y = 0$ and $y = b$. This is a meaningful satisfaction of boundary conditions provided the boundaries are defined by these constant-y values, regardless of how the x-dependence through $X_n(x) = a_n \sinh(n\pi x/b)$ plays in. In addition, these boundary conditions guarantee orthogonality which is needed for satisfying any of the non-homogeneous conditions on boundaries that are perpendicular to $y = 0$ and $y = b$. Now, let us examine the non-homogeneous condition at $x = a$, i.e., $T(a, y) = T_0$. Letting $x = a$ in equation (4.232), we see that

$$T(a, y) = T_0 = \sum_{n=1}^{\infty} a_n \sinh\left(\frac{n\pi a}{b}\right) \sin\left(\frac{n\pi y}{b}\right). \tag{4.233}$$

The function on the left is now a function of only y. Therefore, to satisfy a nonzero condition there in the form of a Fourier expansion, all terms multiplying $\sin(n\pi y/b)$ must be constant as far as the space variables are concerned. This is indeed the case, and the Fourier series expansion in terms of $\sin(n\pi y/b)$ would yield the values of the set of coefficients a_n which are give by equation (4.85). We can see that the method works because on the boundaries, we are able to fix one of the variables while the other one satisfies either homogeneous conditions term-by-term (and as a sum), or non-homogeneous conditions as a Fourier series. If any of the boundaries required two or more variables to define, then we would run into difficulties with this method. Take, for example, the trapezoid shown in Figure 4.13, where Laplace's equation

$$\frac{\partial^2 T}{\partial x^2} + \frac{\partial^2 T}{\partial y^2} = 0 \tag{4.234}$$

is to be satisfied with the boundary conditions as shown. For this system, the solution satisfying the three homogeneous boundary conditions (at $x = 0$, $y = 0$, and $y = b$) can be shown to be

$$T(x, y) = \sum_{n=0}^{\infty} A_n \sinh\left[(n + \tfrac{1}{2})\frac{\pi x}{b}\right] \cos\left[(n + \tfrac{1}{2})\frac{\pi y}{b}\right] \tag{4.235}$$

Since the sloping edge is defined by two variables in the form

$$y = -\frac{b(x - a)}{(a - c)} \qquad \text{or} \qquad x = a - \frac{(a - c)y}{b}$$

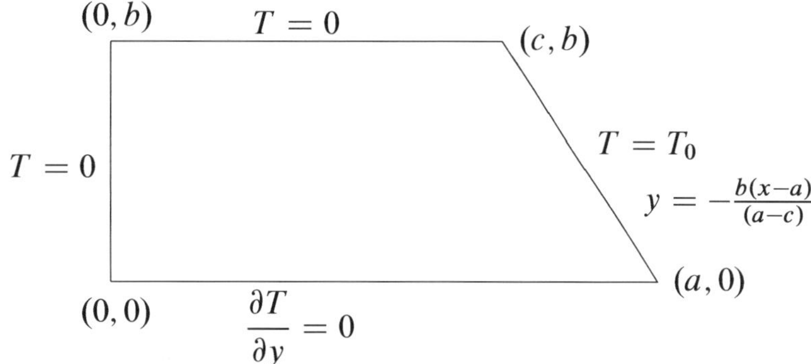

Figure 4.13: Trapezoidal geometry: an illustration of difficulties

an attempt to satisfy the non-homogeneous boundary condition there leads to

$$T_0 = \sum_{n=0}^{\infty} A_n \sinh\left\{(n+\tfrac{1}{2})\frac{\pi\,[a-(a-c)y/b]}{b}\right\}\cos\left[(n+\tfrac{1}{2})\frac{\pi y}{b}\right], \tag{4.236}$$

which does not lend itself to a suitable Fourier series expansion in y since the y-dependence appears in terms besides the cosine-expansion terms. Similar difficulties arise even when the boundary condition on the sloping edge is a homogeneous one. To deal with this class of problems, one needs to transform the variables by applying the conformal mapping technique involving complex variables [4]. A successful mapping would generate a new set of variables which would allow each segment of the boundary to be defined by a single coordinate. While the conformal mapping approach is not discussed here, there are some simple basic geometries for which a straightforward coordinate transformation permits the eigenfunction method to work. These include the cylindrical and spherical coordinate systems which are discussed in the next two chapters, together with the solution techniques.

Chapter 5

Problems in Cylindrical Geometry

As discussed at the end of the Chapter 4, in order to use the method of separation of variables (or eigenfunction expansions), it is necessary to identify each segment of the boundary by a single coordinate. For regions with circular boundaries, the cylindrical coordinate system is therefore quite suitable. In two dimensions, going from the Cartesian (x, y) system to polar (r, θ) coordinates entails the transformation,

$$x = r\cos\theta, \qquad \text{and} \qquad y = r\sin\theta. \tag{5.1}$$

Under this transformation, derivatives with respect to x and y need to be written in terms of r and θ. Following the basic principles of multivariable calculus, differential operators such as the Laplacian, ∇^2 can be stated in terms of r and θ. Specifically,

$$\frac{\partial^2}{\partial x^2} + \frac{\partial^2}{\partial y^2} = \frac{\partial^2}{\partial r^2} + \frac{1}{r}\frac{\partial}{\partial r} + \frac{1}{r^2}\frac{\partial^2}{\partial \theta^2}.$$

The derivatives with respect to z remain unchanged, and the full three-dimensional Laplacian operator has the following form

$$\nabla^2 = \frac{\partial^2}{\partial r^2} + \frac{1}{r}\frac{\partial}{\partial r} + \frac{1}{r^2}\frac{\partial^2}{\partial \theta^2} + \frac{\partial^2}{\partial z^2}. \tag{5.2}$$

We shall first consider cases where there is no z dependence so that for a dependent variable $\psi(r, \theta)$ obeying Laplace's equation would need to satisfy

$$\nabla^2\psi(r,\theta) = \frac{\partial^2\psi}{\partial r^2} + \frac{1}{r}\frac{\partial\psi}{\partial r} + \frac{1}{r^2}\frac{\partial^2\psi}{\partial \theta^2} = 0. \tag{5.3}$$

Depending on the boundary conditions, we would need an eigenfunction expansion in r or θ for a two-variable problem. We shall begin by considering cases requiring expansions in θ.

5.1 Eigenfunction Expansions in θ

Let us now consider a two-dimensional example

Example 5.1

Problem Statement: For the semicircular region shown, solve

$$\nabla^2 T(r, \theta) = 0, \tag{5.4}$$

with boundary conditions as indicated.
Solution
We should first note that the boundary is defined by $\theta = 0$, $\theta = \pi$ and $r = R$, i.e., the constant values of one or the other coordinates specifies the different segments of the boundary. In this case there is no z-dependence, and equation 5.4 reduces to

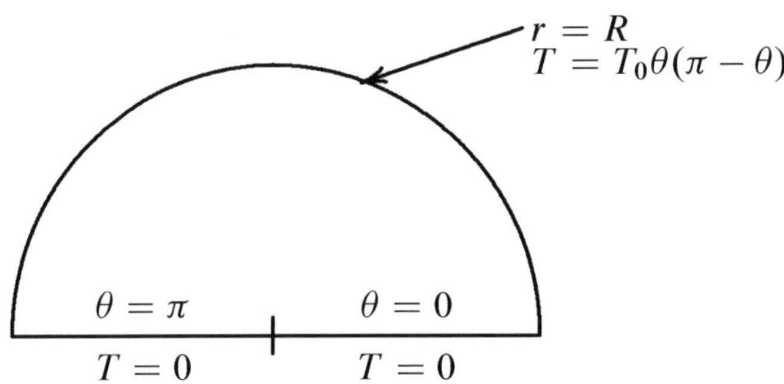

Figure 5.1: Boundary conditions for example 5.1

$$\frac{\partial^2 T}{\partial r^2} + \frac{1}{r}\frac{\partial T}{\partial r} + \frac{1}{r^2}\frac{\partial^2 T}{\partial \theta^2} = 0. \tag{5.5}$$

If we multiply this equation by r^2, we can separate the r and θ derivative operations, i.e.,

$$r^2\frac{\partial^2 T}{\partial r^2} + r\frac{\partial T}{\partial r} + \frac{\partial^2 T}{\partial \theta^2} = 0. \tag{5.6}$$

Based on the procedures developed in Chapter 4, we need to identify a pair of opposite homogeneous boundary conditions to set up a suitable eigenfunction expansion. For this purpose, it will help us if we visualize the problem with r and θ as rectangular coordinates, as shown in figure 5.2. Here we are able to identify a pair of opposite homogeneous boundary conditions at $\theta = 0$ and $\theta = \pi$. This signals us that we should be looking for a set of suitable eigenfunctions in θ. Furthermore, an expansion in θ will also facilitate satisfying the nonzero condition at $r = R$ where θ is a variable along that segment of the boundary. In the differential equation (5.5), all of the θ operations are confined to the last term $\partial^2 T/\partial\theta^2$. Therefore, we should seek eigenfunctions of the operator $d^2/d\theta^2$, i.e., we need to develop a set of solutions of

$$\frac{d^2\Theta(\theta)}{d\theta^2} = -\lambda^2\Theta(\theta). \tag{5.7}$$

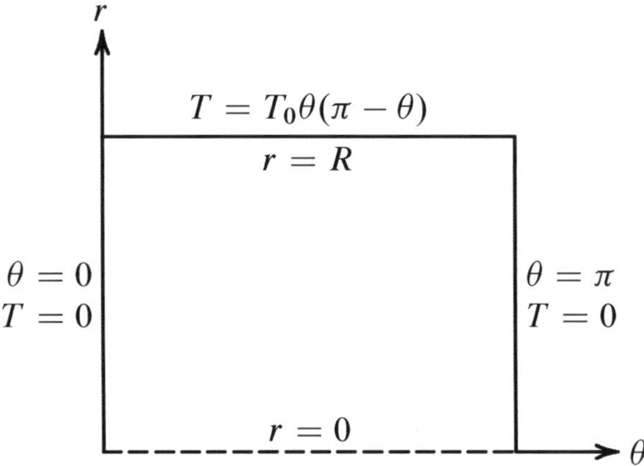

Figure 5.2: Depiction of boundary conditions for example 5.1 in 'rectangularized' coordinates. The point $r = 0$ stretches to a line.

Following the procedures for ordinary differential equations with constant coefficients (see Chapter 1, pages 2-11), it is not difficult obtain the solution,

$$\Theta(\theta) = A \sin \lambda\theta + B \cos \lambda\theta. \tag{5.8}$$

We need to restrict the values of λ so that $\Theta(\theta) = 0$ at $\theta = 0$ and $\theta = \pi$ on order that the two opposite homogeneous boundary conditions for $T(r, \theta)$ are satisfied. Thus, $\Theta(0) = 0$ gives

$$A \cdot 0 + B \cdot 1 = 0, \quad \text{leading to} \quad B = 0,$$

reducing $\Theta(\theta)$ to

$$\Theta(\theta) = A \sin \lambda\theta.$$

With $\Theta(\pi) = 0$ we have

$$A \sin \lambda\pi = 0.$$

We need to satisfy this condition in a nontrivial manner, and therefore, we cannot set $A = 0$. The only choice is

$$\sin \lambda\pi = 0,$$

which is satisfied when we choose

$$\lambda = \lambda_n = n.$$

Thus the set of eigenfunctions $\Theta(\theta)$ of the operator $d^2/d\theta^2$ satisfying $\Theta(0) = \Theta(\pi) = 0$ is given by

$$\Theta_n(\theta) = \sin n\theta, \tag{5.9}$$

where we have adopted the index n, and dropped the coefficient A which can be attached to the radial function in the formal eigenfunction expansion, as we shall see next. We can express $T(r, \theta)$ as the expansion,

$$T(r, \theta) = \sum_{n=1}^{\infty} R_n(r) \sin n\theta. \tag{5.10}$$

Substituting this into the differential equation (5.6) gives

$$\sum_{n=1}^{\infty} r^2 R_n''(r)\sin n\theta + \sum_{n=1}^{\infty} rR_n'(r)\sin n\theta - \sum_{n=1}^{\infty} R_n(r)(-n^2)\sin n\theta = 0. \quad (5.11)$$

We can factor out $\sin n\theta$ are rewrite this as

$$\sum_{n=1}^{\infty} \{r^2 R_n''(r) + rR_n'(r) - n^2 R_n(r)\}\sin n\theta = 0. \quad (5.12)$$

Since this equation should be valid for all values of θ in the range $0 < \theta < \pi$, it is not difficult to see that

$$r^2 R_n''(r) + rR_n'(r) - n^2 R_n(r) = 0. \quad (5.13)$$

This equation is an Euler type (see Chapter 1, pages 24-34) with solutions of the type

$$R_n(r) = r^k.$$

Substitution into equation (5.13) leads to the characteristic equation

$$k(k-1) + k - n^2 = 0,$$

which simplifies to

$$k^2 = n^2,$$

yielding $k = \pm n$. Thus equation (5.13) has the general solution

$$R_n(r) = A_n r^n + B_n r^{-n}, \quad n \neq 0. \quad (5.14)$$

For $n = 0$, we have $k = 0, 0$, and the solution is

$$R_0(r) = A_0 + B_0 \ln r.$$

However, this is not needed since the series expansion in $\sin n\theta$ starts with $n = 1$. Upon using the result (5.14) in equation (5.10), we obtain

$$T(r,\theta) = \sum_{n=1}^{\infty} \left[A_n r^n + B_n r^{-n}\right]\sin n\theta. \quad (5.15)$$

This solution behaves improperly as $r \to 0$ because of the term r^{-n}. Since $r = 0$ is included in the domain of the problem, we have to exclude this term by setting $B_n = 0$, reducing equation (5.15) to

$$T(r,\theta) = \sum_{n=1}^{\infty} A_n r^n \sin n\theta. \quad (5.16)$$

The set of constants A_n can be determined by satisfying the remaining boundary condition,

$$T(R,\theta) = T_0\theta(\pi - \theta).$$

This is a nonhomogeneous condition, and cannot be satisfied term by term. We thus require

$$T_0\theta(\pi - \theta) = \sum_{n=1}^{\infty} \left[A_n R^n\right] \sin n\theta. \tag{5.17}$$

This is a simple Fourier sine series, and following the procedure in Chapter 3, pages 75-87, we obtain the following expression for the coefficients $A_n R^n$

$$\begin{aligned}
A_n R^n &= \frac{2}{\pi}\int_0^{\pi} T_0\theta(\pi-\theta)\sin n\theta \, d\theta \\
&= \frac{2T_0}{\pi}\left[\theta(\pi-\theta)\left(-\frac{\cos n\theta}{n}\right)\Big|_0^{\pi} + \int_0^{\pi}(\pi-2\theta)\left(\frac{\cos n\theta}{n}\right) d\theta\right] \\
&= \frac{2T_0}{\pi}\left[0 + (\pi-2\theta)\left(\frac{\sin n\theta}{n^2}\right)\Big|_0^{\pi} + \int_0^{\pi} 2\left(\frac{\sin n\theta}{n^2}\right) d\theta\right] \\
&= \frac{2T_0}{\pi}\left[0 + 2\left(-\frac{\cos n\theta}{n^3}\right)\Big|_0^{\pi}\right] \\
&= \frac{4T_0\left[1-(1)^n\right]}{n^3\pi} \\
A_n &= \frac{4T_0\left[1-(1)^n\right]}{n^3\pi R^n}
\end{aligned}$$

Using this expression for A_n in equation (5.16) gives us the final result,

$$T(r,\theta) = 4T_0 \sum_{n=1}^{\infty} \frac{\left[1-(1)^n\right]}{n^3\pi}\left(\frac{r}{R}\right)^n \sin n\theta. \tag{5.18}$$

□

The procedure applied in the preceding example is applicable to other shapes (and boundary conditions) that can be defined by constant values of r and θ in polar coordinates. In the next example we consider a semi-circular annulus.

Example 5.2

Problem Statement: For the semicircular region shown, solve

$$\nabla^2 T(r,\theta) = 0,$$

with boundary conditions as indicated.

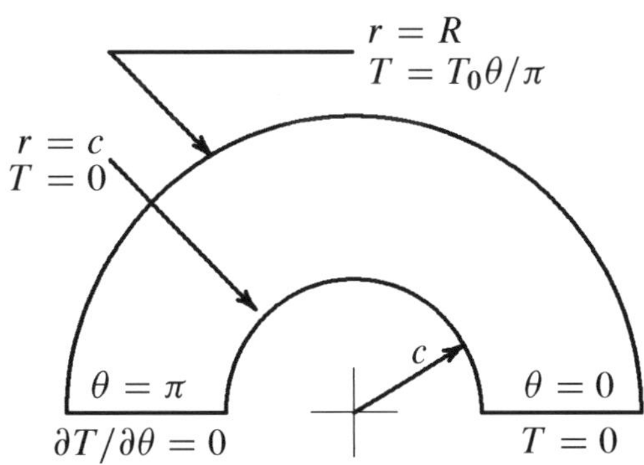

In terms of r and θ, Laplace's equation is the same as that given by equation (5.6) in Example 5.1. Furthermore, we identify a pair of opposite homogeneous boundary conditions at $\theta = 0$ and $\theta = \pi$. Therefore, we need an eigenfunction expansion in θ, i.e., a solution to the differential equation (5.7). The general solution is given by equation (5.8). We can now satisfy the boundary conditions both of which are homogeneous. With $\Theta(0) = 0$, we have

$$A_n \sin\lambda_n 0 + B_n \cos\lambda_n 0 = 0,$$

leading to $B_n = 0$, and reducing $\Theta(\theta)$ to

$$\Theta(0) = A_n \sin\lambda_n x.$$

Next, requiring that $\Theta'(\theta) = 0$ at $\theta = \pi$ gives us

$$A_n\lambda_n \cos\lambda_n\pi = 0.$$

For the solution to be nontrivial we cannot set $A_n = 0$, and therefore require that

$$\cos\lambda_n\pi = 0,$$

resulting in $\lambda_n = \left(n + \frac{1}{2}\right)$. We can thus expand

$$T(r,\theta) = \sum_{n=1}^{\infty} R_n(r)\sin\lambda_n\theta, \tag{5.19}$$

and substitute into the differential equation (5.6). This leads to the ordinary differential equation for $R_n(r)$ similar to equation (5.13),

$$r^2 R_n''(r) + rR_n'(r) - \lambda_n^2 R_n(r) = 0. \tag{5.20}$$

This has the general solution

$$\begin{aligned} R_n(r) &= A_n r^{\lambda_n} + B_n r^{-\lambda_n} \\ &= A_n r^{n+\frac{1}{2}} + B_n r^{-(n+\frac{1}{2})}. \end{aligned} \tag{5.21}$$

At this stage, we can satisfy the boundary condition $T(c,\theta) = 0$ term by term with $R_n(c) = 0$. Again, this term-by-term procedure works because the condition is homogeneous. We therefore have

$$\begin{aligned} R_n(c) = A_n c^{n+\frac{1}{2}} + B_n c^{-(n+\frac{1}{2})} &= 0 \\ B_n &= -A_n c^{2n+1} \\ R_n(r) &= A_n\left[r^{n+\frac{1}{2}} - c^{2n+1} r^{-(n+\frac{1}{2})}\right] \\ &= A_n c^{n+\frac{1}{2}}\left[\left(\frac{r}{c}\right)^{n+\frac{1}{2}} - \left(\frac{c}{r}\right)^{n+\frac{1}{2}}\right] \\ &= a_n\left[\left(\frac{r}{c}\right)^{n+\frac{1}{2}} - \left(\frac{c}{r}\right)^{n+\frac{1}{2}}\right] \end{aligned}$$

Using this expression for $R_n(r)$ in equation (5.19) and noting that $\lambda_n = \left(n + \frac{1}{2}\right)$, we obtain

$$T(r,\theta) = \sum_{n=0}^{\infty} a_n\left[\left(\frac{r}{c}\right)^{n+\frac{1}{2}} - \left(\frac{c}{r}\right)^{n+\frac{1}{2}}\right]\sin\left[\left(n + \tfrac{1}{2}\right)\theta\right], \tag{5.22}$$

What remains now is to satisfy the remaining boundary condition that is non-homogeneous, i.e., $T(R,\theta) = T_0\theta/\pi$. Following through with this, we obtain

$$T_0\frac{\theta}{\pi} = \sum_{n=0}^{\infty} a_n \left[\left(\frac{R}{c}\right)^{n+\frac{1}{2}} - \left(\frac{c}{R}\right)^{n+\frac{1}{2}}\right] \sin\left[\left(n+\tfrac{1}{2}\right)\theta\right].$$

Applying orthogonality, we obtain

$$\begin{aligned}
a_n\left[\left(\frac{R}{c}\right)^{n+\frac{1}{2}} - \left(\frac{c}{R}\right)^{n+\frac{1}{2}}\right] &= \frac{2}{\pi}\int_0^{\pi} T_0\frac{\theta}{\pi}\sin\left[\left(n+\tfrac{1}{2}\right)\theta\right] d\theta \\
&= \frac{2T_0}{\pi^2}\left\{\theta\left[\frac{-\cos\left[\left(n+\frac{1}{2}\right)\theta\right]}{\left(n+\frac{1}{2}\right)}\right]_0^{\pi} + \int_0^{\pi}\frac{\cos\left[\left(n+\frac{1}{2}\right)\theta\right]}{\left(n+\frac{1}{2}\right)}\right\} \\
&= \frac{2T_0}{\pi^2}\left\{0 + \left.\frac{\cos\left[\left(n+\frac{1}{2}\right)\theta\right]}{\left(n+\frac{1}{2}\right)^2}\right|_0^{\pi}\right\} \\
&= \frac{2T_0(-1)^n}{\left(n+\frac{1}{2}\right)^2\pi^2} \\
a_n &= \frac{2T_0(-1)^n}{\left(n+\frac{1}{2}\right)^2\pi^2}\frac{1}{\left[\left(\frac{R}{c}\right)^{n+\frac{1}{2}} - \left(\frac{c}{R}\right)^{n+\frac{1}{2}}\right]}.
\end{aligned}$$

With this expression for a_n, equation (5.22) becomes

$$T(r,\theta) = 2T_0\sum_{n=0}^{\infty}\frac{(-1)^n}{\left(n+\frac{1}{2}\right)^2\pi^2}\frac{\left[\left(\frac{r}{c}\right)^{n+\frac{1}{2}} - \left(\frac{c}{r}\right)^{n+\frac{1}{2}}\right]}{\left[\left(\frac{R}{c}\right)^{n+\frac{1}{2}} - \left(\frac{c}{R}\right)^{n+\frac{1}{2}}\right]}\sin\left[\left(n+\tfrac{1}{2}\right)\theta\right], \quad (5.23)$$

□

The procedure in the preceding two examples is applicable to other geometries conforming to the (r,θ) coordinate system. This includes annular and full sectors of a circle. Some situations will call for expansions in r, and these will be discussed later. Next, we shall examine problems involving a full circle. As an example we consider Laplace's equation for a disk with a specified condition on the circular boundary.

Example 5.3

Solve

$$\nabla^2 T(r,\theta) = 0$$

in the circular region $0 \le r \le R$ boundary conditions as shown in Figure 5.3. The boundary condition at $r = R$ may be stated in piecewise form as

$$T(R,\theta) = f(\theta) = \begin{cases} T_0\theta(\pi-\theta), & 0 \le \theta \le \pi \\ 0 & \pi \le \theta \le 2\pi \end{cases} \quad (5.24)$$

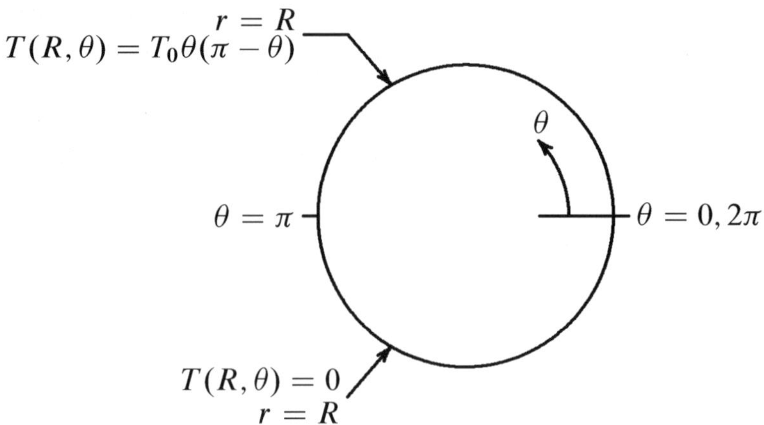

Figure 5.3: Boundary conditions for the circular disk.

In this case there are no boundaries defined by the θ-coordinate, and thus there are no boundary conditions defined by constant θ values. However, with the circular geometry, there is 2π-periodicity. This means that if we take a point (r', θ') within the disk, then

$$\begin{aligned} T(r', \theta') &= T(r', \theta' + 2\pi), & (5.25) \\ \left.\frac{\partial T(r', \theta)}{\partial \theta}\right|_{\theta=\theta'} &= \left.\frac{\partial T(r', \theta)}{\partial \theta}\right|_{\theta=\theta'+2\pi}. & (5.26) \end{aligned}$$

For $T(r, \theta)$, if we had an eigenfunction expansion

$$T(r, \theta) = \sum_{n=0}^{\infty} R_n(r)\Theta_n(\theta),$$

then applying the periodicity condition (5.25), and following with some rearrangement, leads to

$$\begin{aligned} \sum_{n=0}^{\infty} R_n(r)\Theta_n(\theta') &= \sum_{n=0}^{\infty} R_n(r)\Theta_n(\theta' + 2\pi) \\ \sum_{n=0}^{\infty} R_n(r)\left[\Theta_n(\theta') - \Theta_n(\theta' + 2\pi)\right] &= 0. & (5.27) \end{aligned}$$

From here we can see that the periodicity condition (5.25) can be satisfied by requiring that

$$\left[\Theta_n(\theta') - \Theta_n(\theta' + 2\pi)\right] = 0,$$

or

$$\Theta_n(\theta') = \Theta_n(\theta' + 2\pi). \tag{5.28}$$

Similarly, we can show that the condition (5.26) can be satisfied term-by-term with

$$\Theta_n'(\theta') = \Theta_n'(\theta' + 2\pi). \tag{5.29}$$

This set of conditions (5.28)-(5.29) corresponds to the fourth class of conditions that guarantee orthogonality of the eigenfunctions of linear second-order differential operators (see Chapter 3, pages 115-116). Therefore, for the problem under consideration, the differential equation (5.6) is to be solved, and we can construct eigenfunctions of the operator $d^2/d\theta^2$. As with example 5.1, the appropriate ordinary differential equation for these eigenfunctions is (5.7) with the general solution given by (5.8), which is repeated here for convenience

$$\Theta(\theta) = A\sin\lambda\theta + B\cos\lambda\theta.$$

The periodicity conditions (5.28)-(5.29) require that the period is 2π, and

$$\lambda = \lambda_n = n\pi/\pi = n.$$

For this set of values of λ_n, we may index $\Theta(\theta)$ as $\Theta_n(\theta)$ so that

$$\Theta_n(\theta) = A_n\sin n\theta + B_n\cos n\theta.$$

We can now carry out the eigenfunction expansion,

$$T(r,\theta) = \sum_{n=0}^{\infty} R_n(r)\Theta_n(\theta) = \sum_{n=0}^{\infty} R_n(r)\left[A_n\sin n\theta + B_n\cos n\theta\right]. \tag{5.30}$$

Substitution into the differential equation (5.6) leads to

$$\sum_{n=0}^{\infty} r^2R_n''(r)\left[A_n\sin n\theta + B_n\cos n\theta\right] + \sum_{n=0}^{\infty} rR_n'(r)\left[A_n\sin n\theta + B_n\cos n\theta\right]$$
$$+\sum_{n=0}^{\infty} R_n(r)(-n^2)\left[A_n\sin n\theta + B_n\cos n\theta\right] = 0 \quad .$$

As expected, the eigenfunction $\Theta_n(\theta)$ is preserved under the differential operation, and may be factored out so that

$$\sum_{n=0}^{\infty}\left\{r^2R_n''(r) + rR_n'(r) - n^2R_n(r)\right\}\left[A_n\sin n\theta + B_n\cos n\theta\right] = 0.$$

Since this must be valid for all values of θ in the range $[0, 2\pi]$, it follows that

$$r^2R_n''(r) + rR_n'(r) - n^2R_n(r),$$

$$R_n(r) = C_nr^n + D_nr^{-n}.$$

In this solution the term r^{-n} becomes unbounded as $r \to 0$. To exclude this part, we set $D_n = 0$, leaving us with

$$R_n(r) = C_nr^n.$$

Using this in equation (5.30), we have

$$T(r,\theta) = \sum_{n=0}^{\infty} C_n r^n \left[A_n \sin n\theta + B_n \cos n\theta\right], \tag{5.31}$$

where we can absorb C_n into A_n and B_n and write

$$T(r,\theta) = \sum_{n=0}^{\infty} r^n \left[a_n \sin n\theta + b_n \cos n\theta\right]. \tag{5.32}$$

The constants a_n and b_n may be obtained by satisfying the boundary condition (5.24). Applying this leads to

$$\sum_{n=0}^{\infty} R^n \left[a_n \sin n\theta + b_n \cos n\theta\right] = \begin{cases} T_0\theta(\pi-\theta), & 0 \le \theta \le \pi \\ 0 & \pi \le \theta \le 2\pi \end{cases} \tag{5.33}$$

With the orthogonality of the set eigenfunctions, the coefficients a_n and b_n are found to be

$$\begin{aligned}
a_n R^n &= \frac{1}{\pi}\left[T_0 \int_0^{\pi} \theta(\pi-\theta)\sin n\theta \, d\theta + \int_{\pi}^{2\pi} 0 \, \sin n\theta \, d\theta\right] \\
&= \frac{T_0}{\pi}\left[\theta(\pi-\theta)\left(\frac{-\cos n\theta}{n}\right)\Big|_0^{\pi} + \int_0^{\pi} (\pi-2\theta)\frac{\cos n\theta}{n}\, d\theta\right] \\
&= \frac{T_0}{\pi}\left[0 + (\pi-2\theta)\frac{\sin n\theta}{n^2}\Big|_0^{\pi} + \int_0^{\pi} 2\left(\frac{\sin n\theta}{n^2}\right) d\theta\right] \\
&= \frac{T_0}{\pi}\left[0 + 2\left(\frac{-\cos n\theta}{n^3}\right)\Big|_0^{\pi}\right] \\
&= 2T_0\left[\frac{1-(-1)^n}{n^3\pi}\right] \\
a_n &= 2T_0\left[\frac{1-(-1)^n}{n^3\pi}\right]\frac{1}{R^n} \\
a_0 &= 0 \\
b_n R^n &= \frac{1}{\pi}\left[T_0 \int_0^{\pi} \theta(\pi-\theta)\cos n\theta \, d\theta \int_{\pi}^{2\pi} 0 \, \cos n\theta \, d\theta\right] \\
&= \frac{T_0}{\pi}\left[\theta(\pi-\theta)\left(\frac{\sin n\theta}{n}\right)\Big|_0^{\pi} - \int_0^{\pi} (\pi-2\theta)\left(\frac{\sin n\theta}{n}\right) d\theta\right] \\
&= \frac{T_0}{\pi}\left[0 + (\pi-2\theta)\left(\frac{\cos n\theta}{n^2}\right)\Big|_0^{\pi} - \int_0^{\pi} 2\left(\frac{\cos n\theta}{n^2}\right) d\theta\right] \\
&= \frac{T_0}{\pi}\left[\frac{\pi\left[-(-1)^n - 1\right]}{n^2} - 2\left(\frac{\sin n\theta}{n^3}\right)\Big|_0^{\pi}\right] \\
&= -T_0\left[\frac{1+(-1)^n}{n^2}\right] \\
b_n &= -T_0\left[\frac{1+(-1)^n}{n^2}\right]\frac{1}{R^n}
\end{aligned}$$

$$\begin{aligned} b_0 &= \frac{1}{2\pi}\left[T_0\int_0^{\pi}\theta(\pi-\theta)\,d\theta+\int_{\pi}^{2\pi}0\;d\theta\right] \\ &= \frac{1}{2\pi}T_0\left[\tfrac{1}{2}\pi\theta^2-\tfrac{1}{3}\theta^3\right]_0^{\pi} \\ &= \tfrac{1}{12}\pi^2T_0 \end{aligned}$$

With these expressions for a_n and b_n, equation (5.32) takes the final form

$$T(r,\theta)=T_0\left\{\tfrac{1}{12}\pi^2+\sum_{n=1}^{\infty}\left(\frac{r}{R}\right)^n\left[\left(\frac{1-(-1)^n}{n^3\pi}\right)\sin n\theta-\left(\frac{1+(-1)^n}{n^2}\right)\cos n\theta\right]\right\}. \tag{5.34}$$

□

These examples illustrate how we can implement the eigenfunction-expansion procedure for situations where the homogeneous boundary conditions are defined by constant θ values, or if there is periodicity in θ, both cases guaranteeing orthogonality of the θ eigenfunctions. We shall now go on to cases requiring expansion in r.

EXERCISES 5.1

Problem 1

For the $\frac{1}{4}$-circle region shown, solve

$$\nabla^2T(r,\theta)=0,$$

with boundary conditions as shown.

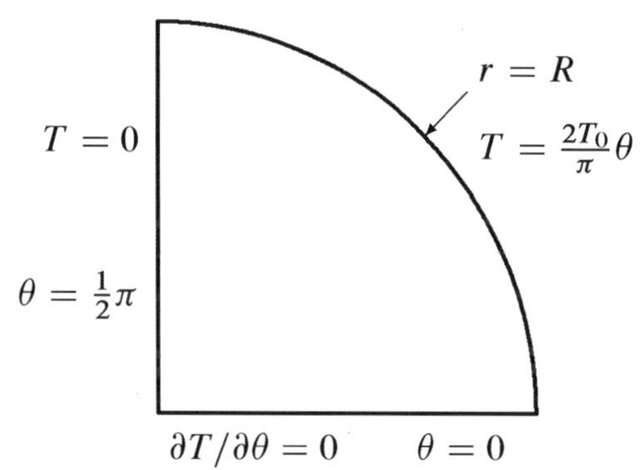

Problem 2

For the $\frac{1}{2}$-circle region shown, solve

$$\nabla^2T(r,\theta)=0,$$

with boundary conditions as shown.

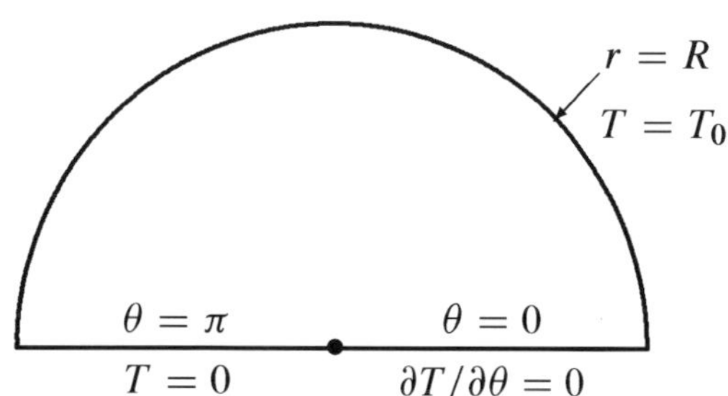

Problem 3

For the semicircular region shown, solve

$$\nabla^2 T(r, \theta) = 0,$$

with boundary conditions as indicated.

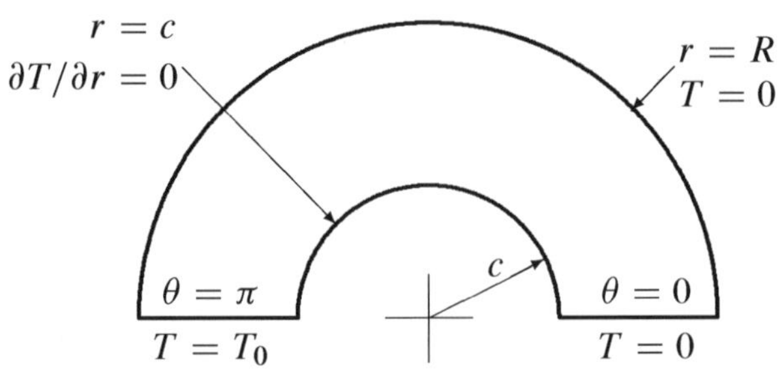

Problem 4

For the semicircular region shown, solve

$$\nabla^2 T(r, \theta) = 0,$$

with boundary conditions as indicated.

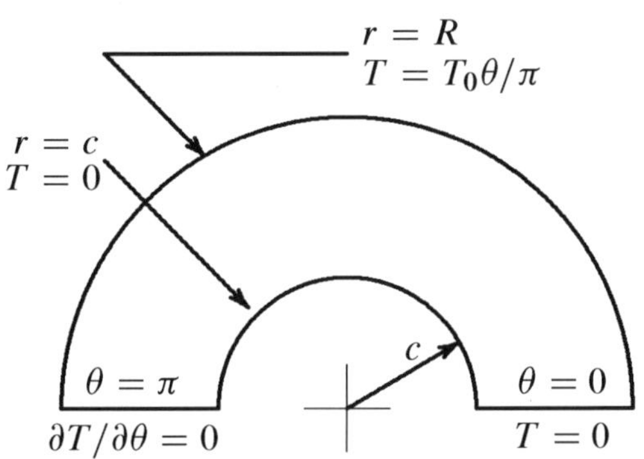

Problem 5

Solve the steady-state laminar flow problem for a circular pipe with a longitudinal fin. Assuming that the flow is unidirectional along the pipe, the velocity $u(r, \theta)$ is given by the solution of the nonhomogeneous differential equation,

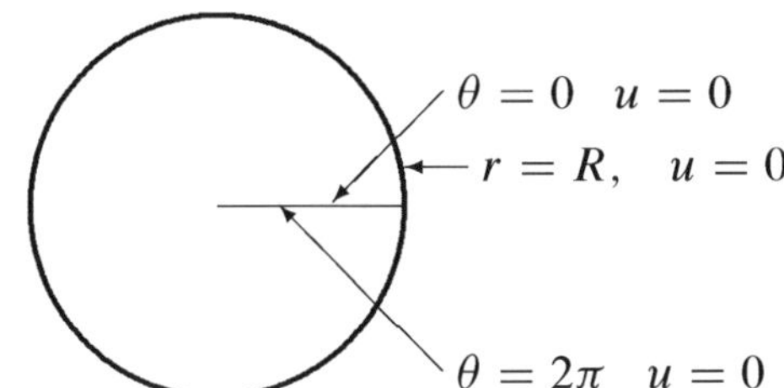

$$\left(\frac{\partial^2 u(r,\theta)}{\partial r^2} + \frac{1}{r}\frac{\partial u(r,\theta)}{\partial r} + \frac{1}{r^2}\frac{\partial^2 u(r,\theta)}{\partial \theta^2}\right) = -\frac{P_0}{\mu_0}, \tag{5.35}$$

where P_0 represents constant pressure gradient, and μ_0 is the fluid viscosity. Proceed as follows:

1. In this case, $\theta = 0$ and $\theta = 2\pi$ are real boundaries. Identify the three homogeneous boundary conditions on rectangularized (r, θ) coordinates. At $r = 0$, we do not have a specific condition but we need not get too concerned. With continuity around the fin, it can be taken to be zero.

2. With homogeneous conditions all around, either r or θ can be taken to be a variable for eigenfunction expansion. Let us choose θ because it will be easier.

3. Multiply the differential equation by r^2, and isolate the $\partial^2 u/\partial\theta^2$ term. Set up the operator as an equation for $\Theta_n(\theta)$ so that the second derivative operation on it yields $-\lambda_n^2\Theta_n(\theta)$.

4. Solve the equation, satisfy the homogeneous conditions at $\theta = 0$ and $\theta = 2\pi$, obtain λ_n values, and establish specific expressions for $\Theta_n(\theta)$.

5. Expand the right-hand side of the differential equation in terms of the eigenfunctions $\Theta_n(\theta)$, i.e.,

$$\frac{P_0}{\mu_0} = \sum_{n=1}^{\infty} p_n \Theta_n(\theta),$$

and determined p_n by orthogonality of the eigenfunctions, using the specific expression for $\Theta_n(\theta)$.

6. Next, express $u(r, \theta)$ as an eigenfunction expansion in θ using the same set of eigenfunctions established in Item 4 and used in Item 5, i.e.,

$$u(r, \theta) = \sum_{n=1}^{\infty} U_n(r)\Theta_n(\theta).$$

7. Substitute both expansions [for $u(r, \theta)$ and P_0/μ_0] into the differential equation (5.35). Keep the p_n notation for the RHS.

8. Equate both sides term-by-term, and obtain an ODE for $U_n(r)$. Multiply the equation by r^2 and note that it is a nonhomogeneous Cauchy-Euler equation.

9. Obtain the general solution to this nonhomogeneous equation (homogeneous plus particular).

10. Eliminate any terms that go to infinity as $r \to 0$. Satisfy the remaining homogeneous condition at $r = R$. This, being homogeneous, can be satisfied term-by-term.

11. Assemble the solution for $u(r, \theta)$. Put back the expression for p_n derived in Item 5. Integrate $u(r, \theta)$ across the pipe cross-section to get the volumetric flowrate Q.

5.2 Eigenfunction Expansions in r

To illustrate cases where an expansion in r is required, we consider another example with a semi-circular annulus.

Example 5.4

Problem Statement: For the semicircular region shown, solve

$$\nabla^2 T(r, \theta) = 0,$$

with boundary conditions as indicated.

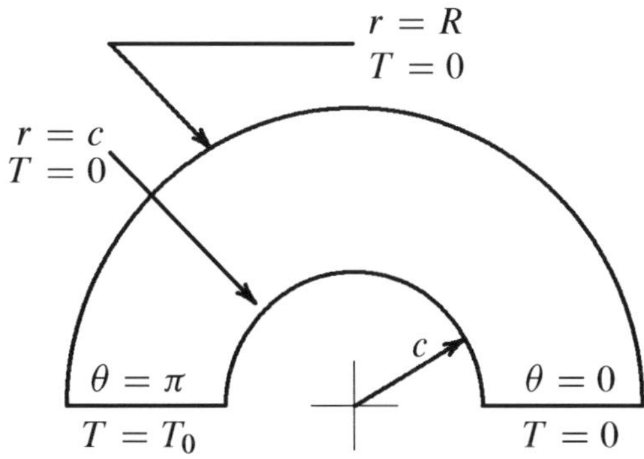

To properly identify the expansion variable, we treat r and θ as rectangular coordinates as shown in the Figure 5.4

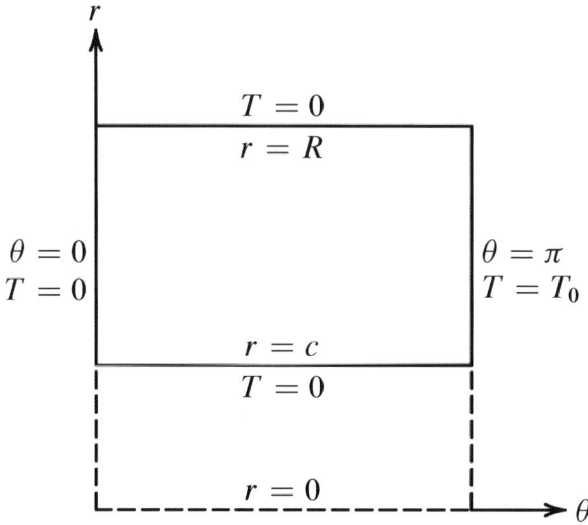

Figure 5.4: Depiction of boundary conditions for example 5.4 in 'rectangularized' coordinates.

In terms of r and θ, Laplace's equation is the same as that given by equation (5.6) in Example 5.1.

In terms of r and θ, Laplace's equation is

$$\frac{\partial^2 T}{\partial r^2} + \frac{1}{r}\frac{\partial T}{\partial r} + \frac{1}{r^2}\frac{\partial^2 T}{\partial \theta^2} = 0.$$

Upon multiplication by r^2, this becomes

$$r^2\frac{\partial^2 T}{\partial r^2} + r\frac{\partial T}{\partial r} + \frac{\partial^2 T}{\partial \theta^2} = 0. \tag{5.36}$$

Based on $T = 0$ at $r = c$ and $r = R$, we need to carry out an eigenfunction expansion in r,

$$T(r,\theta) = \sum \Theta_n(\theta) R_n(r),$$

where $R_n(r)$ is a set of orthogonal eigenfunctions. We therefore need to solve

$$r^2\frac{d^2 R_n(r)}{dr^2} + r\frac{dR_n(r)}{dr} = -\lambda_n^2 R_n(r), \tag{5.37}$$

i.e.,

$$r^2\frac{d^2 R_n(r)}{dr^2} + r\frac{dR_n(r)}{dr} + \lambda_n^2 R_n(r) = 0. \tag{5.38}$$

This is an equation of the Cauchy-Euler type. The solutions are simple powers in r. We let $R_n(r) = r^k$. Upon substitution into the differential equation (5.38), we obtain the characteristic equation

$$k(k-1) + k + \lambda_n^2 = 0,$$

which leads to

$$k^2 + \lambda_n^2 = 0, \quad \text{yielding} \quad k = \pm i\lambda_n.$$

Therefore, the general solution of equation (5.38) is

$$\begin{aligned}
R_n(r) &= a_n^* r^{i\lambda_n} + b_n^* r^{-i\lambda_n} \\
&= a_n^* e^{i\lambda_n \ln r} + b_n^* e^{-i\lambda_n \ln r} \\
&= a_n^* \left[\cos(\lambda_n \ln r) + i \sin(\lambda_n \ln r)\right] + b_n^* \left[\cos(\lambda_n \ln r) + i \sin(\lambda_n \ln r)\right] \\
&= a_n \cos(\lambda_n \ln r) + b_n \sin(\lambda_n \ln r)
\end{aligned}$$

The homogeneous boundary condition,

$$T = 0, \quad \text{at} \quad r = c$$

may be satisfied term by term, i.e.,

$$\begin{aligned}
R_n(c) &= a_n \cos(\lambda_n \ln c) + b_n \sin(\lambda_n \ln c) = 0, \\
a_n &= -b_n \frac{\sin(\lambda_n \ln c)}{\cos(\lambda_n \ln c)}. \\
R_n(r) &= -b_n \cos(\lambda_n \ln r) \left[\frac{\sin(\lambda_n \ln c)}{\cos(\lambda_n \ln c)}\right] + b_n \sin(\lambda_n \ln r) \\
&= b_n \left[\frac{\sin(\lambda_n \ln c)\cos(\lambda_n \ln r) - \sin(\lambda_n \ln r)\cos(\lambda_n \ln c)}{\cos(\lambda_n \ln c)}\right] \\
&= b_n \left[\frac{\sin(\lambda_n \ln r - \lambda_n \ln c)}{\cos(\lambda_n \ln c)}\right] \\
&= \frac{b_n}{\cos(\lambda_n \ln c)} \left[\sin(\lambda_n \ln r - \lambda_n \ln c)\right] \\
&= \frac{b_n}{\cos(\lambda_n \ln c)} \left\{\sin\left[\lambda_n \ln\left(\frac{r}{c}\right)\right]\right\}
\end{aligned}$$

We may now disregard the constant $[b_n/\cos(\lambda_n \ln c)]$ since it may be absorbed into $\Theta_n(\theta)$. Hence,

$$R_n(r) = \sin\left[\lambda_n \ln\left(\frac{r}{c}\right)\right] \tag{5.39}$$

The homogeneous boundary condition at $r = R$ may now be satisfied term-by-term to yield the values of λ_n. Applying this condition on $R_n(r)$ leads to

$$R_n(R) = \sin\left[\lambda_n \ln\left(\frac{R}{c}\right)\right] = 0.$$

This is satisfied if the argument of the sine function is restricted to a set of integer multiples of π, i.e.,

$$\lambda_n \ln\left(\frac{R}{c}\right) = n\pi, \qquad \text{or} \quad \lambda_n = \frac{n\pi}{\ln\left(\frac{R}{c}\right)}$$

Note that if we satisfy $R_n(R) = 0$ first, we would obtain, in place of Equation (5.39), eigenfunctions of the form

$$R_n(r) = \sin\left[\lambda_n \ln\left(\frac{R}{r}\right)\right],$$

and $R_n(c) = 0$ would yield the same λ_n values as above. The radial eigenfunction is completely defined, and may now be used (we will use Equation (5.39)) in the expansion for $T(r, \theta)$, i.e.,

$$T(r,\theta) = \sum_{n=0}^{\infty} \Theta_n(\theta) \sin\left[\lambda_n \ln\left(\frac{r}{c}\right)\right].$$

Upon substitution of this expansion into the partial differential equation (5.36) we obtain

$$\sum_{n=0}^{\infty} \Theta_n(\theta)\left(r^2\frac{d^2}{dr^2} + r\frac{d}{dr}\right)\sin\left[\lambda_n \ln\left(\frac{r}{c}\right)\right] + \sum_{n=0}^{\infty} \Theta_n''(\theta)\sin\left[\lambda_n \ln\left(\frac{r}{c}\right)\right] = 0 \quad (5.40)$$

We need to obtain the first and the second derivatives of the radial eigenfunction $R_n(r)$. This is not absolutely necessary since the derivative operations in r (above) should give us back the eigenfunction. In fact, the first and second derivative operation above may also be written as

$$\left(r^2\frac{d^2}{dr^2} + r\frac{d}{dr}\right) = r\frac{d}{dr}\left(r\frac{d}{dr}\right)$$

Now, let us systematically carry out the derivative operation on $R_n(r)$.

$$\begin{aligned}
\frac{d}{dr}\sin\left[\lambda_n \ln\left(\frac{r}{c}\right)\right] &= \cos\left[\lambda_n \ln\left(\frac{r}{c}\right)\right]\frac{\lambda_n}{r} \\
r\frac{d}{dr}\sin\left[\lambda_n \ln\left(\frac{r}{c}\right)\right] &= \lambda_n \cos\left[\lambda_n \ln\left(\frac{r}{c}\right)\right] \\
\frac{d}{dr}\left(r\frac{d}{dx}\right)\sin\left[\lambda_n \ln\left(\frac{r}{c}\right)\right] &= -\lambda_n\left\{\sin\left[\lambda_n \ln\left(\frac{r}{c}\right)\right]\right\}\frac{\lambda_n}{r} \\
r\frac{d}{dr}\left(r\frac{d}{dx}\right)\sin\left[\lambda_n \ln\left(\frac{r}{c}\right)\right] &= -\lambda_n^2 \sin\left[\lambda_n \ln\left(\frac{r}{c}\right)\right]
\end{aligned}$$

Using this result in (5.40), we find

$$\sum_{n=0}^{\infty} \Theta_n(\theta)\,(-\lambda_n^2)\sin\left[\lambda_n \ln\left(\frac{r}{c}\right)\right] + \sum_{n=0}^{\infty} \Theta_n''(\theta)\sin\left[\lambda_n \ln\left(\frac{r}{c}\right)\right] = 0.$$

$$\sum_{n=0}^{\infty}\left[\Theta_n''(\theta) - \lambda_n^2\Theta_n(\theta)\right]\sin\left[\lambda_n \ln\left(\frac{r}{c}\right)\right] = 0.$$

For this to be valid for all values of r in the range $c \le r \le R$, we have

$$\Theta_n''(\theta) - \lambda_n^2\Theta_n(\theta) = 0.$$

The general solution to this equation is

$$\Theta_n(\theta) = A_n \sinh \lambda_n \theta + B_n \cosh \lambda_n \theta.$$

The boundary condition $T = 0$ at $\theta = 0$, being homogeneous, may be satisfied term-by-term with $\Theta(0) = 0$. Therefore,

$$\Theta(0) = 0 = A_n \sinh \lambda_n 0 + B_n \cosh \lambda_n 0 = B_n.$$

With $B_n = 0$, $\Theta_n(\theta)$ reduces to

$$\Theta_n(\theta) = A_n \sinh \lambda_n \theta,$$

and the solution for $T(r, \theta)$ takes the form

$$T(r, \theta) = \sum_{n=1}^{\infty} A_n \sinh \lambda_n \theta \sin \left[\lambda_n \ln \left(\frac{r}{c}\right)\right].$$

The remaining boundary condition, $T(r, \pi) = T_0$, is a non-homogeneous one, and needs to be satisfied as a series. Thus,

$$T(r, \pi) = 1 = \sum_{n=1}^{\infty} [A_n \sinh \lambda_n \pi] \sin \left[\lambda_n \ln \left(\frac{r}{c}\right)\right]. \tag{5.41}$$

At this point, we need to review the orthogonality of the set of eigenfunctions,

$$\left\{\sin \left[\lambda_n \ln \left(\frac{r}{c}\right)\right], \qquad n = 1, 2, 3, 4, \ldots\right\}.$$

We examine the differential equation (5.38) for $R_n(r)$ which may written as

$$r \frac{d}{dr}\left(r \frac{dR_n(r)}{dr}\right) + \lambda_n^2 R_n(r) = 0. \tag{5.42}$$

Dividing by r, we obtain

$$\frac{d}{dr}\left(r \frac{dR_n(r)}{dr}\right) + \frac{\lambda_n^2}{r} R_n(r) = 0. \tag{5.43}$$

The standard Sturm-Liouville form (given earlier by equation (3.59) on page 110) is

$$\frac{d}{dx}\left[p(x) \frac{dX_n(r)}{dx}\right] + s(x) X_n(x) + \lambda_n^2 q(x) X_n(x) = 0. \tag{5.44}$$

Comparing equation (5.43) with (5.44), we can identify the following:

$$x = r, \quad X_n(x) = R_n(r) = \sin \left[\lambda_n \ln \left(\frac{r}{c}\right)\right],$$

$$s(x) = 0, \quad p(x) = p(r) = r, \quad q(x) = q(r) = \frac{1}{r}.$$

The weighting factor is therefore $1/r$, and the orthogonality relationship is

$$\int_c^R \frac{1}{r}\sin\left[\lambda_n \ln\left(\frac{r}{c}\right)\right]\sin\left[\lambda_m \ln\left(\frac{r}{c}\right)\right]\, dr = 0, \quad m \neq n. \tag{5.45}$$

We may now go ahead and apply this to the boundary condition at $\theta = \pi$ [equation (5.41)]. We multiply both sides of (5.41) by $(1/r)\sin\left[\lambda_m \ln\left(\frac{r}{c}\right)\right]$ and integrate with respect to x over $c \leq r \leq R$. This gives

$$\begin{aligned}
&\int_c^R T_0 \left(\frac{1}{r}\right)\sin\left[\lambda_m \ln\left(\frac{r}{c}\right)\right]\, dr \\
&= \sum_{n=1}^{\infty} [A_n \sinh \lambda_n \pi] \int_c^R \left(\frac{1}{r}\right)\sin\left[\lambda_n \ln\left(\frac{r}{c}\right)\right]\sin\left[\lambda_m \ln\left(\frac{r}{c}\right)\right]\, dr.
\end{aligned}$$

Let us substitute

$$u = \ln\left(\frac{r}{c}\right), \quad \text{with the differential} \quad du = \left(\frac{1}{r}\right) dr.$$

Noting that $u = 0$ when $r = c$ and $u = \ln(R/c)$ when $r = R$, we have

$$\int_0^{\ln(R/c)} T_0 \sin(\lambda_m u)\; dr = \sum_{n=1}^{\infty} [A_n \sinh \lambda_n \pi] \int_0^{\ln(R/c)} \sin(\lambda_n u)\sin(\lambda_m u)\; du.$$

In view of the orthogonality relationship (5.45), the only nonzero term in the summation on the right-hand side is $n = m$ (the term that survives the integration process). Therefore,

$$\begin{aligned}
T_0 \left.\frac{\sin(\lambda_m u)}{\lambda_m}\right|_0^{\ln(R/c)} &= [A_m \sinh \lambda_m \pi] \int_0^{\ln(R/c)} \sin^2(\lambda_m u)\; du \\
T_0 \frac{-\cos(\lambda_m \ln(R/c))}{\lambda_m} &= [A_m \sinh \lambda_m \pi] \int_0^{\ln(R/c)} \tfrac{1}{2}[1 - \cos(2\lambda_m u)]\; du \\
T_0 \frac{1 - \cos m\pi}{\lambda_m} &= [A_m \sinh \lambda_m \pi]\, \tfrac{1}{2} \left[u - \frac{\sin(2\lambda_m u)}{2\lambda_m}\right]\Bigg|_0^{\ln(R/c)} \\
T_0 \frac{1 - (-1)^m}{\lambda_m} &= [A_m \sinh \lambda_m \pi]\, \tfrac{1}{2} \left[\ln\left(\frac{R}{c}\right) - \frac{\sin\left(2\lambda_m \ln\left(\frac{R}{c}\right)\right)}{2\lambda_m}\right] \\
&= [A_m \sinh \lambda_m \pi]\, \tfrac{1}{2} \left[\ln\left(\frac{R}{c}\right) - \frac{\sin(2m\pi)}{2\lambda_m}\right] \\
&= [A_m \sinh \lambda_m \pi]\, \tfrac{1}{2} \ln\left(\frac{R}{c}\right) \\
A_m &= \frac{2T_0[1 - (-1)^m]}{\lambda_m \ln\left(\frac{R}{c}\right)} \left(\frac{1}{\sinh(\lambda_m \pi)}\right) \\
&= \frac{2T_0[1 - (-1)^m]}{m\pi} \left(\frac{1}{\sinh(\lambda_m \pi)}\right)
\end{aligned}$$

The complete solution may now be written as

$$\begin{aligned} T(r,\theta) &= 2T_0 \sum_{n=1}^{\infty} \frac{[1-(-1)^n]}{n\pi} \left(\frac{\sinh(\lambda_n \theta)}{\sinh(\lambda_n \pi)}\right) \sin\left[\lambda_n \ln\left(\frac{r}{c}\right)\right] \\ &= 2T_0 \sum_{n=1}^{\infty} \frac{1}{(2n+1)\pi} \left(\frac{\sinh(\lambda_n \theta)}{\sinh(\lambda_n \pi)}\right) \sin\left[\lambda_n \ln\left(\frac{R}{r}\right)\right] \end{aligned}$$

□

F We now consider another type of expansion in r leading to Bessel's equation. Here, we have an example involving a full cylinder.

Example 5.5

PROBLEM STATEMENT: Here we consider the diffusion equation for a cylinder

$$\frac{1}{\alpha}\frac{\partial T}{\partial t} = \nabla^2 T$$

with initial and boundary conditions as shown in the picture.

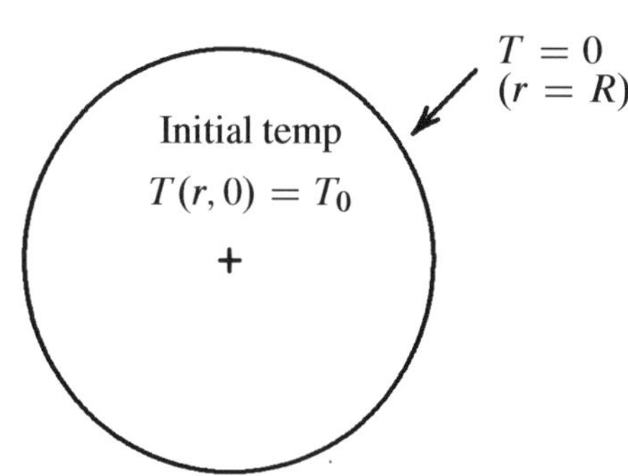

SOLUTION:
On a physical basis, we can see axial symmetry for this problem, and therefore θ and z dependencies do not arise, and T is a function of only r and t . The differential equation above reduces to

$$\frac{1}{\alpha}\frac{\partial T(r,t)}{\partial t} = \frac{\partial^2 T(r,t)}{\partial r^2} + \frac{1}{r}\frac{\partial T(r,t)}{\partial r}. \tag{5.46}$$

It would be useful to examine the problem on rectangularized coordinates as shown in Figure 5.5 We have a pair of opposite homogeneous boundary conditions at $r = 0$ and $r = R$. This will facilitate an eigenfunction expansion in r, and allow us to satisfy the nonzero condition at $t = 0$. We choose to apply the traditional procedure of the separation of variables, as opposed to the eigenfunction expansion method, so that we can see the unified principles of both procedures. Thus, we let

$$T(r,t) = R(r)\Theta(t), \tag{5.47}$$

and substitute into the partial differential Equation (5.46). This leads to

$$\frac{1}{\alpha}R(r)\Theta'(t) = R''(r)\Theta'(t) + \frac{1}{r}R'(r)\Theta(t) \tag{5.48}$$

Division by $R(r)\Theta(t)$ results in

$$\frac{1}{\alpha}\frac{\Theta'(t)}{\Theta(t)} = \frac{R''(r)}{R(r)} + \frac{1}{r}\frac{R'(r)}{R(r)}$$

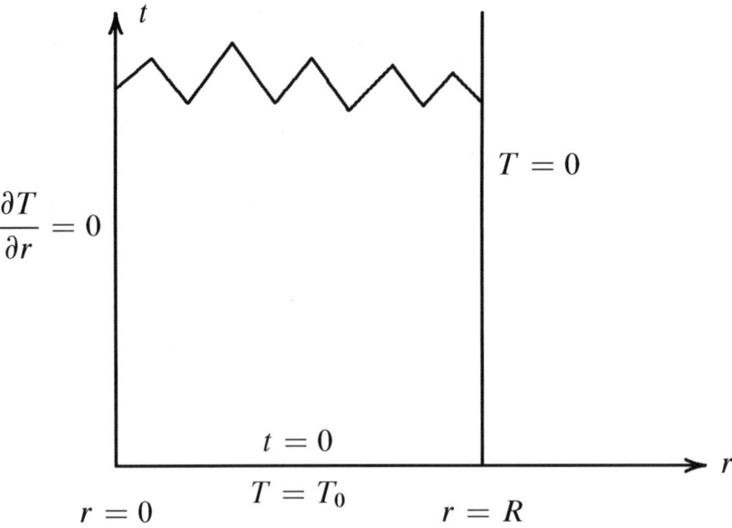

Figure 5.5: Rectangularized view of the cylinder problem in Example 5.5

Here we have a function of t only equal to a function of r only. This is possible only if the function on each side of the equation is a constant. Thus, we set

$$\frac{1}{\alpha}\frac{\Theta'(t)}{\Theta(t)} = \frac{R''(r)}{R(r)} + \frac{1}{r}\frac{R'(r)}{R(r)} = -\lambda^2$$

These are two ordinary differential equations for $R(r)$ and $\Theta(t)$ which may be written as:

$$\frac{1}{\alpha}\frac{\Theta(t)}{\Theta(t)} = -\lambda^2, \quad \text{or} \quad \Theta'(\theta) + \alpha\lambda^2\Theta(\theta) = 0 \tag{5.49}$$

and

$$\frac{R''}{R} + \frac{1}{r}\frac{R'}{R} = -\lambda^2, \quad \text{or} \quad R''(r) + \frac{1}{r}R'(r) + \lambda^2 R(r) = 0. \tag{5.50}$$

Upon multiplying Equation (5.50) by r^2, this equation for $R(r)$ may be written as

$$r^2R'' + rR' + \lambda^2r^2R = 0. \tag{5.51}$$

Here Equation(5.51) is a form of Bessel's equation of order zero. In fact, by letting $x = \lambda r$, it transforms to

$$x^2\frac{d^2f(x)}{dx^2} + \frac{df(x)}{dx} + x^2f(x) = 0, \tag{5.52}$$

where $f(x) = R(r)$. This is a special case of Equation (2.57) with $\nu = 0$ (see Chapter 2, page 62). Therefore, for order zero, we have The solution can be found by the method of Frobenius (see Chapter 2), and is given by

$$f(x) = AJ_0(x) + BY_0(x), \tag{5.53}$$

where expressions for $J_0(x)$ and $Y_0(x)$ are given by Equations (2.69) and (2.70), respectively (see page 68) in Chapter 2). Restoring the equation back to $R(r)$ with $x = \lambda r$, we can write the solution of Equation (5.51) as

$$R(r) = AJ_0(\lambda r) + BY_0(\lambda r). \tag{5.54}$$

From the plots of the two solutions, $J_0(x)$ and $Y_0(x)$ in Figure 2.1 on page 69 in Chapter 2, we see that $Y_0(x) \rightarrow -\infty$ as $x \rightarrow 0$, and obviously, $Y_0(\lambda r) \rightarrow -\infty$ as $r \rightarrow 0$. This is not acceptable (there may be some exceptions when we have a concentrated line source, and we tolerate such logarithmic singular behavior), and we therefore set $B = 0$. This reduces the solution (5.52) to

$$R(r) = AJ_0(\lambda r) \tag{5.55}$$

Next, we work on the boundary condition at $r = R$. The requirement here is that $T(R,t) = 0$. Since we let $T(r,t) = R(r)\Theta(t)$, we can let $R(R)\Theta(t) = 0$. This is satisfied for all values of t if $R(R) = 0$, i.e.,

$$R(R) = AJ_0(\lambda R) = 0. \tag{5.56}$$

To locate the values of λR that satisfy this condition, we examine the roots of $J_0(x)$. This is displayed in Figure 3.3 on page 119 in Chapter 3. Specifically, the function crosses the horizontal axis at $x = \beta_{10}, \beta_{20}, \ldots, \beta_{m0}, \ldots$, where the values of $\beta_{m0},\ m = 1, 2, 3, \ldots$ are tabulated in Table 3.1 on page 123 in Chapter 2. Thus the boundary condition at $r = R$ is satisfied with

$$\lambda_m R = \beta_{m0} \quad \text{or} \quad \lambda_m = \frac{\beta_{m0}}{R}. \tag{5.57}$$

Since there are a multitude of values of λ satisfying this condition, we have chosen to label it as λ_m. Using the same argument, we write the solution (5.55) as

$$R_m(r) = A_m J_0(\lambda_m r) \tag{5.58}$$

The boundary condition at $r = R$ limits to discrete set of values given by Equation (5.57). It is meaningful to graphically examine the solutions, $J_0(\lambda_1 r)$, $J_0(\lambda_2 r)$, $J_0(\lambda_3 r), \ldots$ in the domain $0 \leq r \leq R$. Therefore, we plot the first three of these on the same graph in the range $0 \leq r \leq R$ (see Figure 5.6). We are effectively taking the plots of $J_0(x)$ as given in Figure 3.3 on page 122 successively from $x = 0$ to $x = \beta_{10}, \beta_{20}, \beta_{30}$ and fitting them into the space $0 \leq r \leq R$ by adjusting the argument of $J_0(x)$ with the scaling $x = \beta_{m0} r/R$. With values established as $\lambda = \lambda_m = \beta_{m0}/R$, we may index Equation (5.49) as

$$\Theta_m'(\theta) + \alpha\lambda_m^2 \Theta_m(\theta) = 0, \quad m = 1, 2, 3, \ldots \tag{5.59}$$

Using the techniques for ordinary differential equations with constant coefficients (see Chapter 1, Section 1.2.1, pages 2-11), this may be easily solved to yield

$$\Theta_m(t) = a_m e^{-\alpha\lambda_m^2 t}. \tag{5.60}$$

Considering that we started out with Equation (5.47) that, upon satisfying the differential equation (5.46), gave expressions for $R(r)$ and $\Theta(t)$ by Equations (5.58) and (5.60), respectively. Therefore, the product $R(r)\Theta(t)$, properly indexed with m, may be written as

$$\begin{aligned} T_m(r,t) &= A_m J_0(\lambda_m r) a_m e^{-\alpha\lambda_m^2 t} \\ &= C_m J_0(\lambda_m r) e^{-\alpha\lambda_m^2 t}, \quad m = 1, 2, 3, \ldots \end{aligned} \tag{5.61}$$

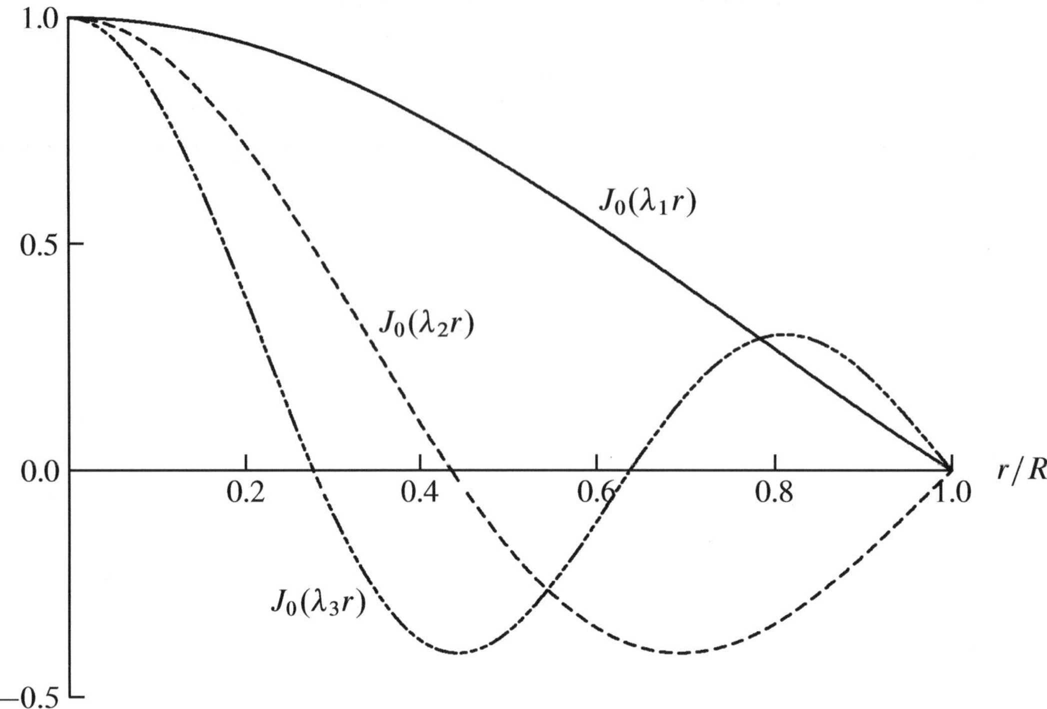

Figure 5.6: Plots for $J_0(\lambda_m r)$ for $m = 1, 2, 3$ within the space $0 \leq r \leq R$.

where we have set $C_m = A_m a_m$ without any loss of generality. Here, $T_m(r,t)$ satisfies the differential equation (5.46) for all values of λ_m. Since the differential equation is linear, any linear combination of the set of solutions (5.61) is also a solution. Thus, we may write

$$T(r,t) = \sum_{m=1}^{\infty} C_m J_0(\lambda_m r) e^{-\alpha \lambda_m^2 t}. \tag{5.62}$$

The set of constants C_m can be determined by the initial condition $T(r,0) = T_0$. Applying this, we obtain

$$T(r,0) = T_0 = \sum_{m=1}^{\infty} C_m J_0(\lambda_m r). \tag{5.63}$$

This is a Fourier-Bessel series that we discussed in Chapter 3, Section3.7.2. For the present case, Example 3.13 on page 119 is relevant. In fact, we have a special case of Equation (3.105) with $f(r) = T_0$. We apply orthogonality, and following that example, it is not difficult to see from Equation (3.107) on page 120 that

$$C_m = \frac{\int_0^R T_0 r J_0(\lambda_m r)\, dr}{\int_0^R r\, [J_0(\lambda_m r)]^2\, dr} \tag{5.64}$$

Following through with the rest of Example 3.13, we can see that the expression for

C_m is similar to that for a_k in Equation (3.113) on page 121, i.e.,

$$C_m = \frac{2T_0}{R\lambda_m} \frac{1}{J_1(\lambda_m R)}. \tag{5.65}$$

Using this result in Equation (5.61), we obtain the final solution

$$T(r,t) = 2T_0 \sum_{m=1}^{\infty} e^{-\alpha\lambda_m^2 t} \frac{J_0(\lambda_m r)}{R\lambda_m J_1(\lambda_m R)} \tag{5.66}$$

□

In the above example, the summation in this result starts with $m = 1$. It should be noted that we indexed the first root of $J_0(\lambda R) = 0$ as $\lambda_1 R = \beta_{01}$, i.e., with $m = 1$. Furthermore, m is just a counter that we define to start with zero or 1 as is convenient. There are cases however when the first root of the eigenfunction turns out to be zero. In such cases, we choose to index that root as $\lambda_0 = 0$, i.e., with $m = 0$, and then the first nonzero root as λ_1. It should be emphasized that the choice of the leading index value is completely a matter of convenience since m is just a counter. The important point however is if $\lambda_0 = 0$ is of any consequence in the solution. In some cases, $\lambda_0 = 0$ simply leads to a zero value of the corresponding term, and is inconsequential. In some instances, $\lambda_0 = 0$ has special significance, and we shall consider an example where this situation arises.

Example 5.6

PROBLEM STATEMENT: Once again we consider a cylinder, as in the previous example, with initial and boundary conditions as shown in the picture. The axial symmetry is apparent, and the differential equation thus reduces to (drop the z and θ dependencies)

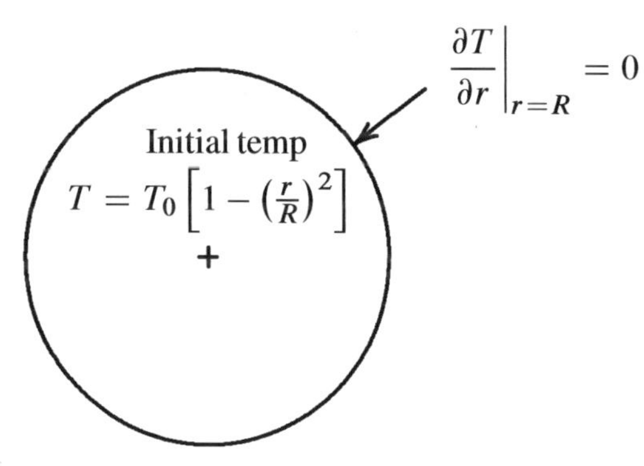

$$\frac{1}{\alpha}\frac{\partial T(r,t)}{\partial t} = \frac{\partial^2 T(r,t)}{\partial r^2} + \frac{1}{r}\frac{\partial T(r,t)}{\partial r}, \tag{5.67}$$

which is the same as Equation (5.46) in Example 5.5.

SOLUTION:

Once again, it would be useful to examine the problem on rectangularized coordinates as shown in Figure 5.7 As in the previous example, we have a pair of opposite homogeneous boundary conditions at $r = 0$ and $r = R$. This facilitates an eigenfunction expansion in r, and allow us to satisfy the nonzero condition at $t = 0$. This time we shall apply the method of eigenfunction expansions, and identify from the differential equation (5.67), the radial operator,

$$\frac{\partial^2}{\partial r^2} + \frac{1}{r}\frac{\partial}{\partial r}, \tag{5.68}$$

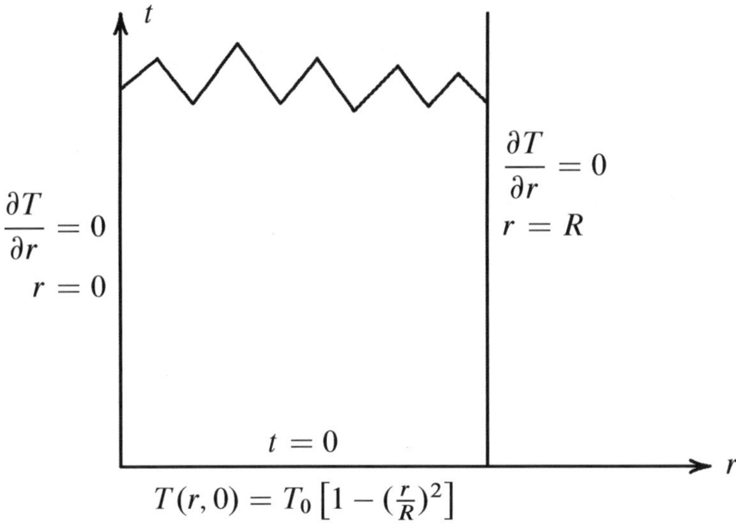

Figure 5.7: Display of the domain $0 \le r \le R$, $t \ge 0$ on a rectangular coordinate system for Example 5.6.

for which we shall determine a set of eigenfunctions. These are given by the solution of the ordinary differential equation

$$\frac{d^2R_m(r)}{dr^2} + \frac{1}{r}\frac{dR_m(r)}{dr} = -\lambda_m^2 R_m(r). \tag{5.69}$$

which is the same as the Equation (5.50), except that Equation (5.69) is indexed with m. We may multiply this equation by r^2 and obtain a form similar to Equation (5.51), i.e.,

$$r^2\frac{d^2R_m(r)}{dr^2} + r\frac{dR_m(r)}{dr} + \lambda_m^2 r^2 R_m(r) = 0. \tag{5.70}$$

Following Equations (5.51)-(5.54) on pages 210-210, we may write

$$R_m(r) = A_m J_0(\lambda_m r) + B_m Y_0(\lambda_m r), \tag{5.71}$$

This represents a set of eigenfunctions of the operator (5.68) with corresponding eigenvalues $-\lambda_m^2$. Once again, the with $Y_0(\lambda_m r)$ going logarthmically to $-\infty$ as $r \to 0$ (see Figure 2.1 on page 69), we eliminate this solution by setting $B_m = 0$. The solution for $R_m(r)$ therefore reduces to

$$R_m(r) = A_m J_0(\lambda_m r), \tag{5.72}$$

which is the same as Equation (5.72), except that the λ_m values will be determined by a different boundary condition at $r = R$. Since this boundary condition is homogeneous, it may be satisfied at this point term-by-term with $R_m'(R) = 0$. However, we shall do this after including the time part in the solution in order to get a better perspective. So, let us go ahead and expand as follows

$$T(r,t) = \sum_{m=0}^{\infty} \Theta_m(t) J_0(\lambda_m r) \tag{5.73}$$

and substitute into the differential equation (5.67). This results in

$$\frac{1}{\alpha}\sum_{m=0}^{\infty}\Theta_m'(t)J_0(\lambda_m r) = \sum_{m=0}^{\infty}\Theta_m(t)\left[\frac{d^2}{dr^2}+\frac{1}{r}\frac{d}{dr}\right]J_0(\lambda_m r). \tag{5.74}$$

Since here $J_0(\lambda_m r)$ is a set of eigenfunctions of the operator in front of it with corresponding eigenvalues $-\lambda_m^2$, we can write

$$\frac{1}{\alpha}\sum_{m=0}^{\infty}\Theta_m'(t)J_0(\lambda_m r) = \sum_{m=0}^{\infty}\Theta_m(t)\left[-\lambda_m^2\right]J_0(\lambda_m r), \tag{5.75}$$

or equivalently,

$$\sum_{m=0}^{\infty}\left[\frac{1}{\alpha}\Theta_m'(t)+\lambda_m^2\Theta_m(t)\right]J_0(\lambda_m r) = 0. \tag{5.76}$$

Since this must be valid for all values of r in the range $0 \le r \le R$, and since the right-hand side is zero, we can satisfy this equation with just

$$\frac{1}{\alpha}\Theta_m'(t)+\lambda_m^2\Theta_m(t) = 0. \tag{5.77}$$

This is the same as Equation (5.49), except that we do not necessarily exclude $m = 0$, and as mentioned before, λ_m values are different. The solution is given by Equation (5.60) which we rewrite as

$$\Theta_m(t) = a_m e^{-\alpha\lambda_m^2 t}. \tag{5.78}$$

Using this in the expansion (5.73), we obtain

$$T(r,t) = \sum_{m=0}^{\infty} a_m e^{-\alpha\lambda_m^2 t}J_0(\lambda_m r), \tag{5.79}$$

where λ_m values will be found by satisfying the pair of opposite boundary conditions at $r = 0$ and $r = R$. The condition at $r = 0$ is already satisfied by radial symmetry, and we just need to work with the one at $r = R$, i.e.,

$$\left.\frac{\partial T}{\partial r}\right|_{r=R} = 0. \tag{5.80}$$

Applying this in Equation (5.79) leads to

$$T(r,t) = \sum_{m=0}^{\infty} a_m e^{-\alpha\lambda_m^2 t}\left.\frac{dJ_0(\lambda_m r)}{dr}\right|_{r=R} = 0. \tag{5.81}$$

Being a homogeneous condition, it can be applied term-by-term, i.e.,

$$\left.\frac{dJ_0(\lambda_m r)}{dr}\right|_{r=R} = \lambda_m J_0'(\lambda_m R) = 0. \tag{5.82}$$

Let us examine $J_0(x)$ graphically and see where we have $J_0'(x) = 0$. A plot of $J_0(x)$ is given in Figure 5.8 Noting that $J_0'(x) = -J_1(x)$, the values of $\lambda_m R$ we are looking

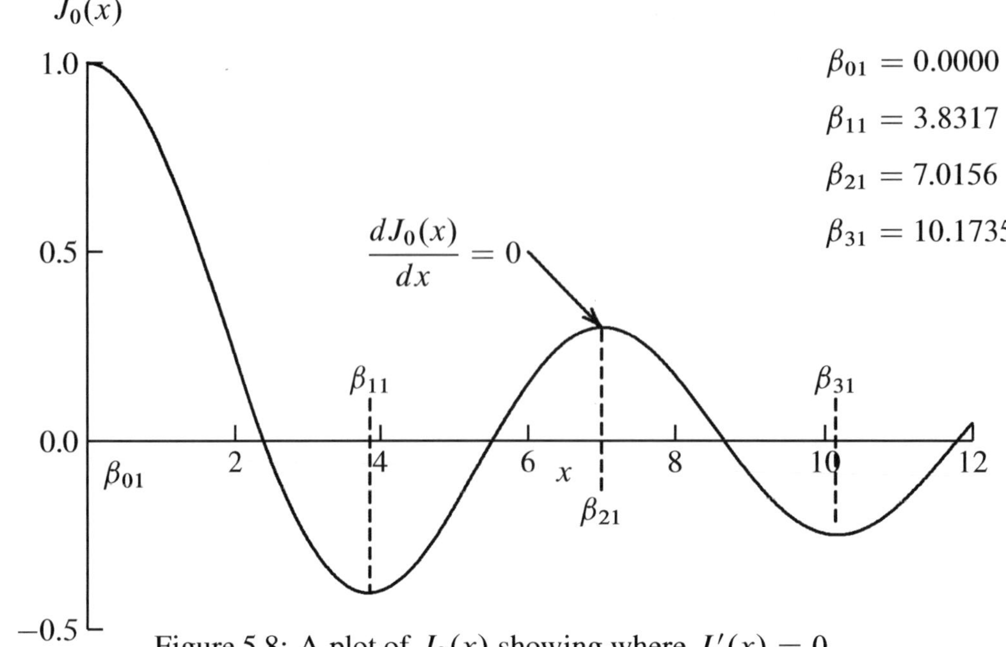

Figure 5.8: A plot of $J_0(x)$ showing where $J_0'(x) = 0$.

for are given by the roots of $J_1(x)$, and represented by β_{m1} (see Table 3.1, page 123 in Chapter 3). Continuing with Equation (5.82), we obtain

$$\lambda_m R = \beta_{m1} \qquad \text{or} \qquad \lambda_m = \frac{\beta_{m1}}{R}. \tag{5.83}$$

At this point, we can satisfy the initial condition,

$$T(r,0) = T_0\left(1 - \frac{r^2}{R^2}\right) = \sum_{n=0}^{\infty} a_m J_0(\lambda_m r) \tag{5.84}$$

Here we invoke the orthogonality of the set $J_0(\lambda_m r), \quad m = 0, 1, 2, 3, \ldots$ which can be seen in detail by following Examples 3.12 and 3.13 on pages 117-121 in Chapter 3. The structural difference between those example and the current one is just in the boundary at $r = R$. Instead of $J_0(\lambda_m R) = 0$, we are requiring $J_0'(\lambda_m R) = 0$ in the present case. The pair of conditions at the end points of the domain $0 \leq r \leq R$ are both Neumann type corresponding to the Sturm-Liouville classification given by Equation (3.69) on page 112. This guarantees orthogonality of the set under discussion in the present example, and the weighting factor r applies as in the previous example as well as Example 3.13. Following Equations (3.106) and (3.107) on page 120, and applying the initial condition (5.84), we obtain the following expression for the set of coefficients, a_m:

$$a_m = \frac{T_0 \int_0^R \left(1 - \frac{r^2}{R^2}\right) r J_0(\lambda_m r) dr}{\int_0^R r[J_0(\lambda_m r)]^2 dr}. \tag{5.85}$$

The denominator can be calculated using the expression (3.130) on page 124 by setting $n = 0$. Noting that for the present example $J_0'(\lambda_m R) = 0$, and replacing the index k

in (3.130) with m, we obtain

$$\int_0^R r\,[J_0(\lambda_m r)]^2\,dr = \tfrac{1}{2}R^2\,[J_0(\lambda_m R)]^2, \qquad (5.86)$$

where λ_m carries just the index m, and we do not use the index for $n = 0$. The expression for the numerator can be calculated through integration by parts, and using the differential equation for $J_0(\lambda_m r)$. We use the form given by Equation (3.110) where again we replace the index k with m, i.e.,

$$\frac{d}{dr}\left[r\frac{dJ_0(\lambda_m r)}{dr}\right] = -\lambda_m^2 r J_0(\lambda_m r) \qquad (5.87)$$

Using this, we now work with the numerator and start integration by parts,

$$\begin{aligned}
&\lambda_m^2 \int_0^R \left(1-\frac{r^2}{R^2}\right) rJ_0(\lambda_m r)\,dr \\
&= -\left(1-\frac{r^2}{R^2}\right)\left[r\frac{dJ_0(\lambda_m r)}{dr}\right]_{r=0}^{r=R} - \int_0^R \left(\frac{2r}{R^2}\right)\left[r\frac{dJ_0(\lambda_m r)}{dr}\right]dr \\
&= 0 - \left(\frac{2r^2}{R^2}\right) J_0(\lambda_m r)\Big|_0^R + \int_0^R 4rJ_0(\lambda_m r)\,dr \\
&= -2J_0(\lambda_m R) - \frac{4}{\lambda_m^2}\left[r\frac{dJ_0(\lambda_m r)}{dr}\right]_0^R \\
&= -2J_0(\lambda_m R)
\end{aligned} \qquad (5.88)$$

Dividing by λ_m^2, we obtain for the numerator in Equation (5.85) as

$$\int_0^R \left(1-\frac{r^2}{R^2}\right) rJ_0(\lambda_m r)dr = -\frac{2}{\lambda_m^2} J_0(\lambda_m R), \quad m \neq 0. \qquad (5.89)$$

This integral is not valid for $m = 0$ because $\lambda_0 = 0$ is in the denominator. For this situation we evaluate the integral separately, noting that $J_0(\lambda_0 r) = J_0(0) = 1$. Thus,

$$\begin{aligned}
\int_0^R \left(1-\frac{r^2}{R^2}\right) rJ_0(\lambda_0 r)\,dr &= \left(1-\frac{r^2}{R^2}\right) r\,dr \\
&= \left(\tfrac{1}{2}r^2 - \frac{r^4}{4R^2}\right)\Big|_0^R \\
&= \tfrac{1}{4}R^2
\end{aligned} \qquad (5.90)$$

Now, using the integrals (5.86) and (5.89) in Equation (5.85), we may write the expression for a_m as

$$\begin{aligned}
a_m &= \frac{-\frac{2T_0}{\lambda_m^2} J_0(\lambda_m R)}{\frac{1}{2}R^2\,[J_0(\lambda_m R)]^2}, \\
&= \frac{-4T_0}{\lambda_m^2 R^2\,[J_0(\lambda_m R)]}, \quad m \neq 0
\end{aligned} \qquad (5.91)$$

For $m = 0$, we use Equation (5.86) along with (5.90) and obtain

$$a_0 = \frac{T_0 \frac{1}{4} R^2}{\frac{1}{2} R^2 \left[J_0(\lambda_0 R)\right]^2}, \qquad (\lambda_0 = 0)$$
$$= \tfrac{1}{2} T_0 \tag{5.92}$$

Using the expressions (5.91) and (5.92) in Equation (5.79), we may write the final solution as

$$T(r,t) = T_0 \left[\frac{1}{2} - 4 \sum_{m=1}^{\infty} \frac{J_0(\lambda_m r) e^{-\alpha \lambda_m^2 t}}{\lambda_m^2 R^2 \left[J_0(\lambda_m R)\right]} \right], \tag{5.93}$$

where the leading term $\frac{1}{2} T_0$ represents the average temperature. □

In the examples we have considered so far, the problems involved only two independent variables $[(r, \theta)$ or $(r, t)]$, and we were able to solve them with expansions in either r or θ. The Bessel-function expansion, as done in Examples 5.5 and 5.6, will also work for Laplaces equation with (r, z) dependence if the problems call for expansions in r. In cylindrical geometry, we often encounter problems with dependence on (r, θ, t) or (r, θ, z). There are of course cases with (r, θ, z, t) dependence but we shall not study those.

EXERCISES 5.2

Problem 1

For the semicircular region shown, solve

$$\nabla^2 T(r, \theta) = 0,$$

with boundary conditions as indicated.

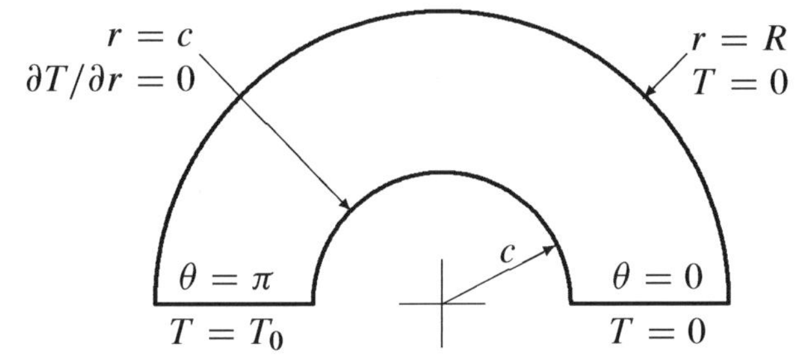

Problem 2

An elastic membrane, mounted on a rigid circular ring, is undergoing vibrations, and the membrane displacement $u(r, \theta, t)$ is given by the wave equation,

$$\frac{1}{c^2} \frac{\partial^2 u(r, \theta, t)}{\partial t^2} = \nabla^2 u(r, \theta, t), \qquad 0 \le r \le R, \quad 0 \le \theta \le 2\pi, \quad t > 0.$$

The initial displacement is given by

$$u(r, \theta, 0) = u_0 \left(1 - \frac{r}{R}\right), \tag{5.94}$$

and the initial velocity is zero, i.e.,

$$\left. \frac{\partial u(r, \theta, t)}{\partial t} \right|_{t=0} = 0. \tag{5.95}$$

Problem 3

Obtain the steady temperature distribution $T(r, z)$ in a circular rod of length l and radius R satisfying

$$\nabla^2 T(r, z) = 0,$$

with boundary conditions,

$$T(R, z) = 0, \qquad T < \infty \quad \text{as} \quad r \to \infty, \qquad -k\frac{\partial T}{\partial z}\bigg|_{z=0} = q_0, \quad T(r, l) = 0.$$

Problem 4

Consider an infinitely long circular pipe of radius R, filled with an incompressible fluid (with kinematic viscosity ν and dynamic viscosity μ). Suddenly, a constant pressure gradient, dp/dz, is imposed to drive the flow along the axis of pipe (in the z-direction), which means that the pressure gradient is zero for $t = 0$, and a constant for $t > 0$. The initial velocity is zero, i.e., $u(r, 0) = 0$. The unsteady-state unidirectional laminar flow in axisymmetric cylindrical coordinates is described by the nonhomogeneous differential equation,

$$\frac{\partial u(r,t)}{\partial t} - \nu\left[\frac{1}{r}\frac{\partial}{\partial r}\left(r\frac{\partial u(r,t)}{\partial r}\right)\right] = \frac{P_0}{\rho_0}, \tag{5.96}$$

where $P_0 = -dp/dz$ denotes the axial pressure gradient (in the z-direction), ν is the kinematic viscosity, and ρ_0 is the density. The velocity at the pipe boundary ($r = R$) is no-slip, i.e., $u(R, t) = 0$. The axial symmetry condition gives $\partial u(r, t)/\partial r|_{r=0} = 0$. Obtain $u(r, t)$, applying the following procedure

1. Decompose the problem into steady-state and time-dependent parts, $u(r, t) = u_1(r) + u_2(r, t)$. Rectangularized domain would be helpful.

2. Let $u_1(r)$ take care of the forcing function P_0/ρ_0. Note that this is an ODE with a straightforward solution.

3. Then $u_2(r, t)$ satisfies a homogeneous PDE. Decide on the variable for eigenfunction expansion. It should be r, and the r-operator is

$$\frac{d^2}{dr^2} + \frac{1}{r}\frac{d}{dr}. \tag{5.97}$$

4. Construct eignfunctions of the operator (5.97). Carry out an eigenfunction expansion in r. You should get Bessels equation of order zero for the radial eigenfunctios $R_m(r)$. Obtain the solution $R_m(r)$, which should be

$$R_m(r) = A_m J_0(\lambda_m r) + B_m Y_0(\lambda_m r).$$

5. Eliminate $Y_0(\lambda_m r)$ on the basis that it goes to $-\infty$ as $r \to 0$. Observe that $u_2(R, t) = 0$ can be satisfied term-by-term with $R_m(R) = A_m J_0(\lambda_m R) = 0$.

6. Require that $J_0(\lambda_m R) = 0$ and establish λ_m values. The roots of $J_0(x) = 0$ are given by $\beta_{m0},\ m = 1, 2, 3, \ldots$ in Table 3.1 on page 123 in Chapter 2.

7. Expand $u_2(r,t)$ in the form

$$u_2(r,t) = \sum_{m=1}^{\infty} T_m(t) J_0(\lambda_m r),$$

and substitute into the differential equation for $u_2(r,t)$. Make use of the fact that $J_0(\lambda_m r)$ is an eigenfunction of the operator (5.97).

8. Construct the ODE for $T_m(t)$ and solve it.

9. Satisfy the initial condition and obtain the unknown set of constants. Be careful that the initial condition for $u_2(r,t)$ should be $u_2(r,0) = 0 - u_1(r) = -u_1(r)$. When applying orthogonality to determine the unknown set of constants, please recall orthogonality relationship in the general Sturm-Liouville form is

$$\int_a^b q(x) X_j(x) X_k(x) dx = 0, \quad j \neq k.$$

The orthogonality relationship of Bessel functions of order 0 should be

$$\int_0^R r J_0(\lambda_j r) J_0(\lambda_k r) dr = 0, \quad j \neq k.$$

See Example 3.12 on pages 117-119 and Example 5.5 on pages 209-213.

10. Combine $u_1(r)$ and $u_2(r,t)$ to obtain $u(r,t)$.

ALTERNATE PROCEDURE: Please note that this problem can also be solved by carrying out an eigenfunction expansion for the pressure term and $u(r,t)$ in terms and of $J_0(\lambda_m r)$ and matching each side term-by-term to obtain a non-homogeneous ODE for $T_m(t)$.

5.3 Eigenfunction Expansions in r and θ

We shall consider an example involving heat diffusion with (r,θ,t) dependence having homogeneous boundary conditions and a nonzero initial condition.

Example 5.7

PROBLEM STATEMENT: Obtain the temperature distribution $T(r, \theta, t)$ in the semi-circular region shown in the figure. The governing equation in this case is

$$\frac{1}{\alpha}\frac{\partial T}{\partial t} = \frac{\partial^2 T}{\partial r^2} + \frac{1}{r}\frac{\partial T}{\partial r} + \frac{1}{r^2}\frac{\partial^2 T}{\partial \theta^2}, \quad (5.98)$$

and the boundary and initial conditions are as shown, i.e.,

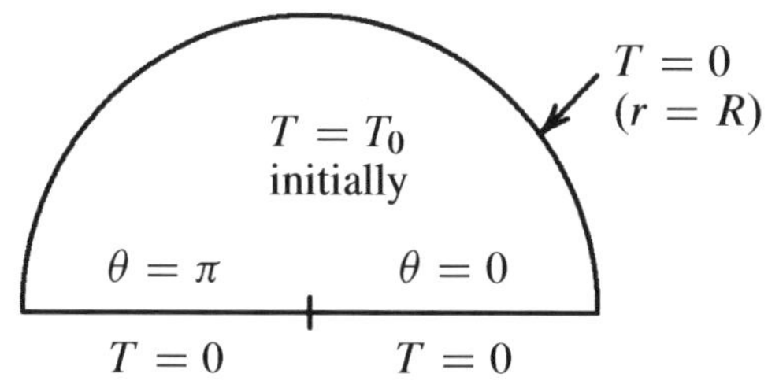

$$T(r, 0, t) = 0, \qquad T(r, \pi, t) = 0, \qquad T(R, \theta, t) = 0. \qquad (5.99)$$

with the initial condition,

$$T(r, \theta, 0) = T_0. \qquad (5.100)$$

SOLUTION:
Once again, it would be useful to examine the problem on rectangularized coordinates as shown in Figure 5.9 Here $r = 0$ which is physically a point in the $r - \theta$ plane,

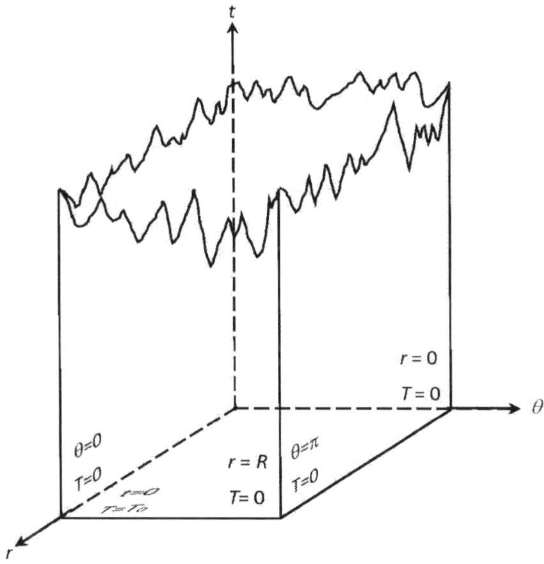

Figure 5.9: A three-dimensional rectangularized depiction of (r, θ, t) coordinates for Example 5.7.

appears as a line. We note from the previous figure that the point $r = 0$ lies where the two boundaries $\theta = 0$ and $\theta = \pi$ meet. At both these boundaries, $T = 0$. Therefore, by continuity, $T = 0$ at $r = 0$. With the initial condition $T(r, \theta, 0) = T_0$, we see that it is formally a function of r and θ. Therefore, we need to expand in these variables. Additionally, we see from Figure 5.9 that we have pairs of opposite homogeneous boundary conditions in both r and θ, thus facilitating eigenfunction expansions in these variables. To carry out such expansions we need to construct the appropriate differential equations that isolate the respective variables. Upon examining Equation (5.98), we see that r^{-2} multiplies the second derivative in θ, and these variables are not isolated from each other in the equation as given. However, if we multiply the equation by r^2, we will isolate the θ-derivative term. Carrying out this operation, we have Multiply equation by r^2

$$r^2 \frac{1}{\alpha} \frac{\partial T}{\partial t} = r^2 \frac{\partial^2 T}{\partial r^2} + r \frac{\partial T}{\partial r} + \frac{\partial^2 T}{\partial \theta^2}, \tag{5.101}$$

where we see that the term, $\partial^2 T/\partial \theta^2$ stands alone while all the other terms have no appearance of θ, neither as a derivative nor as a multiplicative term. Therefore, it is possible to construct suitable eigenfunctions in θ. We shall start with the θ expansion as our only choice. Let us look for eigenfunctions of $d^2/d\theta^2$, i.e., we need solutions to the ordinary differential equation,

$$\frac{d^2 \Theta_n}{d\theta^2} = -\lambda_n^2 \Theta_n,$$

which may also be written as

$$\Theta_n''(\theta) + \lambda_n^2 \Theta_n(\theta) = 0$$

Following Example 5.1, we look at Equation (5.7) on page 192 for which same set of homogeneous boundary conditions are required as the present case. Following the development leading to Equation (5.9), we arrive at the same results, i.e.,

$$\Theta_n(\theta) = \sin n\theta, \quad n = 1, 2, 3, \ldots$$

which satisfies $\Theta_n(0) = 0$ and $\Theta_n(\pi) = 0$. We can now expand $T(r, \theta, t)$ as a series in terms of the eigenfunctions $sin n\theta$, i.e.,

$$T(r, \theta, t) = \sum_{n=1}^{\infty} T_n(r, t) \sin n\theta. \tag{5.102}$$

Upon the substitution of this expansion into Equation (5.101), we obtain

$$r^2 \frac{1}{\alpha} \sum_{n=1}^{\infty} \frac{\partial T_n}{\partial t} \sin n\theta = r^2 \sum_{n=1}^{\infty} \frac{\partial^2 T_n}{\partial r^2} \sin n\theta$$
$$+ r \sum_{n=1}^{\infty} \frac{\partial T_n}{\partial r} \sin n\theta + \sum_{n=1}^{\infty} T_n(r)(-n^2) \sin n\theta. \tag{5.103}$$

We notice that the θ-dependence in the functional form of $\sin n\theta$ is preserved throughout the equation, and it is possible to factor this out so that

$$\sum_{n=1}^{\infty} \left(\frac{r^2}{\alpha} \frac{\partial T_n}{\partial t} - \left[r^2 \frac{\partial^2 T_n}{\partial r^2} + r \frac{\partial T_n}{\partial r} - n^2 T_n \right] \right) \sin n\theta = 0.$$

This is valid for all values of θ in the range $0 \leq \theta \leq \pi$, and therefore, we may set every coefficient of $sin n\theta$ equal to zero. This results in

$$\left(\frac{r^2}{\alpha}\frac{\partial T_n(r,t)}{\partial t} - \left[r^2\frac{\partial^2 T_n(r,t)}{\partial r^2} + r\frac{\partial T_n(r,t)}{\partial r} - n^2 T_n(r,t)\right]\right) = 0.$$

We can now separate the r and the t operations if we divide the equation by r^2, i.e.

$$\frac{1}{\alpha}\frac{\partial T_n(r,t)}{\partial t} - \left[\frac{\partial^2 T_n(r,t)}{\partial r^2} + \frac{1}{r}\frac{\partial T_n(r,t)}{\partial r} - \frac{n^2}{r^2}T_n(r,t)\right] = 0,$$

or equivalently,

$$\frac{1}{\alpha}\frac{\partial T_n(r,t)}{\partial t} - \left[\frac{\partial^2}{\partial r^2} + \frac{1}{r}\frac{\partial}{\partial r} - \frac{n^2}{r^2}\right] T_n(r,t) = 0, \tag{5.104}$$

This is, in fact, a partial differential equation for $T_n(r,t)$. By expanding $T(r,\theta,t)$ in $\sin n\theta$, we have reduced the number of variables from three to two. Equation (5.104) can be solved by suitable eigenfunction expansion in r. For doing so, we need eigenfunctions of the operator

$$\frac{d^2}{dr^2} + \frac{1}{r}\frac{d}{dr} - \frac{n^2}{r^2}, \tag{5.105}$$

that we have identified within the square brackets in Equation (5.104). In other words, we need to solve

$$\left[\frac{d^2}{dr^2} + \frac{1}{r}\frac{d}{dr} - \frac{n^2}{r^2}\right] R(r) = -\lambda^2 R(r). \tag{5.106}$$

Here, for every n, we have a different equation, and we need to index the solution $R(r)$ accordingly. In addition, for every n, we will have a set of values of λ which need to be indexed further. Therefore, we need two indices, n and m. Thus, we set up the above equation as

$$\left[\frac{d^2}{dr^2} + \frac{1}{r}\frac{d}{dr} - \frac{n^2}{r^2}\right] R_{mn}(r) = -\lambda^2 R_{mn}(r). \tag{5.107}$$

Upon multiplying by r^2, and following with some rearrangement, we obtain

$$r^2\frac{d^2 R_{mn}}{dr^2} + r\frac{dR_{mn}}{dr} + \left(-n^2 + \lambda_{mn}^2 r^2\right) R_{mn} = 0. \tag{5.108}$$

Let us compare this with Bessels equation [see Equation (2.57) on page 62] of order n,

$$x^2\frac{d^2 f}{dx^2} + x\frac{df}{dx} + \left(-n^2 + x^2\right) f = 0, \tag{5.109}$$

This is equivalent to Equation 5.108 if we set $x = \lambda_{mn} r$ and $f(x) = R_{mn}(r)$. The solution for Equation (5.109) can be found by the method of Frobenius (see Chapter 2). Such a solution can be expressed as the sum of two linearly independent solutions in the form

$$f(x) = AJ_n(x) + BY_n(x),$$

where $J_n(x)$ and $Y_n(x)$ are given by Equations (2.67) and (2.68) on page 68. With $x = \lambda_{mn} r$, the solution for $R_{mn}(r)$ can be written as

$$R_{mn}(r) = A_{mn} J_n(\lambda_{mn} r) + B_{mn} Y_n(\lambda_{mn} r)$$

The behavior of these functions can be seen on the plots given in Figure 2.1 on page 69. As noted in some earlier examples with order zero Bessel functions, the function $Y_n(x) \to \infty$ as $x \to 0$. Therefore, for regions where $r = 0$ is included in the domain, we usually exclude the solution $Y_n(\lambda_{mn} r)$. This can be done by setting $B_{mn} = 0$. This leaves us with

$$R_{mn}(r) = A_{mn} J_n(\lambda_{mn} r).$$

The function $T_n(r,t)$ can be expanded as a Fourier-Bessel series,

$$T_n(r,t) = \sum_{m=1}^{\infty} T_{mn}(t) J_n(\lambda_{mn} r), \tag{5.110}$$

where the summation starts with $m = 1$ because, as we shall see later, $\lambda_{0n} = 0$ for $n \geq 1$. Also, in this example, $n = 0$ does not come into play because in the sine series in θ, $\sin n\theta$ has no contribution for $n = 0$. The boundary condition $T = 0$ at $r = R$ can be satisfied term-by-term with $T_n(R,t) = 0$ and this can be satisfied term-by-term with $J_n(\lambda_{mn} R) = 0$. Therefore, we need to locate the roots of $J_n(x) = 0$ to determine the values of λ_{mn} so that the homogeneous boundary condition at $r = R$ is satisfied. The roots are given by

$$\lambda_{mn} R = \beta_{mn} \qquad \text{or} \qquad \lambda_{mn} = \frac{\beta_{mn}}{R}, \tag{5.111}$$

and the β_{mn} values are available in Table 3.1 on page 123. Let us now substitute the expansion for $T_n(r,t)$ [Equation (5.110)] into the partial differential equation (5.104). This leads to

$$\frac{1}{\alpha} \sum_{m=1}^{\infty} T'_{mn}(t) J_n(\lambda_{mn} r) = \sum_{m=1}^{\infty} T_{mn}(t) \left[\frac{d^2}{dr^2} + \frac{1}{r}\frac{d}{dr} - \frac{n^2}{r^2} \right] J_n(\lambda_{mn} r). \tag{5.112}$$

Since $J_n(\lambda_{mn} r)$ is a solution of Equation (5.108), it is obviously an eigenfunction of the operator on the right-hand side of Equation (5.112) with eigenvalue $-\lambda_{mn}^2$. Using this property, we obtain

$$\frac{1}{\alpha} \sum_{m=1}^{\infty} T'_{mn}(t) J_n(\lambda_{mn} r) = \sum_{m=1}^{\infty} T_{mn}(t) \left[-\lambda_{mn}^2 \right] J_n(\lambda_{mn} r), \tag{5.113}$$

which may be written as

$$\sum_{m=1}^{\infty} \left[\frac{1}{\alpha} T'_{mn}(t) + \lambda_{mn}^2 T_{mn}(t) \right] J_n(\lambda_{mn} r) = 0. \tag{5.114}$$

With the usual argument that this equation is valid for all values of r in the range $0 \leq r \leq R$, we can require that each coefficient of $J_n(\lambda_{mn}r)$ vanish, i.e.,

$$\left[\frac{1}{\alpha}T'_{mn}(t) + \lambda_{mn}^2 T_{mn}(t)\right] J_n(\lambda_{mn}r) = 0. \tag{5.115}$$

This has the solution

$$T_{mn}(t) = a_{mn}e^{-\alpha\lambda_{mn}^2 t}. \tag{5.116}$$

Let us now assemble the solution backwards, starting with $T_n(r,t)$. Using Equation (5.116) in the expansion (5.110), we obtain

$$T_n(r,t) = \sum_{m=1}^{\infty} a_{mn}e^{-\alpha\lambda_{mn}^2 t} J_n(\lambda_{mn}r), \tag{5.117}$$

Next, we use this expansion for $T_n(r,t)$ in Equation (5.102), leading to

$$T(r,\theta,t) = \sum_{n=1}^{\infty} T_n(r,t)\sin n\theta = \sum_{n=1}^{\infty}\sum_{m=1}^{\infty} a_{mn}e^{-\alpha\lambda_{mn}^2 t} J_n(\lambda_{mn}r)\sin n\theta. \tag{5.118}$$

We can now apply the initial condition, $T(r,\theta,0) = T_0$. Setting $t = 0$ in Equation (5.118), we end up with

$$T_0 = \sum_{n=1}^{\infty}\left[\sum_{m=1}^{\infty} a_{mn}J_n(\lambda_{mn}r)\right]\sin n\theta. \tag{5.119}$$

With the orthogonality of the Fourier sine series, we obtain the coefficients as

$$\begin{aligned}\left[\sum_{m=1}^{\infty} a_{mn}J_n(\lambda_{mn}r)\right] &= \frac{\int_0^{\pi} T_0 \sin n\theta \, d\theta}{\int_0^{\pi} \sin^2 n\theta \, d\theta} \\ &= \frac{2T_0\left[1-(-1)^n\right]}{n\pi}\end{aligned} \tag{5.120}$$

The left-hand side is a Fourier-Bessel series. Applying orthogonality of $J_n(\lambda_{mn}r)$ gives

$$a_{mn} = \frac{\int_0^R \frac{2T_0[1-(-1)^n]}{n\pi} rJ_n(\lambda_{mn}r)dr}{\int_0^R r[J_n(\lambda_{mn}r)]^2 dr} \tag{5.121}$$

The integral in the denominator is given in a general form by Equation (3.130) on page 124. It is repeated here for convenience.

$$\int_0^R r[J_n(\lambda_{mn}r)]^2 dr = \tfrac{1}{2}R^2\left\{[J_n'(\lambda_{mn}R)]^2 + \left(1 - \frac{n^2}{\lambda_{mn}^2 R^2}\right)[J_n(\lambda_{mn}R)]^2\right\}. \tag{5.122}$$

For the present case, $J_n(\lambda_{mn}R) = 0$, and the integral reduces to

$$\int_0^R r[J_n(\lambda_{mn}r)]^2 dr = \tfrac{1}{2}R^2[J_n'(\lambda_{mn}R)]^2, \tag{5.123}$$

which is also given by Equation (3.132) on page 124. The numerator can be evaluated by using the power series of $J_n(x)$, given by Equation (2.67) on page 68 in Chapter 2. Using Equation (2.67), we obtain

$$\int_0^R r[J_n(\lambda_{mn}r)]dr = R^2 \sum_{k=0}^{\infty} \frac{(-1)^k \left(\frac{1}{2}\lambda_{mn}R\right)^{2k+n}}{(k+n)!k!(2k+n+2)}, \quad n = 0, 1, 2, 3 \dots \tag{5.124}$$

Now, using Equations (5.124) and (5.123) in Equation (5.121), we obtain the following expression for a_{mn}:

$$a_{mn} = \left(\frac{4T_0[1-(-1)^n]}{n\pi}\right) \frac{\sum_{k=0}^{\infty} \frac{(-1)^k \left(\frac{1}{2}\lambda_{mn}R\right)^{2k+n}}{(k+n)!k!(2k+n+2)}}{[J_n'(\lambda_{mn}R)]^2}. \tag{5.125}$$

The complete solution as given by Equation (5.118) now takes the form

$$T(r,\theta,t) = 4T_0 \sum_{n=1}^{\infty} \frac{[1-(-1)^n]}{n\pi} \sum_{m=1}^{\infty} \frac{\sum_{k=0}^{\infty} \frac{(-1)^k \left(\frac{1}{2}\lambda_{mn}R\right)^{2k+n}}{(k+n)!k!(2k+n+2)}}{[J_n'(\lambda_{mn}R)]^2} e^{-\alpha\lambda_{mn}^2 t} J_n(\lambda_{mn}r)\sin n\theta. \tag{5.126}$$

□

This was an example where all the boundary conditions were homogeneous, and we were able to carry out eigenfunction expansions in both the spatial variables. Next, we shall consider cases when we have non-homogeneous boundary conditions.

5.3.1 Time-Dependent Problems with Non-Homogeneous Boundary Conditions

In situations where we have nonzero boundary conditions, we can decompose the problem into a steady-state part and a time-dependent part. With this decomposition, the steady-state part takes care of the non-homogeneous boundary conditions while the unsteady part then has to deal with only homogeneous boundary conditions. This approach was taken earlier in Example 4.11 on pages 182-186. This approach works provided the boundary conditions are not time-dependent. The next example will illustrate the procedure.

Example 5.8

PROBLEM STATEMENT: Obtain the temperature distribution $T(r,\theta,t)$ in the semicircular region shown in the figure by solving

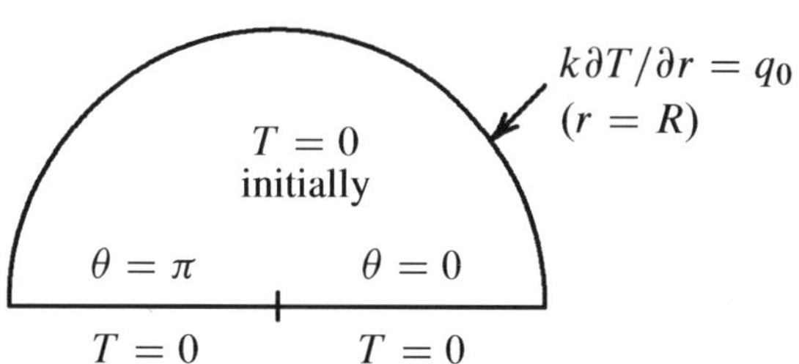

$$\frac{1}{\alpha}\frac{\partial T}{\partial t} = \frac{\partial^2 T}{\partial r^2} + \frac{1}{r}\frac{\partial T}{\partial r} + \frac{1}{r^2}\frac{\partial^2 T}{\partial \theta^2}. \quad (5.127)$$

The boundary conditions as illustrated are

$$T(r,0,t) = 0, \qquad T(r,\pi,t) = 0, \quad (5.128)$$

$$\left.\frac{\partial T(r,\theta,t)}{\partial r}\right|_{r=R} = q_0, \quad (5.129)$$

and initial condition,

$$T(r,\theta,0) = 0. \quad (5.130)$$

SOLUTION:
Here we decompose the problem into a steady and an unsteady part, i.e.,

$$T(r,\theta,t) = T_1(r,\theta) + T_2(r,\theta,t), \quad (5.131)$$

where $T_1(r,\theta)$ is the steady part. This decomposition is illustrated pictorially in Figure 5.10. Using the decomposition (5.131) in Equation (5.127), we obtain the following

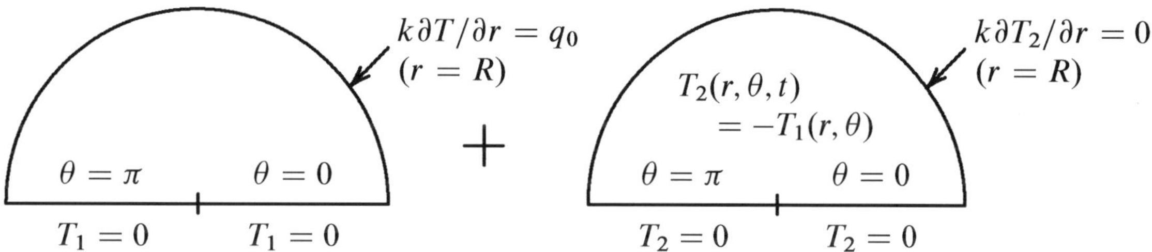

Figure 5.10: Illustration of decomposition of $T(r,\theta,t)$ into steady and unsteady parts.

$$\frac{\partial^2 T_1}{\partial r^2} + \frac{1}{r}\frac{\partial T_1}{\partial r} + \frac{1}{r^2}\frac{\partial^2 T_1}{\partial \theta^2} = 0, \quad (5.132)$$

$$\frac{\partial^2 T_2}{\partial r^2} + \frac{1}{r}\frac{\partial T_2}{\partial r} + \frac{1}{r^2}\frac{\partial^2 T_2}{\partial \theta^2} = \frac{1}{\alpha}\frac{\partial T_2}{\partial t}. \quad (5.133)$$

(5.134)

The steady part $T_1(r,\theta)$ is quite similar to Example 5.1 on pages 192-195. The only difference is in the boundary condition at $r = R$. We can therefore adopt the solution form given by Equation (5.16), i.e.,

$$T_1(r,\theta) = \sum_{n=1}^{\infty} A_n r^n \sin n\theta. \quad (5.135)$$

Here we note that the set of eigenfunctions $\sin n\theta$ satisfies the pair of opposite homogeneous boundary conditions $T_1 = 0$ at $\theta = 0$ and $\theta = \pi$, and the solution given by Equation (5.135) satisfies Equation (5.132). The remaining boundary condition is

$$\left.\frac{\partial T_1(r,\theta)}{\partial r}\right|_{r=R} = \frac{q_0}{k}. \tag{5.136}$$

The application of this boundary condition to the expression (5.135) for $T_1(r,\theta)$ leads to

$$\sum_{n=1}^{\infty} A_n n R^{n-1} \sin n\theta = \frac{q_0}{k}, \tag{5.137}$$

which is a straightforward Fourier sine series. The coefficients are found by orthogonality as

$$\begin{aligned} A_n n R^{n-1} &= \frac{2}{\pi}\int_0^{\pi} \frac{q_0}{k} \sin n\theta \, d\theta, \\ &= \frac{2q_0}{k}\frac{[1-(-1)^n]}{n\pi} \\ A_n &= \left(\frac{2q_0 R}{k}\right)\frac{[1-(-1)^n]}{n^2 \pi R^n} \end{aligned} \tag{5.138}$$

Using this expression for A_n in Equation (5.135) gives us

$$T_1(r,\theta) = \left(\frac{2q_0 R}{k}\right)\sum_{n=1}^{\infty}\frac{[1-(-1)^n]}{n^2\pi}\left(\frac{r}{R}\right)^n \sin n\theta. \tag{5.139}$$

For the time-dependent part, $T_2(r,\theta,t)$, the problem is quite similar to Example 5.7, except for the boundary condition at $r = R$, and the initial condition. Therefore, we can start with the solution form given by Equation (5.118) on page 225, i.e.,

$$T_2(r,\theta,t) = \sum_{n=1}^{\infty}\sum_{m=1}^{\infty} a_{mn} e^{-\alpha\lambda_{mn}^2 t} J_n(\lambda_{mn} r) \sin n\theta. \tag{5.140}$$

This solution satisfies the differential equation (5.133) and the homogeneous boundary conditions $T_2 = 0$ at $\theta = 0$ and $\theta = \pi$. The condition at $r = R$ is different from the previous example and therefore, the λ_{mn} values will be different from those given by Equation (5.111). The boundary condition is

$$\left.\frac{\partial T_2(r,\theta,t)}{\partial r}\right|_{r=R} = 0. \tag{5.141}$$

This, being a homogeneous condition, can be satisfied term-by-term with $J_n'(\lambda_{mn} R) = 0$. The roots of $J_n'(x) = 0$ are given by γ_{mn} in Table 3.2 on page 125. The values of λ_{mn} are determined as

$$\lambda_{mn} R = \gamma_{mn} \qquad \text{or} \qquad \lambda_{mn} = \frac{\gamma_{mn}}{R}. \tag{5.142}$$

The initial condition for the full problem is $T(r,\theta,0)=0$. Taking into consideration the decomposition given by Equation (5.131), we come up with

$$\begin{aligned} T_2(r,\theta,0) &= T(r,\theta,0)-T_1(r,\theta) \\ &= 0-T_1(r,\theta). \end{aligned} \tag{5.143}$$

Using the expressions (5.135) and (5.140) in the initial condition (5.143), we obtain

$$-\sum_{n=1}^{\infty} A_n r^n \sin n\theta = \sum_{n=1}^{\infty}\left[\sum_{m=1}^{\infty} a_{mn} J_n(\lambda_{mn} r)\right]\sin n\theta. \tag{5.144}$$

Equating the coefficients of $\sin n\theta$ term-by-term,

$$-A_n r^n = \left[\sum_{m=1}^{\infty} a_{mn} J_n(\lambda_{mn} r)\right]. \tag{5.145}$$

The right-hand side is a Fourier-Bessel series on which we may apply orthogonality to yield the coefficients,

$$\begin{aligned} a_{mn} &= \frac{\int_0^R -A_n r^n r J_n(\lambda_{mn} r)\,dr}{\int_0^R r\,[J_n(\lambda_{mn} r)]^2\,dr} \\ &= -A_n \frac{\int_0^R r^{n+1} J_n(\lambda_{mn} r)\,dr}{\int_0^R r\,[J_n(\lambda_{mn} r)]^2\,dr}. \end{aligned} \tag{5.146}$$

The integral in the numerator can be found from integral tables (see e.g., Gradshteyn & Ryzhik [2]) as

$$\int_0^R r^{n+1} J_n(\lambda_{mn} r)\,dr = \frac{R^{n+2} J_{n+1}(\lambda_{mn} R)}{\lambda_{mn} R}. \tag{5.147}$$

The denominator in Equation (5.146) is given by Equation (3.130) on page 124. For convenience, we repeat it here,

$$\int_0^R r[J_n(\lambda_{mn} r)]^2 dr = \tfrac{1}{2}R^2\left\{[J_n'(\lambda_{mn} R)]^2 + \left(1-\frac{n^2}{\lambda_{mn}^2 R^2}\right)[J_n(\lambda_{mn} R)]^2\right\}. \tag{5.148}$$

For the present example, $J_n'(\lambda_{mn} R)=0$, and the integral reduces to

$$\int_0^R r\,[J_n(\lambda_{mn} r)]^2\,dr = \tfrac{1}{2}R^2\left\{\left(1-\frac{n^2}{\lambda_{mn}^2 R^2}\right)[J_n(\lambda_{mn} R)]^2\right\}. \tag{5.149}$$

Using these two integrals in Equation (5.146), along with the expression for A_n given by Equation (5.138) we have for a_{mn},

$$\begin{aligned} a_{mn} &= -A_n \frac{\frac{R^{n+2} J_{n+1}(\lambda_{mn} R)}{\lambda_{mn} R}}{\frac{1}{2}R^2\left\{\left(1-\frac{n^2}{\lambda_{mn}^2 R^2}\right)[J_n(\lambda_{mn} R)]^2\right\}} \\ &= -\left(\frac{4q_0 R}{k}\right)\frac{[1-(-1)^n]}{n^2\pi}\frac{\lambda_{mn} R J_{n+1}(\lambda_{mn} R)}{\left(\lambda_{mn}^2 R^2-n^2\right)[J_n(\lambda_{mn} R)]^2}. \end{aligned} \tag{5.150}$$

With the coefficients a_{mn} completely determined, we can substitute the expression (5.150) into Equation (5.140) and together with the expression (5.139) for $T_1(r,\theta)$, and obtain the explicit expression

$$\begin{aligned} T(r,\theta,t) &= T_1(r,\theta) + T_2(r,\theta,t) \\ &= \left(\frac{2q_0 R}{k}\right) \sum_{n=1}^{\infty} \frac{[1-(-1)^n]}{n^2\pi} \left\{ \left(\frac{r}{R}\right)^n \right. \\ & \left. -2\sum_{m=1}^{\infty} \frac{\lambda_{mn} R J_{n+1}(\lambda_{mn}R)}{(\lambda_{mn}^2 R^2 - n^2)\,[J_n(\lambda_{mn}R)]^2} e^{-\alpha\lambda_{mn}^2 t} J_n(\lambda_{mn} r) \right\} \sin n\theta. \end{aligned} \tag{5.151}$$

□

EXERCISES 5.3

Problem 1

For a half-cylinder of length l and radius R, solve

$$\nabla^2 T(r,\theta,z) = 0, \qquad 0 \le r \le R,\ 0 \le \theta \le \pi,\ 0 \le z \le l.$$

with boundary conditions

$$-k\left.\frac{\partial T}{\partial z}\right|_{z=0} = q_0, \quad T(r,\theta,l) = 0.$$

$$T(R,\theta,z) = 0, \quad T(r \to 0,\theta,z) < \infty.$$

Problem 2

For a semi-infinite cylinder of radius R, solve

$$\nabla^2 T(r,\theta,z) = 0, \qquad 0 \le r \le R,\ 0 \le \theta \le 2\pi,\ 0 \le z < \infty.$$

The boundary $z = 0$ is subjected to a uniform heat flux q_0 over half of that end. The lateral surface $r = R$ is kept at $T = 0$.

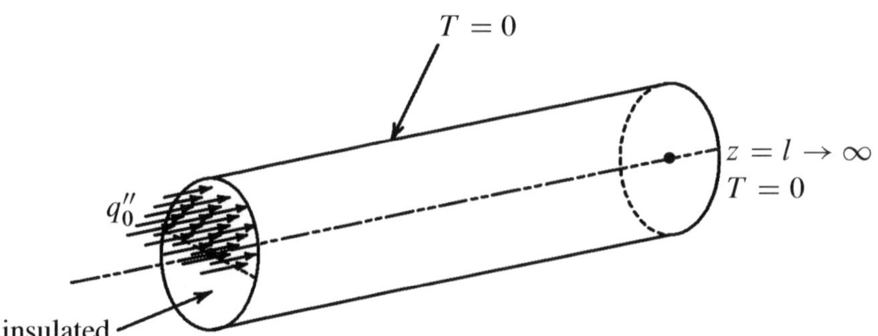

Chapter 6

Problems in Spherical Geometry

The spherical geometry arises in many applications such as the fluid dynamics of spray droplets, and dropwise condensation. Boiling phenomenon requires an understanding of the physics of bubbles which, for sufficiently small sizes, take on spherical shapes. We often model powder particles as approximate spheres. Again, in order to solve partial differential equations for systems with spherical boundaries, we adopt the spherical coordinate system. The transformation from cartesian to spherical is shown schematically in Figure 6.1. Specifically, r is the radial distance of a point, (x, y, z) from the origin. Let us define the position vector $\boldsymbol{r} =< x, y, z >$. The angle that $\boldsymbol{r}$ makes with the z-axis is defined as θ. The angle between the x-z plane and the plane formed by the vectors $\boldsymbol{r}$ and $\hat{z}$ is defined as ϕ. Following the notation in Figure 6.1, we can derive the following relationships between (x, y, z) and (r, θ, ϕ) coordinates:

$$\begin{aligned} z &= r\cos\phi \\ x &= \rho\cos\theta = r\sin\theta\cos\phi \\ y &= \rho\sin\theta = r\sin\theta\sin\phi \end{aligned}$$

where $\rho = r\sin\theta$ represents the distance of (x, y, z) from the z-axis. In the full space, the range of the spherical coordinates $(r; \theta; \phi)$is

$$0 \leq r < \infty,$$
$$0 \leq \theta \leq \pi,$$
$$0 \leq \phi \leq 2\pi.$$

We often encounter the Laplacian operator in many problems of interest. Going from cartesian to spherical coordinates, this may be written as

$$\begin{aligned} \nabla^2 &= \frac{\partial^2}{\partial x^2} + \frac{\partial^2}{\partial y^2} + \frac{\partial^2}{\partial z^2} \\ &= \frac{\partial^2}{\partial r^2} + \frac{2}{r}\frac{\partial}{\partial r} + \frac{1}{r^2\sin\theta}\frac{\partial}{\partial\theta}\left(\sin\theta\frac{\partial}{\partial\theta}\right) + \frac{1}{r^2\sin^2\theta}\frac{\partial^2}{\partial\phi^2}. \end{aligned} \tag{6.1}$$

Let us now consider an example in spherical geometry. We shall begin with cases that require eigenfunction expansions in θ.

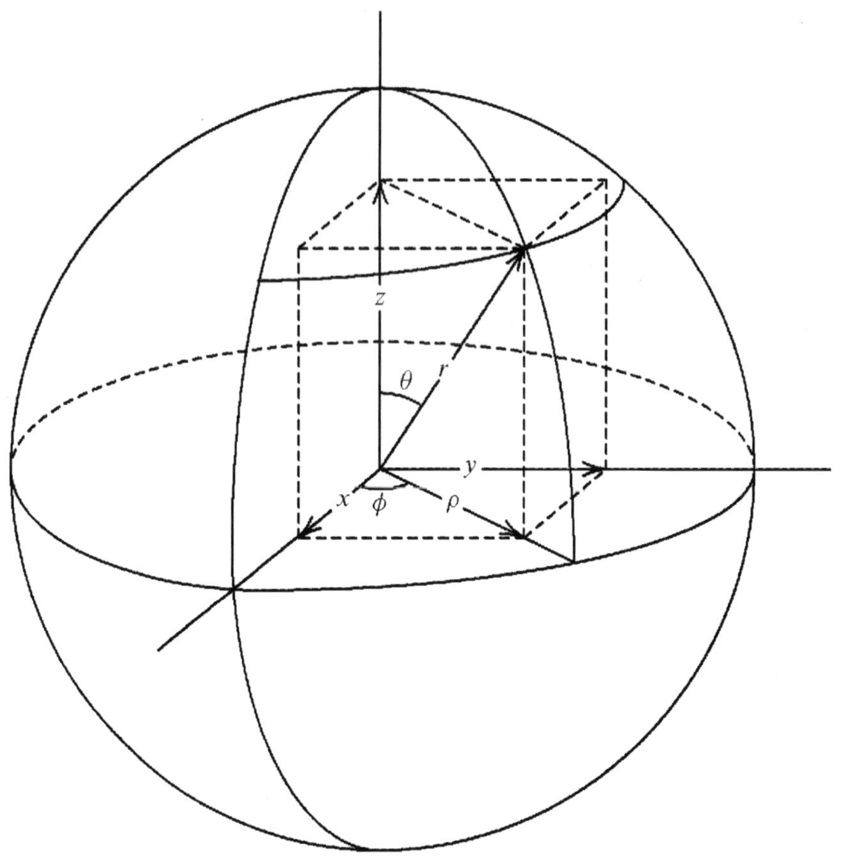

Figure 6.1: Graphical depiction of the spherical coordinate system.

6.1 Eigenfunction Expansions in θ

Example 6.1

PROBLEM STATEMENT: Obtain the electrostatic potential $V(r,\theta)$ inside a sphere. The potential is represented by the solution of Laplaces equation,

$$\nabla^2 V(r,\theta) = 0.$$

On the boundary $r = R$, the potential is given by $V(R,\theta) = f(\theta)$, where

$$f(\theta) = \begin{cases} V_0 \cos\theta, & 0 \le \theta \le \frac{1}{2}\pi \\ 0, & \frac{1}{2}\pi \le \theta \le \pi. \end{cases} \tag{6.2}$$

SOLUTION: In terms of r and θ, Laplaces equation for $V(r, \theta)$ is

$$\frac{\partial^2 V}{\partial r^2}+\frac{2}{r}\frac{\partial V}{\partial r}+\frac{1}{r^2 \sin\theta}\frac{\partial}{\partial\theta}\left(\sin\theta\frac{\partial V}{\partial\theta}\right)=0, \tag{6.3}$$

where we have dropped the ϕ dependence in the Laplacian operator given by Equation (6.1). For the present problem, in order to satisfy the boundary condition $V(R, \theta) = f(\theta)$, we would need an eigenfunction expansion in θ. Let us multiply Equation (6.3) by r^2 in order to isolate the θ operations. This results in

$$\left(r^2\frac{\partial^2 V}{\partial r^2}+2r\frac{\partial V}{\partial r}\right)+\frac{1}{\sin\theta}\frac{\partial}{\partial\theta}\left(\sin\theta\frac{\partial V}{\partial\theta}\right) \tag{6.4}$$

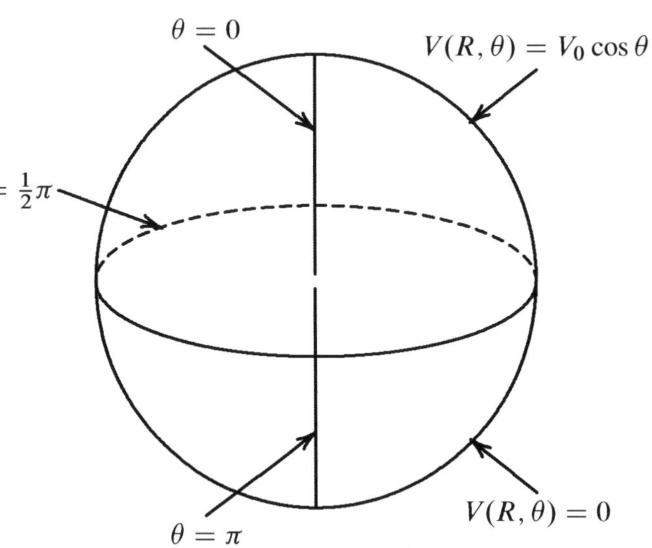

For expansion in θ, we need eigenfunctions of the operator

$$\frac{1}{\sin\theta}\frac{\partial}{\partial\theta}\left(\sin\theta\frac{\partial}{\partial\theta}\right), \tag{6.5}$$

i.e we need to solve the ordinary differential equation,

$$\frac{1}{\sin\theta}\frac{d}{d\theta}\left(\sin\theta\frac{d\Theta_n(\theta)}{d\theta}\right)=-\lambda_n^2\Theta_n(\theta). \tag{6.6}$$

This equation reduces to one of the standard equations with the change of variable $\mu = \cos\theta$. Let us implement this change in the sequence of derivative operations

$$\begin{aligned}
\frac{d\Theta_n}{d\theta} &= \frac{d\Theta_n}{d\mu}\frac{d\mu}{d\theta} \\
&= \frac{d\Theta_n}{d\mu}(-\sin\theta) \\
&= -\sin\theta\frac{d\Theta_n}{d\mu} \\
\left(\sin\theta\frac{d\Theta_n}{d\theta}\right) &= -\sin^2\theta\frac{d\Theta_n}{d\mu} \\
&= -(1-\mu^2)\frac{d\Theta_n}{d\mu} \\
\frac{d}{d\theta}\left(\sin\theta\frac{d\Theta_n}{d\theta}\right) &= -\frac{d}{d\theta}\left[(1-\mu^2)\frac{d\theta_n}{d\mu}\right] \\
&= -\frac{d}{d\mu}\left[(1-\mu^2)\frac{d\theta_n}{d\mu}\right]\frac{d\mu}{d\theta} \\
&= -\frac{d}{d\mu}\left[(1-\mu^2)\frac{d\theta_n}{d\mu}\right]-\sin\theta
\end{aligned}$$

$$
\begin{aligned}
&= -\frac{d}{d\mu}\left[(1-\mu^2)\frac{d\theta_n}{d\mu}\right]\frac{d\mu}{d\theta} \\
&= \sin\theta\frac{d}{d\mu}\left[(1-\mu^2)\frac{d\Theta_n}{d\mu}\right] \\
\frac{1}{\sin\theta}\frac{d}{d\theta}\left(\sin\theta\frac{d\Theta_n}{d\theta}\right) &= \frac{d}{d\mu}\left[(1-\mu^2)\frac{d\Theta_n}{d\mu}\right]
\end{aligned}
$$

Here, it is easy to see that the differential operator (6.5) may be expressed as

$$
\frac{1}{\sin\theta}\frac{d}{d\theta}\left(\sin\theta\frac{d}{d\theta}\right) = \frac{d}{d\mu}\left[(1-\mu^2)\frac{d}{d\mu}\right] \tag{6.7}
$$

The differential equation (6.6) for $\Theta_n(\theta)$ therefore becomes

$$
\frac{d}{d\mu}\left[(1-\mu^2)\frac{d\Theta_n(\mu)}{d\mu}\right] + \lambda_n^2\Theta_n(\mu) = 0, \tag{6.8}
$$

which may also be written as

$$
(1-\mu^2)\frac{d^2\Theta_n}{d\mu^2} - 2\mu\frac{d\Theta_n}{d\mu} + \lambda_n^2\Theta_n(\mu) = 0, \tag{6.9}
$$

where we have changed the argument of Θ from θ to $\cos\theta = \mu$. The differential equation given by Equation (6.4) becomes

$$
r^2\frac{\partial^2 V}{\partial r^2} + 2r\frac{\partial V}{\partial r} + \frac{\partial}{\partial\mu}\left[(1-\mu^2)\frac{\partial V}{\partial\mu}\right] = 0. \tag{6.10}
$$

Let us start using μ as a coordinate, with $-1 \leq \mu \leq 1$. The value $\mu = 1$ corresponds to the positive z axis and $\mu - 1$ defines the negative z axis. The equatorial plane is defined by $\mu = 0$. Equation (6.9) may be solved by the power series method (see Chapter 2). When λ_n^2 takes on integer values according to $\lambda_n^2 = n(n+1)$, one of the power-series solutions for each n value truncates to a polynomial of order n. These polynomials are called Legendre polynomials, and have been discussed in Section 2.4.1 of Chapter 2 on pages 58-62. For problems involving full spheres, the λ_n^2 values will remain restricted to $n(n+1)$. This is somewhat analogous to the periodicity condition for cylinders which leads to $\lambda_n = n$ for $\sin\lambda_n\theta$ and/or $\cos\lambda_n\theta$ (see Example 5.3 on pages 197-201 in Chapter 5). Also, for hemisphere problems with homogeneous boundary conditions of the first or second kind on base plane ($\theta = \frac{1}{2}\pi$ or $\mu = 0$), this type of restriction applies with further limitation to odd or even n. We shall therefore stay restricted to the the following form of Legendres equation:

$$
(1-\mu^2)\frac{d^2\Theta_n}{d\mu^2} - 2\mu\frac{d\Theta_n}{d\mu} + n(n+1)\Theta_n = 0. \tag{6.11}
$$

The general solution consisting of two linearly independent functions is

$$
\Theta_n(\mu) = A_n P_n(\mu) + B_n Q_n(\mu), \tag{6.12}
$$

As mentioned above, the detailed discussion on these functions is given in Section 2.4.1 on pages 58-62. Specific expressions for for $P_n(\mu)$ and $Q_n(\mu)$ are given by Equations (2.53) and (2.54) on page 61. The set $\{P_n(\mu),\ n = 0, 1, 2, 3, \ldots\}$ is orthogonal in the range $-1 \leq \mu \leq 1$, and one can expand a function $f(\mu)$ as a Fourier-Legendre series. Please see Section 3.7.5, pages 134-142 in Chapter 3 for the discussion on the orthogonality of Legendre polynomials. The orthogonality relationship,

$$\int_{-1}^{1} P_n(\mu) P_m(\mu) d\mu = \begin{cases} 0, & m \neq n, \\ \frac{2}{2n+1}, & m = n, \end{cases}$$

is repeated here for convenience. It should be noted here that the weighting factor for this orthogonality relationship is unity when using μ as the variable in the argument of P_n. Let us now examine the solution for $\Theta_n(\theta)$ given by Equation (6.12). As mentioned earlier, $P_n(\theta)$ represents polynomials of order n, and the values are finite in the range $-1 \leq \mu \leq 1$. The solution $Q_n(\mu)$, however, has logarithmic singularities at $\mu = \pm 1$ $(\theta = 0, \pi)$. Specifically, $Q_n(\mu)$ expressions contain the term

$$\ln\left(\frac{1+\mu}{1-\mu}\right).$$

which goes to infinity as $\mu \to \pm 1$ (see Equation (2.54) on page 61). The appropriate eigenfunctions in μ for this problem are therefore

$$\Theta_n(\mu) = P_n(\mu), \quad n = 0, 1, 2, 3, \ldots. \tag{6.13}$$

We can now go ahead and expand $V(r, \mu)$ as

$$V(r, \mu) = \sum_{n=0}^{\infty} V_n(r) P_n(\mu). \tag{6.14}$$

Let us now substitute this expansion into the differential equation (6.10). This results in

$$\sum_{n=0}^{\infty} V_n''(r) P_n(\mu) + \frac{2}{r} \sum_{n=0}^{\infty} V_n'(r) P_n(\mu) + \frac{1}{r^2} \sum_{n=0}^{\infty} V_n(r) \frac{d}{d\mu}\left[(1-\mu^2)\frac{dP_n(\mu)}{d\mu}\right] = 0. \tag{6.15}$$

Noting that $P_n(\mu)$ satisfies Equation(6.11), or for that matter,

$$\frac{d}{d\mu}\left[(1-\mu^2)\frac{dP_n(\mu)}{d\mu}\right] = -n(n+1)P_n(\mu), \tag{6.16}$$

Equation (6.15) reduces to

$$\sum_{n=0}^{\infty} V_n''(r) P_n(\mu) + \frac{2}{r} \sum_{n=0}^{\infty} V_n'(r) P_n(\mu) + \frac{1}{r^2} \sum_{n=0}^{\infty} V_n(r) \left[-n(n+1)P_n(\mu)\right] = 0. \tag{6.17}$$

We see that the eigenfunction $P_n(\mu)$ is preserved, and taking it as a common factor in the summations, we may write Equation (6.17) as

$$\sum_{n=0}^{\infty} \left[V_n''(r) + \frac{2}{r} V_n'(r) - \frac{n(n+1)}{r^2} V_n(r)\right] P_n(\mu) = 0, \tag{6.18}$$

from where it is easy to see that the ordinary differential equation for $V_n(r)$ is

$$\left[V_n''(r) + \frac{2}{r}V_n'(r) - \frac{n(n+1)}{r^2}V_n(r)\right] = 0. \tag{6.19}$$

Upon multiplying by r^2, we obtain

$$r^2V_n''(r) + rV_n'(r) - n(n+1)V_n(r) = 0. \tag{6.20}$$

This is the Cauchy-Euler equation, and we expect solutions of the for $V_n(r) = r^k$. Assuming this form leads to the characteristic equation,

$$\begin{aligned} k(k-1) + 2k - n(n+1) &= 0, \\ k^2 + k - n(n+1) &= 0. \end{aligned}$$

This quadratic equation has solutions

$$k = n, -(n+1),$$

and thus, we may write the general solution for Equation (6.20) as

$$V_n(r) = A_n r^n + B_n r^{-(n+1)}. \tag{6.21}$$

Since $r = 0$ is included in the problem domain, and since $r^{-(n+1)} \to \infty$ as $r \to 0$, we set $B_n = 0$, leaving us with

$$V_n(r) = A_n r^n. \tag{6.22}$$

The expansion for $V(r, \mu)$ given by Equation (6.14) becomes

$$V(r, \mu) = \sum_{n=0}^{\infty} A_n r^n P_n(\mu). \tag{6.23}$$

The set of coefficients A_n can be found by satisfying the boundary condition (6.2) which may be written in terms of μ as

$$V(R, \mu) = F(\mu) = \begin{cases} 0, & -1 \le \mu \le 0 \\ V_0\mu, & 0 \le \mu \le 1 \end{cases} \tag{6.24}$$

Implementing this boundary condition in the expansion (6.18) leads to the Fourier-Legendre series

$$V(R, \mu) = F(\mu) = \sum_{n=0}^{\infty} A_n R^n P_n(\mu), \tag{6.25}$$

and the application of the orthogonality of the set $\{P_n(\mu), \quad n = 0, 1, 2, 3, \ldots\}$ gives us (see the development in Equations (3.179)-(3.185) on pages 134-135 in Chapter 3)

$$\begin{aligned} A_n R^n &= \left(n + \tfrac{1}{2}\right) \int_{-1}^{1} F(\mu) P_n(\mu) d\mu \\ &= \left(n + \tfrac{1}{2}\right) \left[\int_{-1}^{0} 0 \ P_n(\mu) d\mu + \int_{0}^{1} V_0 \mu P_n(\mu) \, d\mu\right] \\ &= \left(n + \tfrac{1}{2}\right) V_0 \int_{0}^{1} \mu P_n(\mu) \, d\mu. \end{aligned} \tag{6.26}$$

To evaluate the integral here, let us examine the differential equation for $P_n(\mu)$ by writing Equation (6.16) as

$$\begin{aligned}
\left[(1-\mu^2)P_n'(\mu)\right]' &= -n(n+1)P_n(\mu), \\
P_n(\mu) &= -\frac{\left[(1-\mu^2)P_n'(\mu)\right]'}{n(n+1)}
\end{aligned} \tag{6.27}$$

With the use of this expression, the integral on the right-hand side of Equation (6.26) may be evaluated as follows

$$\begin{aligned}
I_n = \int_0^1 \mu P_n(\mu)\, d\mu &= -\frac{1}{n(n+1)} \int_0^1 \mu[(1-\mu^2)P_n'(\mu)]' d\mu, n \neq 0 \\
&= -\frac{1}{n(n+1)} \left[\mu[(1-\mu^2)P_n'(\mu)]|_0^1 - \int_0^1 (1-\mu^2)P_n'(\mu) d\mu\right] \\
&= -\frac{1}{n(n+1)} \left[0 - (1-\mu^2)P_n(\mu)|_0^1 - \int_0^1 2\mu P_n(\mu) d\mu\right] \\
I_n &= -\frac{1}{n(n+1)}[P_n(0)] + \frac{2}{n(n+1)} I_n \\
I_n\left(1 - \frac{2}{n(n+1)}\right) &= -\frac{P_n(0)}{n(n+1)} \\
I_n \frac{n^2+n-2}{n(n+1)} &= -\frac{P_n(0)}{n(n+1)} \\
I_n(n+2)(n-1) &= -P_n(0) \\
I_n &= -\frac{P_n(0)}{(n+2)(n-1)}, \quad n \neq 1 \\
I_0 &= \int_0^1 \mu \cdot 1\, d\mu, \quad (\text{using } P_0(\mu) = 1) \\
&= \tfrac{1}{2}\mu^2\big|_0^1 = \\
I_1 &= \int_0^1 \mu P_1(\mu) d\mu \\
&= \int_0^1 \mu^2 d\mu, \quad (\text{using } P_1(\mu) = \mu) \\
&= \tfrac{1}{3}\mu^3\big|_0^1 = \tfrac{1}{3}
\end{aligned}$$

Using this set of expressions for I_n in Equation (6.26), we obtain

$$\begin{aligned}
A_n R^n &= \left(n + \tfrac{1}{2}\right) V_0 I_n \\
&= \left(n + \tfrac{1}{2}\right) V_0 \left[\frac{-P_n(0)}{(n+2)(n-1)}\right], \quad n \neq 1 \\
A_1 R^1 &= \tfrac{3}{2} \cdot \tfrac{1}{3} V_0 = \tfrac{1}{2} V_0 \\
A_0 R^0 &= A_0 = \tfrac{1}{2} I_0 V_0 = \tfrac{1}{4} V_0 \\
V(r,\mu) &= V_0 \left[\tfrac{1}{4} + \tfrac{1}{2}\left(\frac{r}{R}\right)\mu - \sum_{n=2}^{\infty} \frac{P_n(0)}{(n+2)(n-1)} \left(\frac{r}{R}\right)^n P_n(\mu)\right].
\end{aligned} \tag{6.28}$$

At specific μ values $P_n(\mu)$ can be calculated using the recurrence relationships.

- $P'_{n+1}(\mu) - P'_{n-1}(\mu) = (2n+1)P_n(\mu)$
- $(n+1)P_{n+1}(\mu) + nP_{n-1}(\mu) = (2n+1)\mu P_n(\mu)$

where $P_0(\mu) = 1$ and $P_1(\mu) = \mu$ may be used for startup.

□

In the next example, we shall obtain the electrostatic potential around a grounded sphere with a point charge placed near it.

Example 6.2

PROBLEM STATEMENT: Obtain the electrostatic potential due to a point charge of strength q_0 outside a grounded sphere of radius R. The point charge is s distance c from the center of the sphere. We can choose the line joining the center of the sphere and the point charge as the z-axis, and this represents an axis of symmetry. The potential therefore depends only on r and θ is represented by the solution of Laplaces equation,

$$\nabla^2 V(r,\theta) = 0.$$

On the boundary $r = R$, the potential is given by $V(R,\theta) = 0$.

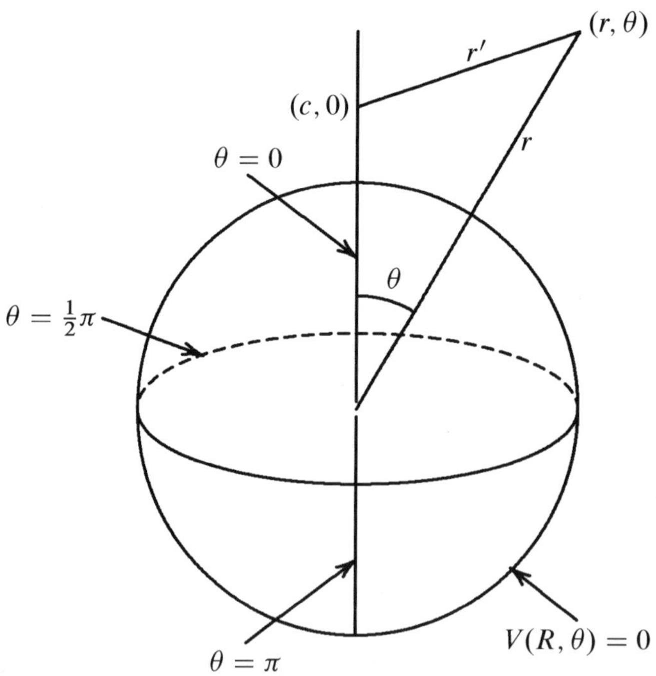

SOLUTION: As in the previous example, in terms of r and θ, Laplaces equation for $V(r,\theta)$ is

$$\frac{\partial^2 V}{\partial r^2} + \frac{2}{r}\frac{\partial V}{\partial r} + \frac{1}{r^2 \sin\theta}\frac{\partial}{\partial\theta}\left(\sin\theta\frac{\partial V}{\partial\theta}\right) = 0, \tag{6.29}$$

Let us consider a field point (r, θ) a distance r' from the point charge. For an isolated charge, without the presence of the sphere, the electrostatic potential takes the spherically symmetric form,

$$V_0(r, \theta) = \frac{q_0}{r'}, \tag{6.30}$$

where $r' = \sqrt{r^2 + c^2 - 2rc\cos\theta}$ is the radial distance from the point charge. With the presence of a grounded sphere, this potential is disrupted. Let us take the center of the coordinate system to coincide with the center of the sphere. We can consider the potential $V(r, \theta)$ to be given as a sum of undisturbed potential (6.30) and a 'correction' $V_1(r, \theta)$ due to the presence of the sphere, i.e.,

$$V(r, \theta) = \frac{q_0}{\sqrt{r^2 + c^2 - 2rc\cos\theta}} + V_1(r, \theta), \tag{6.31}$$

The 'correction potential' $V_1(r, \theta)$ satisfies the same differential equation as $V(r, \theta)$, i.e.,

$$\frac{\partial^2 V_1}{\partial r^2} + \frac{2}{r}\frac{\partial V_1}{\partial r} + \frac{1}{r^2 \sin\theta}\frac{\partial}{\partial\theta}\left(\sin\theta \frac{\partial V_1}{\partial\theta}\right) = 0, \tag{6.32}$$

With the requirement on the grounded sphere that $V(R, \theta) = 0$, we obtain from Equation (6.31)

$$V_1(R, \theta) = -\frac{q_0}{\sqrt{R^2 + c^2 - 2Rc\cos\theta}}. \tag{6.33}$$

In addition, we require that the total potential $V(r, \theta)$ vanish as $r \to \infty$. Again, from Equation (6.31), it is not difficult to see that

$$V_1(r, \theta) \to 0 \quad \text{as} \quad r \to \infty. \tag{6.34}$$

For the present problem, in order to satisfy the boundary condition (6.33), we would need an eigenfunction expansion in θ. As in the previous example, let us multiply Equation (6.32) by r^2 in order to isolate the θ operations. This results in

$$\left(r^2 \frac{\partial^2 V_1}{\partial r^2} + 2r\frac{\partial V_1}{\partial r}\right) + \frac{1}{\sin\theta}\frac{\partial}{\partial\theta}\left(\sin\theta \frac{\partial V_1}{\partial\theta}\right). \tag{6.35}$$

Following the development in Example 6.1, Equations (6.4)-(6.21) on pages 233-236, we can write the solution to Equation (6.35) as

$$V_1(r, \mu) = \sum_{n=0}^{\infty} V_{1n}(r) P_n(\mu), \tag{6.36}$$

where $\mu = \cos\theta$, and

$$V_{1n}(r) = A_n r^n + B_n r^{-(n+1)}. \tag{6.37}$$

Since $r = 0$ is not included in the problem domain, and since $r^n \to \infty$ as $r \to 0$, we set $A_n = 0$, leaving us with

$$V_{1n}(r) = B_n r^{-(n+1)}. \tag{6.38}$$

The expansion for $V_1(r,\mu)$ given by Equation (6.36) becomes

$$V_1(r,\mu) = \sum_{n=0}^{\infty} B_n r^{-(n+1)} P_n(\mu). \tag{6.39}$$

The set of coefficients B_n can be found by satisfying the boundary condition (6.33) which may be written in terms of μ as

$$V_1(R,\mu) = \sum_{n=0}^{\infty} B_n R^{-(n+1)} P_n(\mu) = -\frac{q_0}{\sqrt{R^2 + c^2 - 2Rc\mu}} \tag{6.40}$$

Here we can express the right-hand side as a Legendre series using the generating function given by Equation (3.216) on page 142 which is repeated here,

$$\frac{1}{\sqrt{1 - 2xt + t^2}} = \sum_{n=0}^{\infty} t^n P_n(x). \tag{6.41}$$

Setting $t = R/c$ along with $x = \mu$, and multiplying both sides by q_0, we obtain

$$\frac{q_0 c}{\sqrt{c^2 - 2cR\mu + R^2}} = q_0 \sum_{n=0}^{\infty} \left(\frac{R}{c}\right)^n P_n(\mu), \tag{6.42}$$

leading to

$$\frac{-q_0}{\sqrt{c^2 - 2cR\mu + R^2}} = -\frac{q_0}{c} \sum_{n=0}^{\infty} \left(\frac{R}{c}\right)^n P_n(\mu), \tag{6.43}$$

Upon using this result in Equation (6.40), we obtain

$$\sum_{n=0}^{\infty} B_n R^{-(n+1)} P_n(\mu) = -\frac{q_0}{c} \sum_{n=0}^{\infty} \left(\frac{R}{c}\right)^n P_n(\mu). \tag{6.44}$$

Matching each side term-by-term yields

$$B_n R^{-(n+1)} = -\frac{q_0}{c} \left(\frac{R}{c}\right)^n. \tag{6.45}$$

With this expression for B_n, we can write the potential $V_1(r,\mu)$ in Equation (6.39) as

$$V_1(r,\mu) = -\frac{q_0}{c} \sum_{n=0}^{\infty} \left(\frac{R}{r}\right)^{n+1} \left(\frac{R}{c}\right)^n P_n(\mu). \tag{6.46}$$

This may now be used in Equation (6.31) to give the total potential as

$$V(r,\mu) = \frac{q_0}{c} \left[\frac{c}{\sqrt{r^2 + c^2 - 2rc\mu}} - \sum_{n=0}^{\infty} \left(\frac{R}{r}\right)^{n+1} \left(\frac{R}{c}\right)^n P_n(\mu) \right], \tag{6.47}$$

In this particular case, the Legendre series for $V_1(r, \mu)$ can expressed in compact form using the generating function (3.216). We can write the right-hand side of Equation (6.46) as follows

$$V_1(r, \mu) = -\frac{q_0}{c}\left(\frac{R}{r}\right)\sum_{n=0}^{\infty}\left(\frac{R^2}{rc}\right)^n P_n(\mu), \tag{6.48}$$

and identify $t = \left(R^2/rc\right)$ in Equation (3.216) which gives us

$$\begin{aligned} V_1(r, \mu) &= -\frac{q_0}{c}\left(\frac{R}{r}\right)\sum_{n=0}^{\infty}\left(\frac{R^2}{rc}\right)^n P_n(\mu), \\ &= -\frac{q_0}{c}\left(\frac{R}{r}\right)\frac{1}{\sqrt{1 + \left(\frac{R^2}{rc}\right)^2 - 2\mu\left(\frac{R^2}{rc}\right)}} \\ &= -\frac{q_0 R}{\sqrt{(rc)^2 + R^4 - 2\mu R^2 rc}}, \end{aligned} \tag{6.49}$$

whereupon Equation (6.50) becomes

$$V(r, \mu) = q_0\left[\frac{1}{\sqrt{r^2 + c^2 - 2rc\mu}} - \frac{1}{\sqrt{(rc/R)^2 + R^2 - 2rc\mu}}\right]. \tag{6.50}$$

□

We just did some examples where the only physical boundary was at the surface of the sphere ($r = R$). The Fourier-Legendre series expansion method also works hemispheres where the equatorial plane ($\mu = 0$ or $\theta = \frac{1}{2}\pi$) defines the base boundary, provided the boundary condition on the base is homogeneous first or the second kind. Certain cases with some special types of non-homogeneous boundary conditions can also be treated. Next, we consider an example for Laplaces equation in a hemispherical region.

Example 6.3

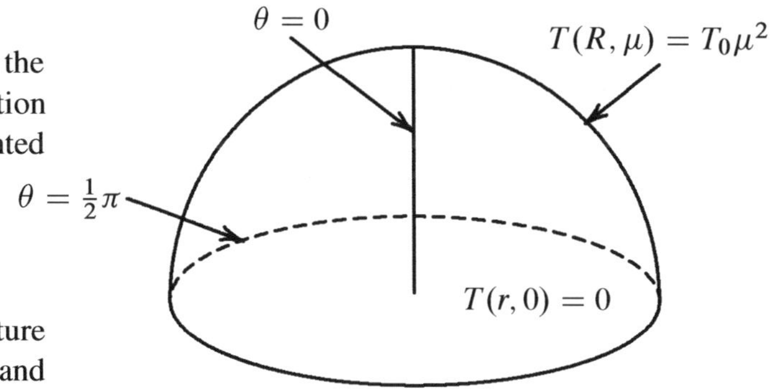

PROBLEM STATEMENT: Obtain the steady-state temperature distribution $T(r, \mu)$ inside a hemisphere, represented by the solution of Laplaces equation,

$$\nabla^2 T(r, \mu) = 0.$$

On the boundary $r = R$, the temperature is given by $T(R, \mu) = F(\mu) = T_0\mu^2$, and at the base defined by $\mu = 0$, the we have $T(r, 0) = 0$. On the axis, due to the symmetry, we require $\partial T/\theta = 0$ (or equivalently, $\partial T/\mu = 0$).

$\theta = 0$

$T(R, \mu) = T_0\mu^2$

$\theta = \frac{1}{2}\pi$

$T(r, 0) = 0$

SOLUTION: The partial differential equation for this problem is the same as the one for the previous examples. In terms of r and θ, Equations (6.3) and (6.4) apply. Upon making the change of variable $\mu = \cos\theta$, these take the form given by Equation (6.10) which is repeated here with a change of the dependent variable to $T(r, \mu)$,

$$r^2 \frac{\partial^2 T}{\partial r^2} + 2r \frac{\partial T}{\partial r} + \frac{\partial}{\partial \mu}\left[(1-\mu^2)\frac{\partial T}{\partial \mu}\right] = 0. \tag{6.51}$$

In rectangularized domain, as shown in Figure 6.2, we can see that with a nonzero condition at $r = R$, we need an eigenfunction expansion in μ to satisfy it. Examining

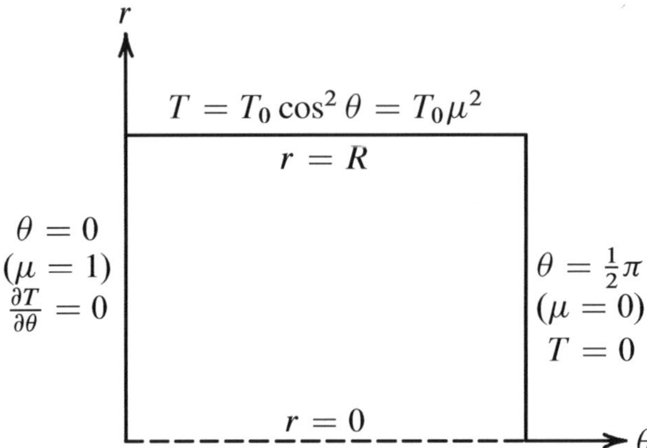

Figure 6.2: Illustration of boundary conditions for Example 6.3 in 'rectangularized' coordinates. As in some previous cases (such as Example 5.1, pages 192-195, the point $r = 0$ stretches to a line.

the differential equation (6.51), and following the previous example, and we see that we need to set up eigenfunctions of the operator (6.7), i.e., we need to solve the the ordinary differential equation (6.11). This equation has the relevant solution given by Equation (6.13) but we need to further restrict the eigenvalues selection so that the boundary condition $T(r, 0) = 0$ is satisfied. To do so, let us examine the symmetry properties of $P_n(\mu)$. From the expressions of Legendre polynomials given by Equation (2.53) on page 61 in Chapter 2, we see that $P_0(\mu)$, $P_2(\mu)$, $P_4(\mu)$,... are even functions and $P_n'(0) = 0$ for $n = 0, 2, 4, \ldots$. We also notice that $P_1(\mu)$, $P_3(\mu)$, $P_5(\mu)$,... are odd functions and $P_n(0) = 0$ for $n = 1, 3, 5, \ldots$. Therefore, if we selected

$$\Theta_n(\mu) = P_n(\mu), \quad n = 1, 3, 5, \ldots, \tag{6.52}$$

as a set of eigenfunctions for expansion of $T(r, \mu)$ in the form

$$T(r, \mu) = \sum_{n=1,3,5,\ldots}^{\infty} T_n(r) P_n(\mu), \tag{6.53}$$

the boundary condition $T(r, 0) = 0$ at the base is satisfied. Let us now substitute this into the differential equation (6.51). We now follow the procedure in the previous example, starting with Equation (6.14), substitution into (6.10), leading to (6.18) and resulting in (6.20). Thus, it is not difficult to see that the ordinary differential equation for $T_n(r)$ is

$$\frac{d^2T_n}{dr^2} + \frac{2}{r}\frac{dT_n}{dr} - \frac{n(n+1)}{r^2}T_n = 0$$

This is the Cauchy-Euler equation with the general solution the same as Equation (6.20), i.e.,

$$T_n(r) = A_n r^n + B_n r^{-(n+1)}. \tag{6.54}$$

Again, we see that $r^{-(n+1)} \to \infty$ as $r \to 0$. Since $r = 0$ is included in the problem domain, we need to exclude the singular solution. We therefore set $B_n = 0$, and end up with

$$T_n(r) = A_n r^n,$$

which, when used in Equation (6.53), leads to

$$T(r, \mu) = \sum_{n=1,3,5,\ldots}^{\infty} A_n r^n P_n(\mu) \tag{6.55}$$

where $\{P_n(\mu), ;\ n = 1, 3, 5, \ldots\}$ is an orthogonal set in the range $0 \le \mu \le 1$. At $r = R$,

$$T(R, \mu) = T_0\mu^2 = \sum_{n=1,3,5}^{\infty} A_n R^n P_n(\mu)$$

By orthogonality,

$$A_n R^n = \frac{\int_0^1 T_0\mu^2 P_n(\mu) d\mu}{\int_0^1 [P_n(\mu)]^2 d\mu}.$$

Since

$$\int_{-1}^{1} [P_n(\mu)]^2 d\mu = \frac{2}{(2n+1)},$$

it is not difficult to see that

$$\int_{0}^{1} [P_n(\mu)]^2 d\mu = \frac{1}{(2n+1)}.$$

Following Example 3.16, on pages 136-138, the expression given by Equation (3.192) is applicable. This expression is not valid for $n = 0, 2$ but that does not concern us since we are taking only the odd values of n. Thus,

$$\begin{aligned} A_n R^n &= (2n+1)\int_0^1 T_0\mu^2 P_n(\mu) d\mu \\ &= -T_0 \frac{2(2n+1)P_n'(0)}{n(n+1)(n+3)(n-2)}, \quad n = 1, 2, 3, \ldots && (6.56) \\ A_n &= -2T_0 \frac{(2n+1)P_n'(0)}{n(n+1)(n+3)(n-2)R^n}, \quad n = 1, 2, 3, \ldots && (6.57) \end{aligned}$$

Using this result in Equation (6.55), we obtain the final form of the solution as

$$T(r, \mu) = -2T_0 \sum_{n=1,3,5,\ldots}^{\infty} \frac{(2n+1)P_n'(0)}{n(n+1)(n+3)(n-2)} \left(\frac{r}{R}\right)^n P_n(\mu) \tag{6.58}$$

□

Up to now, we have considered examples where center of the sphere or the hemisphere is included in the problem domain. With spherical annuli, $r = 0$ is excluded, and terms such as $r^{-(n+1)}$ are not singular, and may be admitted in the solution. Let us consider an example dealing with a hemispherical annulus.

Example 6.4

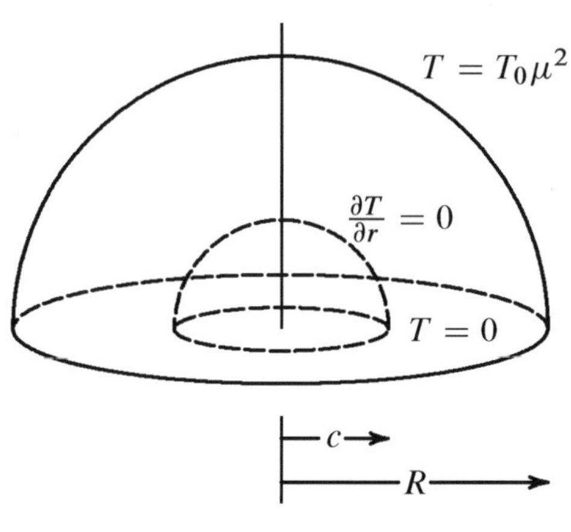

PROBLEM STATEMENT: Calculate the steady-state temperature distribution $T(r, \mu)$ inside a hemispherical annulus, by solving Laplaces equation,

$$\nabla^2 T(r, \mu) = 0.$$

The boundary conditions are as follows

$$T = 0 \quad \text{at} \quad \theta = \tfrac{1}{2}\pi \tag{6.59}$$

$$T = T_0\mu^2 \quad \text{at} \quad r = R \tag{6.60}$$

$$\frac{\partial T}{\partial r} = 0 \quad \text{at} \quad r = c \tag{6.61}$$

SOLUTION:

In this example, as in the previous two, we are dealing with the axisymmetric form of Laplace's equation in spherical coordinates. As a standard operating procedure we use the transformation $\mu = \cos\theta$. The differential equation we need to solve is given by Equation (6.51), i.e.,

$$r^2\frac{\partial^2 T}{\partial r^2} + 2r\frac{\partial T}{\partial r} + \frac{\partial}{\partial \mu}\left[(1-\mu^2)\frac{\partial T}{\partial \mu}\right] = 0. \tag{6.62}$$

As with previous examples, let us set up the problem on a rectangular (r, μ) domain as shown in Figure 6.3. Again, we see that we have a non-homogeneous boundary condition at $r = R$ that is a function of μ. This will require an eigenfunction expansion in μ. As in the previous example, we need eigenfunctions of the operator given by Equation (6.7),

$$\frac{1}{\sin\theta}\frac{d}{d\theta}\left(\sin\theta\frac{d}{d\theta}\right) = \frac{d}{d\mu}\left[(1-\mu^2)\frac{d}{d\mu}\right]. \tag{6.63}$$

The eigenfunctions are given by the solution of Equation (6.11) in Example 6.1. As in Example 6.3, Equation (6.52), we can select

$$\Theta_n(\mu) = P_n(\mu), \quad n = 1, 3, 5, \ldots. \tag{6.64}$$

Here, this set of solutions represents eigenfunctions of the operator (6.63), and satisfies $\Theta_n(0) = 0$. Therefore, as in Example 6.3, Equation (6.53),

$$T(r, \mu) = \sum_{n=1,3,5,\ldots}^{\infty} T_n(r)P_n(\mu), \tag{6.65}$$

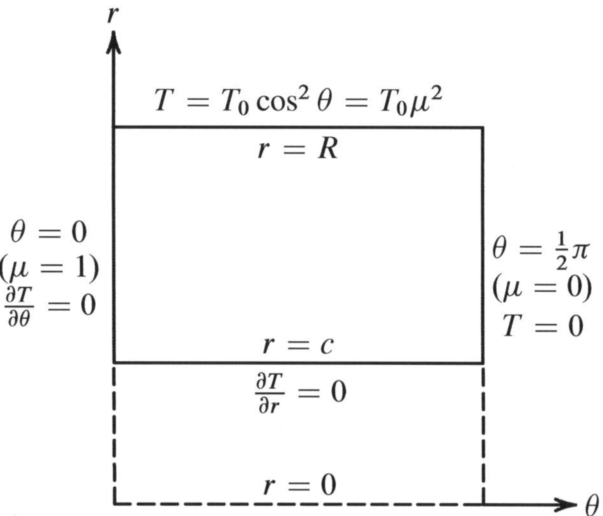

Figure 6.3: Display of boundary conditions for Example 6.4 in 'rectangularized' coordinates.

represents a suitable eigenfunction expansion satisfying the homogeneous the boundary condition $T(r, 0) = 0$ at the base of the hemispherical annulus. Again, following Example 6.3 up to Equation (6.54), it is easy to show that

$$T_n(r) = A_n r^n + B_n r^{-(n+1)}. \tag{6.66}$$

Replacing the expression for $T_n(r)$ in Equation (6.65) with (6.66), we obtain

$$T(r, \mu) = \sum_{n=1,3,5,..}^{\infty} [A_n r^n + B_n r^{-(n+1)}] P_n(\mu). \tag{6.67}$$

Here, the radial function given by Equation (6.66) stays finite throughout the range $c \le r \le R$. Therefore, we keep both r^n and $r^{-(n+1)}$. The boundary condition (6.61) at $r = c$ will give a relationship between A_n and B_n. This boundary condition can be satisfied term-by-term by requiring $T_n'(c) = 0$, i.e.,

$$\begin{aligned} T_n'(c) &= A_n n c^{(n-1)} - B_n (n+1) c^{-(n+2)} = 0, \\ B_n &= \frac{n}{(n+1)} A_n c^{(2n+1)}. \end{aligned} \tag{6.68}$$

Thus, the expression for $T_n(r)$ given by Equation (6.66) may be written as

$$T_n(r) = A_n \left[r^n + \frac{n}{(n+1)} c^{(2n+1)} r^{-(n+1)} \right], \tag{6.69}$$

which, after some rearrangement, may be written as

$$T_n(r) = a_n \left[(n+1) \left(\frac{r}{c}\right)^n + n \left(\frac{c}{r}\right)^{(n+1)} \right], \tag{6.70}$$

where

$$a_n\left(\frac{n+1}{c^n}\right) = A_n.$$

The solution (6.67) therefore becomes

$$T(r,\mu) = \sum_{n=1,3,5,..}^{\infty} a_n\left[(n+1)\left(\frac{r}{c}\right)^n + n\left(\frac{c}{r}\right)^{(n+1)}\right]P_n(\mu). \tag{6.71}$$

The boundary condition (6.60) at $r = R$ will determine the remaining set of constants a_n. Carrying this out gives us

$$T(r,\mu) = \sum_{n=1,3,5,..}^{\infty} a_n\left[(n+1)\left(\frac{R}{c}\right)^n + n\left(\frac{c}{R}\right)^{n+1}\right]P_n(\mu) = T_0\mu^2. \tag{6.72}$$

Making use of the orthogonality of odd Legendre polynomials in the range $0 \leq \mu \leq 1$,

$$\begin{aligned} a_n\left[(n+1)\left(\frac{R}{c}\right)^n + n\left(\frac{c}{R}\right)^{n+1}\right] &= \frac{\int_0^1 T_0\mu^2 P_n(\mu)\,d\mu}{\int_0^1 [P_n(\mu)]^2\,d\mu} \\ &= (2n+1)T_0\int_0^1 \mu^2 P_n(\mu)\,d\mu \end{aligned} \tag{6.73}$$

The integral here is the same as in the previous example which refers to Example 3.16, on pages 136-138. The expression given by Equation (3.192) as well as Equation (6.56) is applies here. Thus,

$$\begin{aligned} & a_n\left[(n+1)\left(\frac{R}{c}\right)^n + n\left(\frac{c}{R}\right)^{n+1}\right] \\ &= -T_0\frac{2(2n+1)P_n'(0)}{n(n+1)(n+3)(n-2)}, \quad n = 1,2,3,\ldots, \\ a_n &= -2T_0\frac{\frac{(2n+1)P_n'(0)}{n(n+1)(n+3)(n-2)}}{(n+1)\left(\frac{R}{c}\right)^n + n\left(\frac{c}{R}\right)^{n+1}}, \quad n = 1,2,3,\ldots, \end{aligned} \tag{6.74}$$

which is not valid for $n = 0, 2$, but that does not concern us since only odd terms are needed for the Legendre-series expansion. Using this expression for a_n in Equation (6.71), we end up with the following form of the final solution

$$\begin{aligned} T(r,\mu) = -2T_0 & \sum_{n=1,3,5,..}^{\infty} \frac{(2n+1)P_n'(0)}{n(n+1)(n+3)(n-2)} \\ & \times\left[\frac{(n+1)\left(\frac{r}{c}\right)^n + n\left(\frac{c}{r}\right)^{n+1}}{(n+1)\left(\frac{R}{c}\right)^n + n\left(\frac{c}{R}\right)^{n+1}}\right]P_n(\mu). \end{aligned} \tag{6.75}$$

□

EXERCISES 6.1

1. Solve $\nabla^2 T = 0$ for the hemisphere shown. The boundary conditions are

$$\left.\frac{\partial T}{\partial r}\right|_{r=R} = \begin{cases} q_0/k, & 0 < \theta < \theta_0 \\ 0, & \theta_0 < \theta < \frac{1}{2}\pi. \end{cases}$$

and

$$T|_{\theta=\frac{1}{2}\pi} = 0.$$

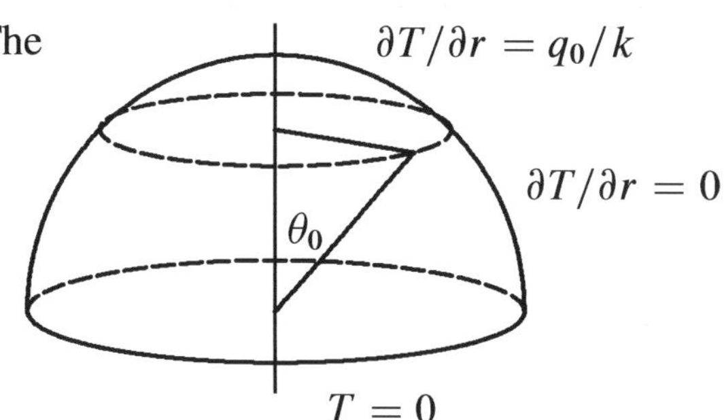

2. Solve $\nabla^2 V = 0$ for the region around the sphere defined by $r = R$. The boundary conditions are

$$V|_{r=R} = f(\theta) = \begin{cases} V_0 \cos\theta, & 0 \le \theta \le \frac{1}{2}\pi \\ 0 & \frac{1}{2}\pi \le \theta \le \pi \end{cases}$$

and

$$V|_{r\to\infty} = 0.$$

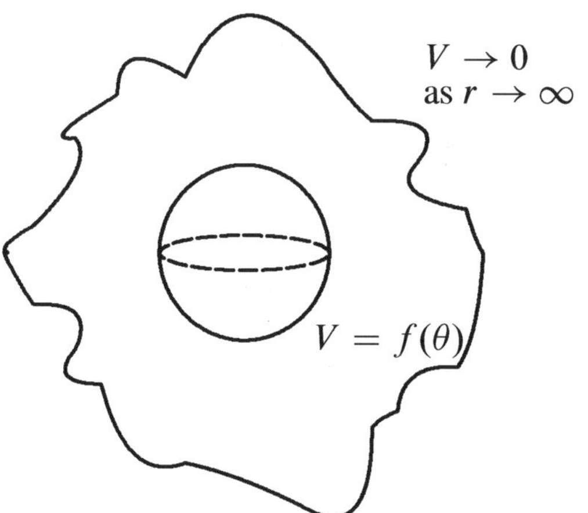

3. Find the solution for $\nabla^2 T = 0$ in the hemispherical shell region. The boundary conditions are given by

$$\begin{aligned} T &= 0 \quad \text{at} \quad \theta = \tfrac{1}{2}\pi \\ T &= T_0 \quad \text{at} \quad r = R \\ T &= 0 \quad \text{at} \quad r = c \end{aligned}$$

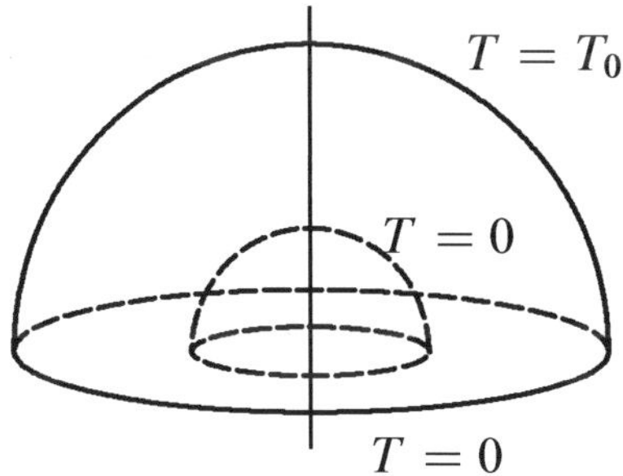

We have considered four examples spherical geometry, all to do with Laplace's equation. Let us now examine the time-dependent diffusion equation. We shall begin with spherically-symmetric examples involving dependence on just r and t. The first one among these examples is a simple case of heat diffusion in a sphere which requires an eigenfunction expansion in r.

6.2 Eigenfunction Expansions in r

Example 6.5

PROBLEM STATEMENT: Calculate the temperature distribution $T(r,t)$ inside a sphere $0 \le r \le R$, initially at a temperature T_0. Starting at $t = 0$, the boundary $r = R$ is kept at $T = 0$. The differential equation in this case is

$$\frac{1}{\alpha}\frac{\partial T}{\partial t} = \frac{\partial^2 T}{\partial r^2} + \frac{2}{r}\frac{\partial T}{\partial r}. \tag{6.76}$$

The initial condition can be formally expressed as

$$T(r,0) = T_0, \tag{6.77}$$

and the boundary condition at $r = R$ is

$$T(R,t) = 0. \tag{6.78}$$

By radial symmetry, we can argue that

$$\left.\frac{\partial T}{\partial r}\right|_{r=0} = 0. \tag{6.79}$$

However, as we shall see later, this condition will not need to be applied.

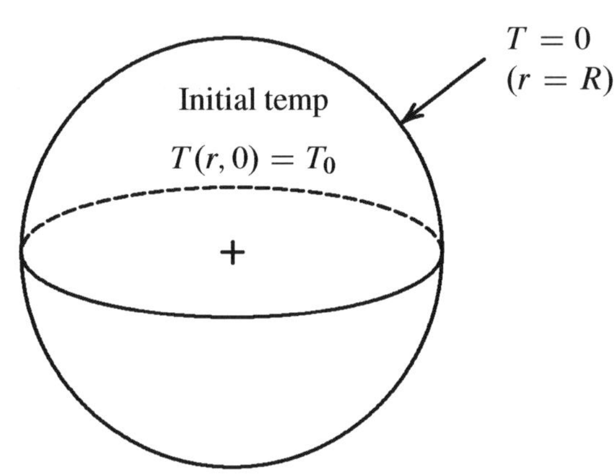

SOLUTION: As we have done in several previous examples, it would be useful to examine the problem on rectangularized coordinates as shown in Figure 6.4 Based on

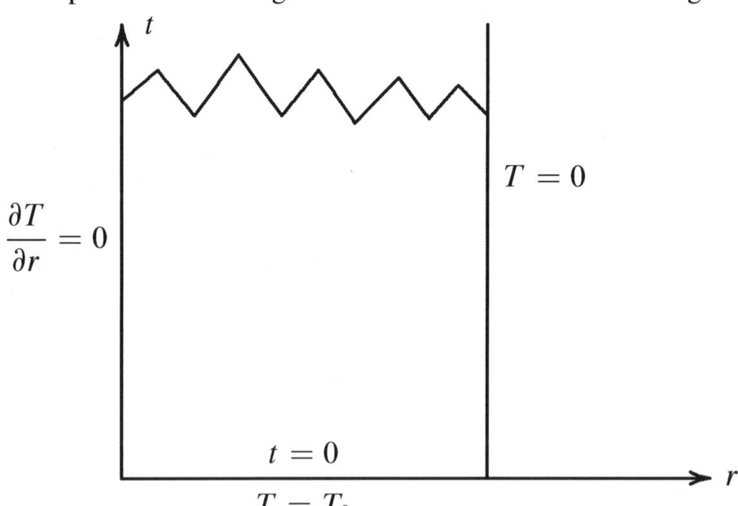

Figure 6.4: Rectangularized view of the sphere problem in Example 6.5

the initial condition (6.77) being at $t = 0$ and along the r coordinate as shown in

Figure 6.4, we need an eigenfunction expansion in r. From Equation (6.76), we can see that the radial operator is

$$\frac{d^2}{dr^2} + \frac{2}{r}\frac{d}{dr}. \tag{6.80}$$

Eigenfunctions of this operator satisfy the ordinary differential equation

$$\frac{d^2 R_m(r)}{dr^2} + \frac{2}{r}\frac{dR_m(r)}{dr} = -\lambda_m^2 R_m(r), \tag{6.81}$$

which, upon multiplying by r^2 and some rearrangement, becomes

$$r^2\frac{d^2 R_m(r)}{dr^2} + 2r\frac{dR_m(r)}{dr} + \lambda_m^2 r^2 R_m(r) = 0. \tag{6.82}$$

This is the spherical Bessel's equation of order zero . The order-n equation is given by Equation (2.74) on page 70 in Chapter 2. We repeat the equation here for convenience,

$$x^2 f''(x) + 2xf'(x) + \left[x^2 - n(n+1)\right] f(x) = 0. \tag{6.83}$$

The general solution can be found by the method of Frobenius, and is given as

$$f(x) = Aj_n(x) + By_n(x), \tag{6.84}$$

where power-series expressions for $j_n(x)$ and $y_n(x)$ are given by Equations (2.76) and (2.77) on page 70. What we need to do next is to construct a general solution for Equation (6.82). If carry out the change of variables,

$$x = \lambda_m r, \quad f(x) = R_m(r),$$

and take $n = 0$, it is not difficult to show that Equation (6.84) transforms to (6.82). Thus, we may write the solution to Equation (6.82) as

$$R_m(r) = A_m j_0(\lambda_m r) + B_m y_0(\lambda_m r), \tag{6.85}$$

As a convenient alternative, this solution can be written in terms of trigonometric functions using the expressions given by Equations (2.79) and (2.82) on page (72),

$$R_m(r) = A_m\frac{\sin\lambda_m r}{\lambda_m r} + B_m\frac{\cos\lambda_m r}{\lambda_m r}, \tag{6.86}$$

Within the domain $0 \leq r \leq R$,

$$\frac{\cos\lambda_m r}{\lambda_m r} \to \infty \quad \text{as} \quad r \to 0,$$

while

$$\frac{\sin\lambda_m r}{\lambda_m r} \to 1.$$

Therefore, we set $B_m = 0$ to require the solution to be finite, leaving us with

$$R_m(r) = A_m\frac{\sin\lambda_m r}{\lambda_m r}. \tag{6.87}$$

Since the boundary condition at $r = R$ (Equation (6.78)) is homogeneous, we can satisfy it term-by-term with $R_m(R) = 0$, i.e.,

$$R_m(R) = A_m \frac{\sin \lambda_m R}{\lambda_m R} = 0. \tag{6.88}$$

This requirement can be met with

$$\sin \lambda_m R = 0, \quad m = 1, 2, 3, \ldots, \tag{6.89}$$

which is satisfied with $\lambda_m = m\pi/R$, $m = 1, 2, 3, \ldots$. We can now carry out an expansion in terms of the radial eigenfunctions in the form,

$$T(r,t) = \sum_{m=1}^{\infty} \Theta_m(t) \frac{\sin \lambda_m r}{\lambda_m r}. \tag{6.90}$$

Substituting this expansion into the differential equation (6.76) leads to

$$\begin{aligned} \frac{1}{\alpha} \sum_{m=1}^{\infty} \Theta'_m(t) \frac{\sin \lambda_m r}{\lambda_m r} &= \sum_{m=1}^{\infty} \Theta_m(t) \left[\frac{d^2}{dr^2} + \frac{2}{r}\frac{d}{dr} \right] \frac{\sin \lambda_m r}{\lambda_m r} \\ &= \sum_{m=1}^{\infty} \Theta_m(t) \left[-\lambda_m^2 \right] \frac{\sin \lambda_m r}{\lambda_m r}, \end{aligned} \tag{6.91}$$

which may be written as

$$\sum_{m=1}^{\infty} \left[\frac{1}{\alpha} \Theta'_m(t) + \lambda_m^2 \Theta_m(t) \right] \frac{\sin \lambda_m r}{\lambda_m r} = 0, \tag{6.92}$$

from where we can identify the separated ordinary differential equation for the time part as

$$\frac{1}{\alpha} \Theta'_m(t) + \lambda_m^2 \Theta_m(t). \tag{6.93}$$

The solution to this equation is

$$\Theta'_m(t) = a_m e^{-\alpha \lambda_m^2 t}. \tag{6.94}$$

Using this result in the expansion (6.90) gives us the solution

$$T(r,t) = \sum_{m=1}^{\infty} a_m e^{-\alpha \lambda_m^2 t} \frac{\sin \lambda_m r}{\lambda_m r}. \tag{6.95}$$

The set of constants $\{a_m\}$ can be found by satisfying the initial condition (6.77). Thus,

$$T_0 = \sum_{m=1}^{\infty} a_m \frac{\sin \lambda_m r}{\lambda_m r} = \sum_{m=1}^{\infty} a_m j_0(\lambda_m r). \tag{6.96}$$

Here, we can invoke the orthogonality of the spherical Bessel functions,

$$j_0(\lambda_m r), \quad m = 1, 2, 3, \ldots.$$

An example of eigenfunction expansions in $j_n(\lambda_{mn}r)$ was considered in Chapter 3 (see Example 3.14, pages 128-129). While the present case is very similar, we shall independently establish the orthogonality. We first rewrite the differential equation (6.82) for $j_0(\lambda_m r)$ as

$$r^2 \frac{d^2 j_0(\lambda_m r)}{dr^2} + 2r \frac{d j_0(\lambda_m r)}{dr} + \lambda_m^2 r^2 j_0(\lambda_m r) = 0. \tag{6.97}$$

By combining the first two terms, we obtain

$$\frac{d}{dr}\left(r^2 \frac{d j_0(\lambda_m r)}{dr}\right) + \lambda_m^2 r^2 j_0(\lambda_m r) = 0. \tag{6.98}$$

Let us compare this with the general Sturm-Liouville form given by Equation (3.59),

$$\frac{d}{dx}\left[p(x)\frac{dX_m(x)}{dx}\right] - s(x)X_m(x) + \lambda_m^2 q(x)X_m(x) = 0, \tag{6.99}$$

where n has been changed to n. We identify the variables as follows

$$x = r, \quad p(x) = p(r) = r^2, \quad q(x) = q(r) = r^2, \quad s(x) = 0, \quad X_m(x) = j_0(\lambda_m r).$$

Thus the general orthogonality relationship (3.67) on page 111,

$$\int_a^b q(x)X_j(x)X_k(x)\,dx = 0, \quad j \neq k, \tag{6.100}$$

for the present case is

$$\int_0^R r^2 j_0(\lambda_j r) j_0(\lambda_k r)\,dr = 0, \quad j \neq k. \tag{6.101}$$

This relationship is valid on the basis of the boundary conditions (6.79) and (6.78). These can be satisfied term-by-term with $R_m'(0) = 0$ and $R_m(R) = 0$, i.e., $j_0'(0) = 0$ and $j_0(\lambda_m R) = 0$. We satisfied the latter condition earlier (see Equations (6.88)-(6.89)). We did not formally apply the condition at $r = 0$ but this is not necessary because $j_0'(0) = 0$ satisfied by the symmetry of the function about $r = 0$. In fact, $j_0(x) = \sin x/x$ is an even power series about $x = 0$. Therefore, the boundary conditions that the set of eigenfunctions satisfy at $r = 0$ and $r = R$, and the second and the first types (Neumann and Dirichlet), respectively, and fall within the classification that guarantees orthogonality (see discussion on pages 111-116 in Chapter 3). Let us now apply orthogonality to Equation (6.96). As usual, we multiply both sides by $r^2 j_0(\lambda_k r)$, and integrate over $[0, R]$, i.e.,

$$\begin{aligned}\int_0^R T_0 r^2 j_0(\lambda_k r)\,dr &= \sum_{m=1}^{\infty} a_m \int_0^R r^2 j_0(\lambda_m r) j_0(\lambda_k r)\,dr \\ &= a_1 0 + a_2 0 + \cdots + a_k \int_0^R r^2 \left[j_0(\lambda_k r)\right]^2 dr \\ &\quad + a_{k+1} 0 + \cdots\end{aligned}$$

$$
\begin{aligned}
T_0 \int_0^R r^2 \left(\frac{\sin\lambda_k r}{\lambda_k r}\right) dr &= a_k \int_0^R r^2 \left(\frac{\sin\lambda_k r}{\lambda_k r}\right)^2 dr \\
\frac{T_0}{\lambda_k}\int_0^R r\sin\lambda_k r\, dr &= \frac{a_k}{\lambda_k^2}\int_0^R \sin^2\lambda_k r\, dr \\
T_0\lambda_k\left[r\frac{-\cos\lambda_k r}{\lambda_k}\Big|_0^R - \int_0^R \frac{-\cos\lambda_k r}{\lambda_k} dr\right] &= a_k \int_0^R \tfrac{1}{2}[1-\cos(2\lambda_k r)]\, dr \\
T_0\lambda_k\left[R\frac{[1-\cos\lambda_k R]}{\lambda_k} + \frac{\sin\lambda_k R}{\lambda_k^2}\right] &= a_k \tfrac{1}{2}\left[r - \frac{\cos(2\lambda_k r)}{2\lambda_k}\right]_0^R \quad (6.102)
\end{aligned}
$$

With $\lambda_k R = k\pi$, as shown by Equation (6.89), the above result becomes

$$
\begin{aligned}
T_0\frac{k\pi}{R}\left(R^2\frac{[1-(-1)^k]}{k\pi} + 0\right) &= a_k\tfrac{1}{2}\left(R - \frac{1-1}{2k\pi/R}\right) \\
a_k &= -2T_0[1-(-1)^k], \quad k = 1,2,3,\ldots (6.103)
\end{aligned}
$$

Using this expression for a_k in Equation (6.95), together with $\lambda_k R = k\pi$, the final solution can be written as

$$
\begin{aligned}
T(r,t) &= -2T_0 \sum_{m=1}^{\infty} [1-(-1)^m] e^{-\alpha(m\pi/R)^2 t}\frac{\sin\left(\frac{m\pi r}{R}\right)}{\frac{m\pi r}{R}} \\
&= -2T_0 \sum_{m=1}^{\infty} \frac{[1-(-1)^m]}{m\pi} e^{-\alpha(m\pi/R)^2 t}\left(\frac{R}{r}\right)\sin\left(\frac{m\pi r}{R}\right). \quad (6.104)
\end{aligned}
$$

In this example, we took the formal Sturm-Liouville approach of applying orthogonality of the spherical Bessel functions. It is possible to take a simplified approach (while still consistent with the Sturm-Liouville formalism) of considering the orthogonality of the Fourier sine series in Equation (6.96),

$$
T_0 = \sum_{m=1}^{\infty} a_m \left(\frac{\sin\lambda_m r}{\lambda_m r}\right). \quad (6.105)
$$

Multiplying both sides by r, and noting that $\lambda_m R = m\pi$, yields

$$
\begin{aligned}
T_0 r &= \sum_{m=1}^{\infty} a_m \left(\frac{\sin\lambda_m r}{\lambda_m}\right) \\
&= \sum_{m=1}^{\infty} \frac{a_m}{m\pi/R}\sin\left(\frac{m\pi r}{R}\right), \quad (6.106)
\end{aligned}
$$

where it is possible to apply the orthogonality of the set $\{\sin(m\pi r/R),\ m = 1,2,3,\ldots\}$ in the range $0 \le r \le R$, and obtain the same result as (6.103). □

For the next example, we again consider the heat diffusion equation in a spherically-symmetric domain. The difference here is that instead of a whole sphere, we consider the annular region $c \le r \le R$.

Example 6.6

PROBLEM STATEMENT: For the spherical annulus, $c \le r \le R$, calculate the temperature distribution $T(r,t)$ described by the heat-diffusion equation,

$$\frac{1}{\alpha}\frac{\partial T}{\partial t} = \frac{\partial^2 T}{\partial r^2} + \frac{2}{r}\frac{\partial T}{\partial r}. \tag{6.107}$$

The initial condition is

$$T(r,0) = T_0, \tag{6.108}$$

and the boundary condition at $r = c$ is that for an insulated boundary,

$$\left.\frac{\partial T(r,t)}{\partial r}\right|_{r=c} = 0. \tag{6.109}$$

At the outer boundary $r = R$, we have

$$T(R,t) = 0. \tag{6.110}$$

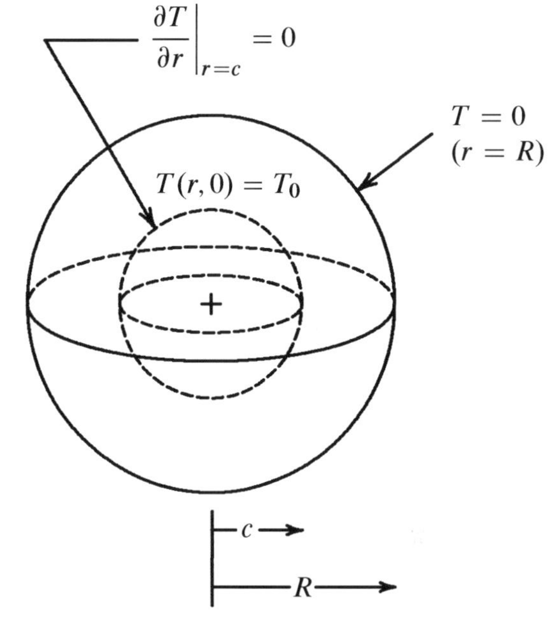

SOLUTION: Following the last example, let us view the problem in rectangularized coordinates as shown in Figure 6.5 Based on this figure, we can identify two 'opposite' homogeneous boundary conditions at $r = c$ and $r = R$. This will facilitate an eigenfunction expansion in r, and also allow us to satisfy the nonhomogeneous initial condition (6.108). The differential equation (6.107) for the current problem is the

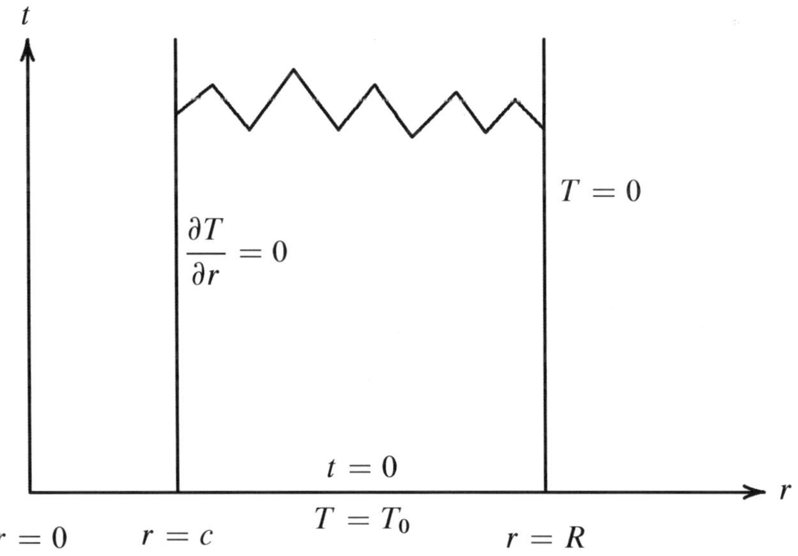

Figure 6.5: Sphere problem in Example 6.6 viewed in rectangularized coordinates.

same as the one in the previous example, and since we will be carrying out an eigenfunction expansion in r, we need to work with the differential operator (6.80). The eigenfunctions of this operator have already been set up and are given by Equations (6.85) and (6.86). For convenience, we rewrite these

$$R_m(r) = A_m j_0(\lambda_m r) + B_m y_0(\lambda_m r), \quad (6.111)$$
$$= \frac{A_m \sin\lambda_m r + B_m \cos\lambda_m r}{\lambda_m r}, \quad (6.112)$$

We can satisfy the two homogeneous boundary conditions (6.109) and (6.110) term-by-term with $R'_m(c) = 0$ and $R_m(R) = 0$. It will be easier to satisfy $R_m(R) = 0$ first. Applying this condition,

$$R_m(R) = 0 = \frac{A_m \sin\lambda_m R + B_m \cos\lambda_m R}{\lambda_m R}, \quad (6.113)$$

This can be satisfied with

$$A_m \sin\lambda_m R + B_m \cos\lambda_m R = 0, \quad (6.114)$$

leading us to the following relationship between A_m and B_m:

$$B_m = -A_m\left(\frac{\sin\lambda_m R}{\cos\lambda_m R}\right). \quad (6.115)$$

Using this in Equation (6.112), we obtain

$$\begin{aligned} R_m(r) &= \frac{A_m}{\lambda_m r}\left[\sin\lambda_m r - \frac{\cos\lambda_m r}{\lambda_m r}\left(\frac{\sin\lambda_m R}{\cos\lambda_m R}\right)\right] \\ &= \frac{A_m}{\lambda_m r\cos\lambda_m R}\left[\sin\lambda_m r\cos\lambda_m R - \cos\lambda_m r\sin\lambda_m R\right] \\ &= a_m\frac{\sin\left[\lambda_m(R-r)\right]}{\lambda_m r}, \end{aligned} \quad (6.116)$$

where

$$a_m = -\frac{A_m}{\cos\lambda_m R}.$$

We may now satisfy the homogeneous boundary condition at $r = c$ with $R'_m(c) = 0$, i.e.,

$$R'_m(c) = a_m \left.\frac{-r\lambda_m\cos\left[\lambda_m(R-r)\right] - \sin\left[\lambda_m(R-r)\right]}{\lambda_m r^2}\right|_{r=c} = 0, \quad (6.117)$$

which leads to the transcendental equation

$$-c\lambda_m\cos\left[\lambda_m(R-c)\right] - \sin\left[\lambda_m(R-c)\right], \quad (6.118)$$

which may also be written as

$$\tan\left[\lambda_m(R-c)\right] = -\lambda_m c. \quad (6.119)$$

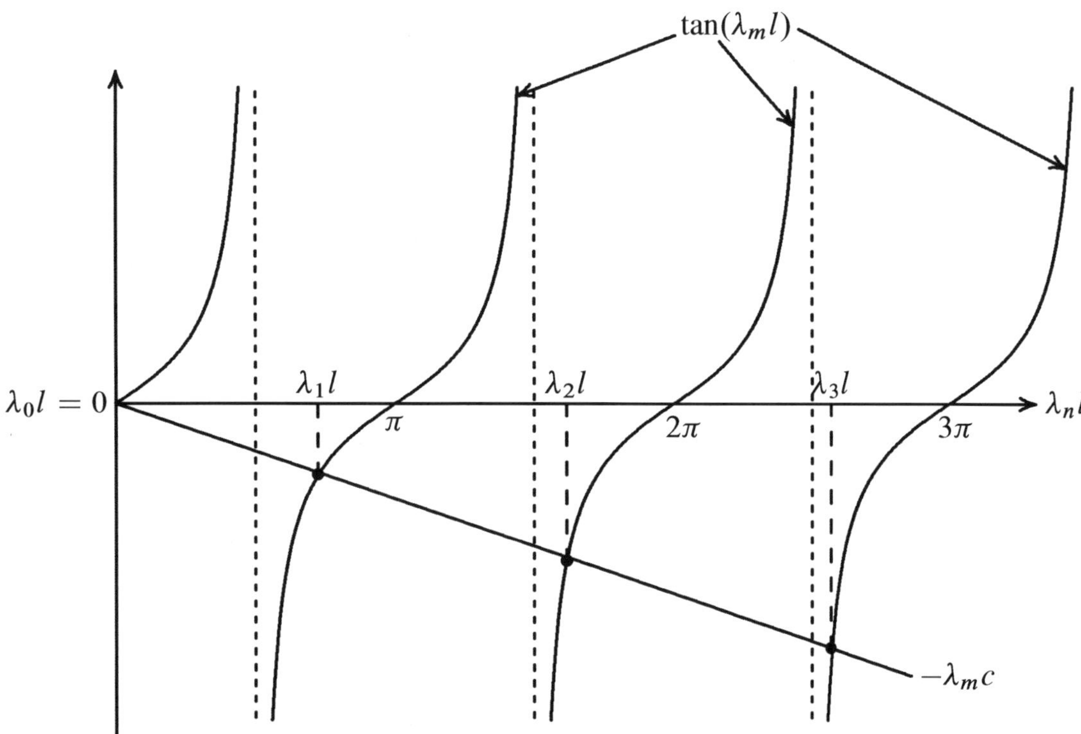

Figure 6.6: Graphical representation of the roots of equation (6.119). Here, we let $l = (R - c)$.

The roots of this equation give the values of λ_m. We may graphically visualize the roots of this equation by plotting both the left-hand side and the right-hand side on the same graph, as shown in Figure 6.6 With λ_m values established, we can write the radial eigenfunction as that given by Equation (6.116). We can therefore express $T(r,t)$ as an eigenfunction expansion,

$$T(r,t) = \sum_{m=1}^{\infty} \Theta_m(t) \frac{\sin[\lambda_m(R-r)]}{\lambda_m r}. \tag{6.120}$$

Substituting this into the differential equation (6.107) gives

$$\begin{aligned} \frac{1}{\alpha} \sum_{m=1}^{\infty} \Theta'_m(t) \frac{\sin[\lambda_m(R-r)]}{\lambda_m r} &= \sum_{m=1}^{\infty} \Theta_m(t) \left[\frac{d^2}{dr^2} + \frac{2}{r}\frac{d}{dr} \right] \frac{\sin[\lambda_m(R-r)]}{\lambda_m r} \\ &= \sum_{m=1}^{\infty} \Theta_m(t) \left[-\lambda_m^2 \right] \frac{\sin[\lambda_m(R-r)]}{\lambda_m r}, \end{aligned} \tag{6.121}$$

which may be written as

$$\sum_{m=1}^{\infty} \left[\frac{1}{\alpha} \Theta'_m(t) + \lambda_m^2 \Theta_m(t) \right] \frac{\sin[\lambda_m(R-r)]}{\lambda_m r} = 0. \tag{6.122}$$

As with the last example (see Equations (6.92-(6.94) on page 250), we can identify the separated ordinary differential equation for the time part as

$$\frac{1}{\alpha}\Theta'_m(t) + \lambda_m^2\Theta_m(t), \tag{6.123}$$

which has the solution

$$\Theta'_m(t) = a_m e^{-\alpha\lambda_m^2 t}. \tag{6.124}$$

Making use of this result in the expansion (6.120) gives us the solution

$$T(r,t) = \sum_{m=1}^{\infty} a_m e^{-\alpha\lambda_m^2 t}\frac{\sin[\lambda_m(R-r)]}{\lambda_m r}. \tag{6.125}$$

We may now obtain the set of constants $\{a_m\}$ which can be found by satisfying the initial condition (6.108). Therefore,

$$T_0 = \sum_{m=1}^{\infty} a_m \frac{\sin[\lambda_m(R-r)]}{\lambda_m r} \tag{6.126}$$

Once again, on the basis of the boundary conditions $R'_m(c) = 0$ and $R_m(R) = 0$, we can count on the set of eigenfunctions $\{R_m(r),\ m = 1, 2, 3, \ldots\}$ to be orthogonal in the range $c \le r \le R$. The weighting factor can be established as r^2 on the basis of the discussion in the previous example (see Equations (6.98)-(6.101) on page 251). The type of reasoning used to establish orthogonality in Example 6.5 will work for the present example as well, and we can be assured of the orthogonality relationship,

$$\int_c^R r^2 \frac{\sin[\lambda_m(R-r)]}{\lambda_m r}\cdot\frac{\sin[\lambda_k(R-r)]}{\lambda_k r}\,dr = 0, \quad k \ne m. \tag{6.127}$$

Let us now multiply both sides of Equation (6.126) by

$$r^2\frac{\sin[\lambda_k(R-r)]}{\lambda_k r},$$

and integrate over $[c, R]$. This leads to

$$\begin{aligned}
T_0\int_c^R r^2\left(\frac{\sin[\lambda_k(R-r)]}{\lambda_k r}\right)dr &= a_1\cdot 0 + a_2\cdot 0 + \cdots + a_{k-1}\cdot 0 \\
&\quad + a_k\int_c^R r^2\left(\frac{\sin[\lambda_k(R-r)]}{\lambda_k r}\right)^2 dr + a_{k+1}\cdot 0 + \cdots \\
T_0\lambda_k\int_c^R r\sin[\lambda_k(R-r)]\,dr &= a_k\int_c^R \sin^2[\lambda_k(R-r)]\,dr
\end{aligned}$$

$$\begin{aligned}
T_0\lambda_k\left[r\frac{\cos[\lambda_k(R-r)]}{\lambda_k}\bigg|_c^R - \int_c^R\frac{\cos[\lambda_k(R-r)]}{\lambda_k}\,dr\right] \\
= a_k\int_c^R \tfrac{1}{2}\{1-\cos[2\lambda_k(R-r)]\}\,dr
\end{aligned}$$

$$T_0\lambda_k\left[\frac{R-c\cos[\lambda_k(R-c)]}{\lambda_k}+\frac{\sin[\lambda_k(R-r)]}{\lambda_k^2}\bigg|_c^R\right]$$
$$=a_k\frac{1}{2}\left[r+\frac{\sin[2\lambda_k(R-r)]}{2\lambda_k}\right]_c^R,$$
$$2T_0\left\{\lambda_kR-\lambda_kc\cos[\lambda_k(R-c)]-\sin[\lambda_k(R-c)]\right\}$$
$$=a_k\left\{\lambda_k(R-c)-\tfrac{1}{2}\sin[2\lambda_k(R-c)]\right\}.\quad(6.128)$$

Using the transcendental equation (6.119), we can eliminate some terms, leaving us with

$$\begin{aligned}2T_0\lambda_kR &= a_k\left\{\lambda_k(R-c)-\sin[\lambda_k(R-c)]\cos[\lambda_k(R-c)]\right\}\\ &= a_k\left\{\lambda_kR-\lambda_kc\left(1-\cos^2[\lambda_k(R-c)]\right)\right\}\\ &= a_k\left\{\lambda_kR-\lambda_kc\sin^2[\lambda_k(R-c)]\right\}\\ a_k &= \frac{2T_0}{\left\{1-(c/R)\sin^2[\lambda_k(R-c)]\right\}},\qquad k=1,2,3,\ldots.\end{aligned}\quad(6.129)$$

By inserting this expression for a_k in Equation (6.125), we obtain the final form of the solution as

$$T(r,t)=2T_0\sum_{m=1}^{\infty}\frac{e^{-\alpha\lambda_m^2t}\sin[\lambda_m(R-r)]}{\lambda_mr\left\{1-(c/R)\sin^2[\lambda_m(R-c)]\right\}},\quad(6.130)$$

where the values of λ_m have to be determined computationally from the transcendental equation (6.119). □

EXERCISES 6.2

1. Obtain the solution for
$$\frac{1}{\alpha}\frac{\partial T}{\partial t}=\frac{\partial^2T}{\partial r^2}+\frac{2}{r}\frac{\partial T}{\partial r}$$
in the solid spherical region, $0<r<R$.

   The initial/boundary conditions are
$$\begin{array}{lll}T=0 & \text{at} & r=R\\ T<\infty & \text{as} & r\to 0\\ T=T_0\left[1-\left(\frac{r}{R}\right)^2\right] & \text{at} & t=0.\end{array}$$

2. Solve
$$\frac{1}{\alpha}\frac{\partial T}{\partial t}=\frac{\partial^2T}{\partial r^2}+\frac{2}{r}\frac{\partial T}{\partial r}$$
in the spherical shell region $c\le r\le R$.

   The initial/boundary conditions are
$$\begin{array}{lll}T=0 & \text{at} & r=c\\ \frac{\partial T}{\partial r}=0 & \text{at} & r=R\\ T=T_0 & \text{at} & t=0.\end{array}$$

3. A small spherical container of radius R is filled with a quiescent liquid. At time $t = 0$, a chemical solution with solute concentration c_0 is deposited at the center covering the region $0 \le r \le r_0$ so that the initial concentration in the entire container may be described piecewise as

$$c(r,0) = \begin{cases} c_0, & 0 \le r \le r_0, \\ 0, & r_0 < r \le R. \end{cases}$$

The solute spreads into the entire container according to the mass diffusion equation for $c(r,t)$,

$$\frac{1}{D}\frac{\partial c}{\partial t} = \frac{\partial^2 c}{\partial r^2} + \frac{2}{r}\frac{\partial c}{\partial r}, \quad 0 \le r \le R,$$

where D is the mass diffusion coefficient. The container wall may be considered impenetrable to the solute, i.e.,

$$\left.\frac{\partial c}{\partial r}\right|_{r=R} = 0.$$

Obtain the concentration $c(r,t)$ within the shell.

We can now go on to problems in three independent variables. We shall stay confined to axisymmetric problems with spatial dependence on r and θ, and in addition have time dependence. We shall begin with a class of problems requiring eigenfunction expansions in r and μ where $\mu = \cos\theta$.

6.3 Eigenfunction Expansions in r and μ

In understanding the solution procedure, thermal or mass diffusion problems are quite instructive and we shall deal with some examples of this type. We begin with an example of a hemisphere, initially at a temperature T_0 while all the boundaries are maintained at zero temperature for subsequent time.

Example 6.7

PROBLEM STATEMENT: Obtain the temperature distribution $T(r,\theta,t)$ in the hemispherical region shown in the figure. The governing equation in this case is

$$\frac{1}{\alpha}\frac{\partial T}{\partial t} = \frac{\partial^2 T}{\partial r^2} + \frac{2}{r}\frac{\partial T}{\partial r} + \frac{1}{r^2 \sin\theta}\frac{\partial}{\partial \theta}\left(\sin\theta \frac{\partial T}{\partial \theta}\right), \tag{6.131}$$

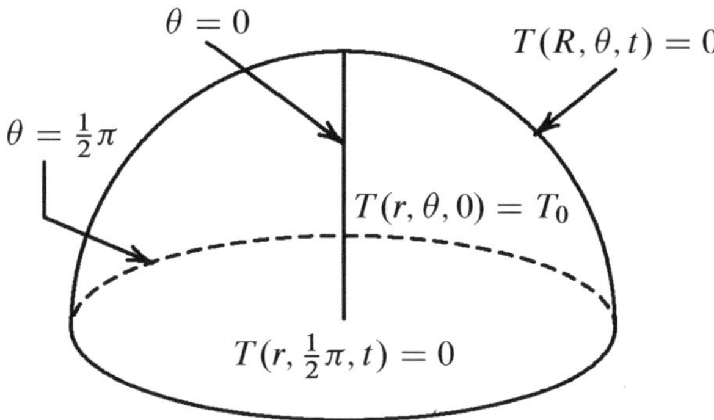

and the boundary and initial conditions for $T(r, \theta, t)$ are as shown, i.e.,

$$\left.\frac{\partial T(r, \theta, t)}{\partial \theta}\right|_{\theta=0} = 0, \qquad T(r, \tfrac{1}{2}\pi, t) = 0, \qquad T(R, \theta, t) = 0, \tag{6.132}$$

and

$$T(r, \theta, 0) = T_0. \tag{6.133}$$

SOLUTION:
This problem is similar in structure to Example 5.7 in several respects, and we shall closely follow the same procedure. However, we first transform $\mu = \cos\theta$ whereupon the differential operator in θ becomes

$$\frac{1}{\sin\theta}\frac{d}{d\theta}\left(\sin\theta\frac{d}{d\theta}\right) = \frac{d}{d\mu}\left[(1-\mu^2)\frac{d}{d\mu}\right], \tag{6.134}$$

as derived earlier (see the procedure preceding Equation (6.7)). The differential equation (6.131) then becomes

$$\frac{1}{\alpha}\frac{\partial T}{\partial t} = \frac{\partial^2 T}{\partial r^2} + \frac{2}{r}\frac{\partial T}{\partial r} + \frac{1}{r^2}\frac{\partial}{\partial \mu}\left[(1-\mu^2)\frac{\partial T}{\partial \mu}\right]. \tag{6.135}$$

We can now rewrite the boundary and initial conditions (6.132) and (6.133) for $T(r, \mu, t)$. The first of the boundary conditions in Equation (6.132) is a symmetry condition that becomes

$$\left.\frac{\partial T(r, \mu, t)}{\partial \mu}(-\sin\theta)\right|_{\theta=0} = 0, \tag{6.136}$$

which, owing to the fact the $\sin\theta$ vanishes as $\theta \to 0$, only requires that

$$\left.\frac{\partial T(r, \mu, t)}{\partial \mu}\right|_{\mu=1} < \infty, \tag{6.137}$$

The second boundary condition (on the base plane) may be written as

$$T(r, \mu, t)|_{\mu=0} = 0, \tag{6.138}$$

and the one at the dome of the hemisphere may be stated as

$$T(r, \mu, t)|_{r=R} = 0. \tag{6.139}$$

The initial condition (6.133) becomes

$$T(r, \mu, t)|_{t=0} = 0. \tag{6.140}$$

Let us now examine the problem on rectangularized coordinates as shown in Figure 6.7 Here, since the base plane of the hemisphere is at $T = 0$, by continuity, at the

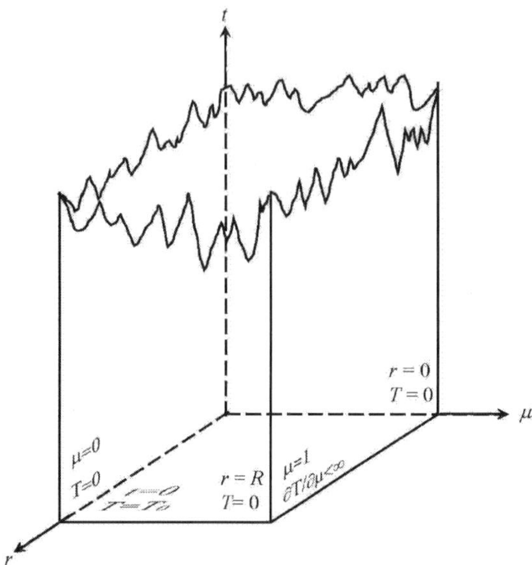

Figure 6.7: A three-dimensional rectangularized depiction of (r, μ, t) coordinates for Example 6.7.

point $r = 0$, we have $T = 0$. We now see that we have a pair of opposite homogeneous boundary conditions in r, at $r = 0$ and $r = R$. This will guarantee orthogonal eigenfunction in r. In μ, we have a homogeneous boundary condition at $\mu = 0$ and a requirement that T remain finite as $\mu \to 1$. While the latter condition is not homogeneous in μ, it is sufficient to generate orthogonal eigenfunctions in μ, as seen in Examples 6.3 and 6.4. With the initial condition $T(r, \mu, 0) = T_0$, we see that it is formally a function of r and μ. Therefore, we need to expand in these variables. Upon examining Equation (6.135), we see that r^{-2} multiplies the derivative term in μ, and these variables (r and μ) are not isolated from each other in the equation as given. Let us multiply the equation by r^2, we will isolate the μ-derivative term. Carrying out this operation, we have

$$\frac{r^2}{\alpha}\frac{\partial T}{\partial t} = r^2\frac{\partial^2 T}{\partial r^2} + 2r\frac{\partial T}{\partial r} + \frac{\partial}{\partial \mu}\left[(1-\mu^2)\frac{\partial T}{\partial \mu}\right]. \tag{6.141}$$

where we see that the last term on the right-hand side stands alone while all the other terms have no appearance of μ, either as a derivative or as a multiplicative term. Therefore, it is possible to construct suitable eigenfunctions in μ, and we shall start with the μ expansion as our only choice. Let us look for eigenfunctions of the operator

$$\frac{d}{d\mu}\left[(1-\mu^2)\frac{d}{d\mu}\right], \tag{6.142}$$

i.e., we need solutions of the ordinary differential equation,

$$\frac{d}{d\mu}\left[(1-\mu^2)\frac{d\Theta_n(\mu)}{d\mu}\right]+\lambda_n^2\Theta_n(\mu)=0. \tag{6.143}$$

This has already been dealt with in detail in Example 6.1. There, it was argued that for full spheres and hemispheres, we need to specify $\lambda_n^2 = n(n+1)$. With such eigenvalues, one of the solutions consists of integer powers in μ. Following Equations (6.8)-(6.12) on pages 234-234, the general solution for Equation (6.143) is

$$\Theta_n(\mu) = A_n P_n(\mu) + B_n Q_n(\mu), \tag{6.144}$$

where expressions for for $P_n(\mu)$ and $Q_n(\mu)$ are given by Equations (2.53) and (2.54) on page 61. Since $Q_n(\mu) \to \infty$ with logarithmic behavior as $\mu \to \pm 1$, this solution is unacceptable. Therefore, we set $B_n = 0$. As with several previous examples in this chapter, we have

$$\Theta_n(\mu) = A_n P_n(\mu). \tag{6.145}$$

Next, we consider the boundary condition (6.138). Based the steady-state hemisphere problems (Examples 6.3 and 6.4), we take only odd $P_n(\mu)$ polynomials since $P_n(0) = 0$ for $n = 1, 3, 5, \ldots$. We thus set

$$\Theta_n(\mu) = P_n(\mu), \quad n = 1, 3, 5 \ldots, \tag{6.146}$$

where we have dropped the set of constants A_n. The set $\{P_n(\mu),\ n = 1, 3, 5\ldots\}$ is orthogonal in the range $0 \le \mu \le 1$, and one can expand a function $f(\mu)$ as a Fourier-Legendre series. We can now expand $T(r, \mu, t)$ as a series in terms of the eigenfunctions $P_n(\mu)$, i.e.,

$$T(r,\theta,t) = \sum_{n=1,3,5,\ldots}^{\infty} T_n(r,t)P_n(\mu). \tag{6.147}$$

Upon the substitution of this expansion into Equation (6.141), we obtain

$$\begin{aligned} r^2\frac{1}{\alpha}&\sum_{n=1,3,5,\ldots}^{\infty}\frac{\partial T_n}{\partial t}P_n(\mu) \\ &= r^2\sum_{n=1}^{\infty}\frac{\partial^2 T_n}{\partial r^2}P_n(\mu) + 2r\sum_{n=1}^{\infty}\frac{\partial T_n}{\partial r}P_n(\mu) + \sum_{n=1}^{\infty}T_n(r)\frac{d}{d\mu}\left[(1-\mu^2)\frac{d}{d\mu}\right]P_n(\mu) \\ &= r^2\sum_{n=1}^{\infty}\frac{\partial^2 T_n}{\partial r^2}P_n(\mu) + 2r\sum_{n=1}^{\infty}\frac{\partial T_n}{\partial r}P_n(\mu) + \sum_{n=1}^{\infty}T_n(r)\left[-n(n+1)\right]P_n(\mu). \end{aligned} \tag{6.148}$$

We can now factor out $P_n(\mu)$ so that

$$\sum_{n=1,3,5,\ldots}^{\infty}\left(\frac{r^2}{\alpha}\frac{\partial T_n}{\partial t} - \left[r^2\frac{\partial^2 T_n}{\partial r^2} + 2r\frac{\partial T_n}{\partial r} - n(n+1)T_n\right]\right)P_n(\mu) = 0.$$

This is valid for all values of μ in the range $0 \le \mu \le 1$, and therefore, we may set the coefficients of $P_n(\mu)$ equal to zero. This results in

$$\left(\frac{r^2}{\alpha}\frac{\partial T_n(r,t)}{\partial t} - \left[r^2\frac{\partial^2 T_n(r,t)}{\partial r^2} + 2r\frac{\partial T_n(r,t)}{\partial r} - n(n+1)T_n(r,t)\right]\right) = 0.$$

We can now separate the r and the t operations if we divide the equation by r^2, i.e.

$$\frac{1}{\alpha}\frac{\partial T_n(r,t)}{\partial t} - \left[\frac{\partial^2 T_n(r,t)}{\partial r^2} + \frac{2}{r}\frac{\partial T_n(r,t)}{\partial r} - \frac{n(n+1)}{r^2}T_n(r,t)\right] = 0,$$

or equivalently,

$$\frac{1}{\alpha}\frac{\partial T_n(r,t)}{\partial t} - \left[\frac{\partial^2}{\partial r^2} + \frac{2}{r}\frac{\partial}{\partial r} - \frac{n(n+1)}{r^2}\right]T_n(r,t) = 0, \tag{6.149}$$

This represents the partial differential equation for $T_n(r,t)$. By expanding $T(r,\mu,t)$ in $P_n(\mu)$, the number of independent variables have been reduced from three to two. We may solve Equation (6.149) by suitable eigenfunction expansion in r. Therefore, we need eigenfunctions of the operator identified within the square brackets in Equation (6.149),

$$\frac{d^2}{dr^2} + \frac{2}{r}\frac{d}{dr} - \frac{n(n+1)}{r^2}, \tag{6.150}$$

We can set up the ordinary differential equation for the eigenfunctions of this operator as

$$\left[\frac{d^2}{dr^2} + \frac{2}{r}\frac{d}{dr} - \frac{n(n+1)}{r^2}\right]R(r) = -\lambda^2 R(r). \tag{6.151}$$

As in Example 5.7 on pages 221-226, for every n, we have a different equation, and we need to index the solution $R(r)$ accordingly. In addition, for every n, we will have a set of values of λ which need to be indexed further. Therefore, we need two indices, n and m. Thus, we set up the above equation as

$$\left[\frac{d^2}{dr^2} + \frac{2}{r}\frac{d}{dr} - \frac{n(n+1)}{r^2}\right]R_{mn}(r) = -\lambda^2 R_{mn}(r). \tag{6.152}$$

Next, we multiply this equation by r^2 and rearrange some terms to obtain

$$r^2\frac{d^2 R_{mn}}{dr^2} + 2r\frac{dR_{mn}}{dr} + \left[-n(n+1) + \lambda_{mn}^2 r^2\right]R_{mn} = 0. \tag{6.153}$$

Let us now compare this with spherical Bessels equation given by (2.74) on page 70],

$$x^2\frac{d^2 f}{dx^2} + 2x\frac{df}{dx} + \left[-n(n+1) + x^2\right]f(x) = 0, \tag{6.154}$$

This is equivalent to Equation 6.153 if we set $x = \lambda_{mn}r$ and $f(x) = R_{mn}(r)$. The solution for Equation (5.109) can be found by the method of Frobenius (see Chapter 2), and is given by the sum of two linearly independent solutions in the form

$$f(x) = Aj_n(x) + By_n(x).$$

The expressions for $j_n(x)$ and $y_n(x)$ are given by Equations (2.76) and (2.77) on page 70. In addition, these solutions can be conveniently expressed as combinations of sine and cosine functions [see Equations (2.81) and (2.84)]. With $x = \lambda_{mn} r$, the solution for $R_{mn}(r)$ can be written as

$$R_{mn}(r) = A_{mn} y_n(\lambda_{mn} r) + B_{mn} y_n(\lambda_{mn} r)$$

The behavior of these functions is quite similar to the cylindrical Bessel functions $J_n(\lambda_{mn} r)$ and $Y_n(\lambda_{mn} r)$ the plots of which are given in Figure 2.1 page 69. The behavior of $y_n(x)$ is asymptotically like $\sim x^{-(n+1)}$ as $x \to 0$. Therefore, for problems where $r = 0$ is included in the domain, we exclude the solution $y_n(\lambda_{mn} r)$ when the solution tending to infinity cannot be tolerated. We therefore set $B_{mn} = 0$, leaving us with

$$R_{mn}(r) = A_{mn} j_n(\lambda_{mn} r).$$

The function $T_n(r, t)$ can be expanded as a spherical Bessel series,

$$T_n(r, t) = \sum_{m=1}^{\infty} T_{mn}(t) j_n(\lambda_{mn} r), \tag{6.155}$$

where the summation begins with $m = 1$ because, as we shall see later, $\lambda_{0n} = 0$ for $n \geq 1$. Also, in this example, $n = 0$ does not come into play because in the Legendre series in μ, $P_n(\mu)$ has no contribution for $n = 0$ since we have constructed the series in only odd $P_n(\mu)$. The boundary condition $T = 0$ at $r = R$ can be satisfied term-by-term with $T_n(R, t) = 0$, and this in turn can be satisfied term-by-term with $j_n(\lambda_{mn} R) = 0$. Therefore, we need to locate the roots of $j_n(x) = 0$ to determine the values of λ_{mn} so that the homogeneous boundary condition at $r = R$ is satisfied. The roots are given by

$$\lambda_{mn} R = \alpha_{mn} \qquad \text{or} \qquad \lambda_{mn} = \frac{\alpha_{mn}}{R}, \tag{6.156}$$

and the α_{mn} values are available in mathematical handbooks. Let us now substitute the expansion for $T_n(r, t)$ [Equation (6.155)] into the partial differential equation (6.149). This leads to

$$\frac{1}{\alpha} \sum_{m=1}^{\infty} T'_{mn}(t) j_n(\lambda_{mn} r) = \sum_{m=1}^{\infty} T_{mn}(t) \left[\frac{d^2}{dr^2} + \frac{2}{r}\frac{d}{dr} - \frac{n(n+1)}{r^2} \right] j_n(\lambda_{mn} r). \tag{6.157}$$

Since $j_n(\lambda_{mn} r)$ is a solution of Equation (6.153), it is obviously an eigenfunction of the operator on the right-hand side of Equation (6.157) with eigenvalue $-\lambda_{mn}^2$. Using this property, we obtain

$$\frac{1}{\alpha} \sum_{m=1}^{\infty} T'_{mn}(t) j_n(\lambda_{mn} r) = \sum_{m=1}^{\infty} T_{mn}(t) \left[-\lambda_{mn}^2 \right] j_n(\lambda_{mn} r), \tag{6.158}$$

which may be written as

$$\sum_{m=1}^{\infty} \left[\frac{1}{\alpha} T'_{mn}(t) + \lambda_{mn}^2 \right] j_n(\lambda_{mn} r) = 0. \tag{6.159}$$

Here we invoke the usual argument that this equation is valid for all values of r in the range $0 \le r \le R$, and require that each coefficient of $j_n(\lambda_{mn} r)$ vanish, i.e.,

$$\left[\frac{1}{\alpha} T'_{mn}(t) + \lambda_{mn}^2\right] j_n(\lambda_{mn} r) = 0. \tag{6.160}$$

This has the solution

$$T_{mn}(t) = a_{mn} e^{-\alpha\lambda_{mn}^2 t}. \tag{6.161}$$

Let us now assemble the solution backwards, starting with $T_n(r,t)$. Using Equation (6.161) in the expansion (6.155), we obtain

$$T_n(r,t) = \sum_{m=1}^{\infty} a_{mn} e^{-\alpha\lambda_{mn}^2 t} j_n(\lambda_{mn} r), \tag{6.162}$$

Next, we use this expansion for $T_n(r,t)$ in Equation (6.147), leading to

$$T(r,\mu,t) = \sum_{n=1,3,5,\ldots}^{\infty} T_n(r,t) P_n(\mu) = \sum_{n=1}^{\infty}\sum_{m=1}^{\infty} a_{mn} e^{-\alpha\lambda_{mn}^2 t} j_n(\lambda_{mn} r) P_n(\mu). \tag{6.163}$$

We can now apply the initial condition, $T(r,\theta,0) = T_0$. Setting $t = 0$ in Equation (6.163), we end up with

$$T_0 = \sum_{n=1,3,5,\ldots}^{\infty} \left[\sum_{m=1}^{\infty} a_{mn} j_n(\lambda_{mn} r)\right] P_n(\mu). \tag{6.164}$$

We may now apply the orthogonality of the half-range Fourier-Legendre series. See pages 138-142 for a discussion on even and odd Legendre series. Example 3.17 is particularly relevant here, and following it, we obtain the coefficients as

$$\begin{aligned}\left[\sum_{m=1}^{\infty} a_{mn} j_n(\lambda_{mn} r)\right] &= \frac{\int_0^1 T_0 P_n(\mu)\, d\mu}{\int_0^1 [P_n(\mu)]^2\, d\mu} \\ &= \frac{T_0(2n+1)}{n(n+1)} P'_n(0),\end{aligned} \tag{6.165}$$

where we have used Equation (3.207) on page 140. The left-hand side is a Fourier-Bessel series. We may now apply the orthogonality of $j_n(\lambda_{mn} r)$. Here, we can make use of the calculations in Example 3.14 on pages 128-129. With the weighting factor r^2, we obtain the following expression for a_{mn}:

$$\begin{aligned} a_{mn} &= \frac{\int_0^R \frac{T_0(2n+1)}{n(n+1)} P'_n(0) r^2 j_n(\lambda_{mn} r) dr}{\int_0^R r^2 [j_n(\lambda_{mn} r)]^2 dr} \\ &= \frac{2T_0(2n+1)}{n(n+1)} P'_n(0) \sum_{k=0}^{\infty} \frac{(-1)^k (k+n)! 2^n (\lambda_{mn} R)^{2k+3+n}}{k!(2k+2n+1)!(2k+n+3)\left[j'_n(\lambda_{mn} R)\right]^2}, \end{aligned} \tag{6.166}$$

where we have used the set of Equations (3.152)-(3.155) on page 129. The complete solution as given by Equation (6.163) now takes the form

$$T(r,\theta,t) = 2T_0 \sum_{n=1,3,5,...}^{\infty} \frac{(2n+1)}{n(n+1)} P_n'(0) \times \sum_{m=1}^{\infty} \left(\sum_{k=0}^{\infty} \frac{(-1)^k (k+n)! 2^n \, (\lambda_{mn} R)^{2k+3+n}}{k!(2k+2n+1)!(2k+n+3) \left[j_n'(\lambda_{mn} R)\right]^2} \right) \times e^{-\alpha \lambda_{mn}^2 t} j_n(\lambda_{mn} r) P_n(\mu) \tag{6.167}$$

□

EXERCISES 6.3

1. Two solid hemispheres, one initially at a temperature T_0, and the other at zero temperature, are joined together at their equatorial planes. Both hemispheres are of the same material, and we assume perfect thermal contact so that the system may be treated as one seamless solid sphere with initial temperature given piecewise as

$$T(r,\mu,0) = \begin{cases} 0, & -1 \le \mu \le 0, \\ T_0, & 0 < \mu \le 1, \end{cases}$$

The outer surface of this whole sphere is kept insulated so that

$$\left. \frac{\partial T}{\partial r} \right|_{r=R} = 0.$$

Obtain the temperature distribution $T(r,\mu,t)$ by solving

$$\frac{1}{\alpha} \frac{\partial T}{\partial t} = \frac{\partial^2 T}{\partial r^2} + \frac{2}{r} \frac{\partial T}{\partial r} + \frac{1}{r^2} \frac{\partial}{\partial \mu} \left[\left(1-\mu^2\right) \frac{\partial T}{\partial \mu} \right]$$

in the solid spherical region, $0 \le r \le R$, with initial and boundary conditions as indicated.

6.3.1 Non-Homogeneous Boundary Conditions

The time-dependent diffusion problem we just did had homogeneous boundary conditions all around. For situations when the boundary conditions are non-homogeneous, decomposition is necessary. We shall consider a case somewhat similar to Example 5.8 in cylindrical coordinates on pages 227-230.

Example 6.8

PROBLEM STATEMENT: Obtain the temperature distribution $T(r, \theta, t)$ in the hemispherical region shown in the figure by solving the equation

$$\frac{1}{\alpha}\frac{\partial T}{\partial t} = \frac{\partial^2 T}{\partial r^2} + \frac{2}{r}\frac{\partial T}{\partial r} + \frac{1}{r^2 \sin\theta}\frac{\partial}{\partial \theta}\left(\sin\theta \frac{\partial T}{\partial \theta}\right), \tag{6.168}$$

and the boundary and initial conditions for $T(r, \theta, t)$ are as shown, i.e.,

$$T(r, \tfrac{1}{2}\pi, t) = 0, \tag{6.169}$$

$$k\frac{\partial T(r,\theta,t)}{\partial r}\bigg|_{r=R} = q_0, \tag{6.170}$$

and initial condition,

$$T(r, \theta, 0) = 0. \tag{6.171}$$

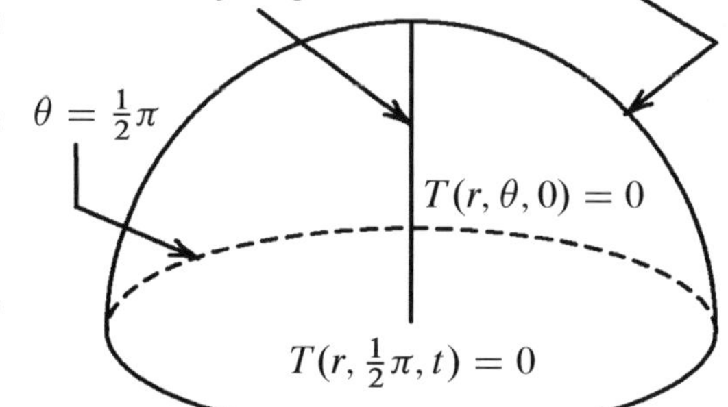

SOLUTION:
We first transform $\mu = \cos\theta$, and rewrite the set of Equations (6.168)-(6.171) as the following for $T(r, \mu, t)$:

$$\frac{1}{\alpha}\frac{\partial T}{\partial t} = \frac{\partial^2 T}{\partial r^2} + \frac{2}{r}\frac{\partial T}{\partial r} + \frac{1}{r^2}\frac{\partial}{\partial \mu}\left[(1-\mu^2)\frac{\partial T}{\partial \mu}\right]. \tag{6.172}$$

which is the same as Equation (6.135) in Example 6.7. The boundary conditions (6.169) and (6.170) become

$$T(r, 0, t) = 0, \tag{6.173}$$

and

$$k\frac{\partial T(r,\mu,t)}{\partial r}\bigg|_{r=R} = q_0, \tag{6.174}$$

and initial condition,

$$T(r, \mu, 0) = 0. \tag{6.175}$$

Here we decompose the problem into a steady and an unsteady part, i.e.,

$$T(r, \mu, t) = T_1(r, \mu) + T_2(r, \mu, t), \tag{6.176}$$

where $T_1(r, \mu)$ is the steady part. This decomposition is illustrated pictorially in Figure 6.8. Using the decomposition (6.176) in Equation (6.168), we obtain the following

$$0 = \frac{\partial^2 T_1}{\partial r^2} + \frac{2}{r}\frac{\partial T_1}{\partial r} + \frac{1}{r^2}\frac{\partial}{\partial \mu}\left[(1-\mu^2)\frac{\partial T_1}{\partial \mu}\right], \tag{6.177}$$

$$\frac{1}{\alpha}\frac{\partial T_2}{\partial t} = \frac{\partial^2 T_2}{\partial r^2} + \frac{2}{r}\frac{\partial T_2}{\partial r} + \frac{1}{r^2}\frac{\partial}{\partial \mu}\left[(1-\mu^2)\frac{\partial T_2}{\partial \mu}\right], \tag{6.178}$$

(6.179)

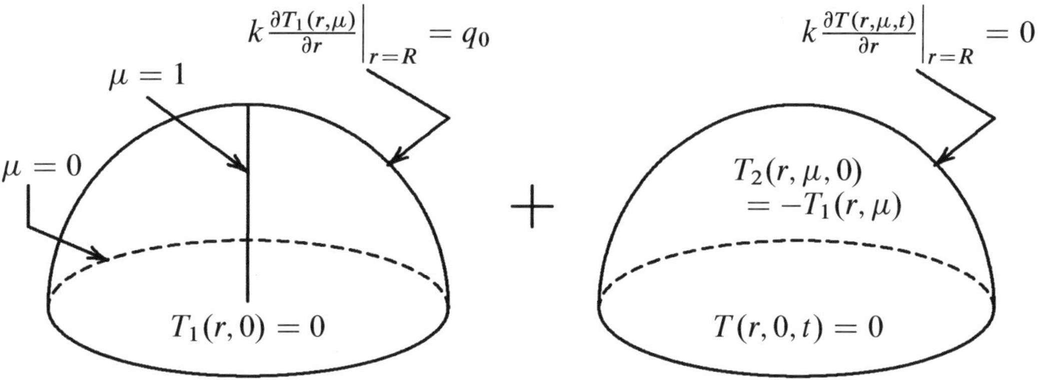

Figure 6.8: Decomposition of $T(r, \mu, t)$ into steady and unsteady parts.

The steady part $T_1(r, \mu)$ is quite similar to Example 6.3 on pages 241-243. The only difference is in the boundary condition at $r = R$. We can therefore adopt the solution form given by Equation (6.55) on page 243, i.e.,

$$T_1(r,\mu) = \sum_{n=1,3,5,\ldots}^{\infty} A_n r^n P_n(\mu). \tag{6.180}$$

The set of coefficients $\{A_n,\ n = 1, 3, 5, \ldots\}$ can be determined upon satisfying the boundary condition,

$$\frac{\partial T_1(r,\mu)}{\partial r}\bigg|_{r=R} = \frac{q_0}{k}, \tag{6.181}$$

which leads to

$$\sum_{n=1,3,5,\ldots}^{\infty} A_n n R^{n-1} P_n(\mu) = \frac{q_0}{k}, \tag{6.182}$$

which is a Fourier-Legendre series. The coefficients are found by orthogonality, following Example 3.17 (see page 140), as

$$\begin{aligned} A_n n R^{n-1} &= (2n+1)\int_0^1 \frac{q_0}{k} P_n(\mu)\,d\mu, \\ &= \left(\frac{q_0}{k}\right)\frac{(2n+1)P_n'(0)}{n(n+1)} \\ A_n &= \left(\frac{q_0 R}{k}\right)\frac{(2n+1)P_n'(0)}{n^2(n+1)R^n} \end{aligned} \tag{6.183}$$

Using this expression for A_n in Equation (6.180) gives us

$$T_1(r,\theta) = \left(\frac{q_0 R}{k}\right)\sum_{n=1,3,5,\ldots}^{\infty} \frac{(2n+1)P_n'(0)}{n^2(n+1)}\left(\frac{r}{R}\right)^n P_n(\mu). \tag{6.184}$$

For the time-dependent part, $T_2(r, \mu, t)$, the problem is quite similar to Example 6.7, except for the boundary condition at $r = R$, and the initial condition. Therefore, we can start with the solution form given by Equation (6.163) on page 264, i.e.,

$$T_2(r,\mu,t) = \sum_{n=1,3,5,\ldots}^{\infty} \sum_{m=1}^{\infty} a_{mn} e^{-\alpha\lambda_{mn}^2 t} j_n(\lambda_{mn} r) P_n(\mu). \tag{6.185}$$

This solution satisfies the differential equation (6.178) and the homogeneous boundary condition $T_2 = 0$ at $\mu = 0$. The condition at $r = R$ is different from the previous example and therefore, the λ_{mn} values will be different from those given by Equation (6.156). The boundary condition is

$$\left.\frac{\partial T_2(r,\theta,t)}{\partial r}\right|_{r=R} = 0. \tag{6.186}$$

This, being a homogeneous condition, can be satisfied term-by-term with $j_n'(\lambda_{mn} R) = 0$. The roots of $j_n'(x) = 0$ are denoted by γ_{mn} and can be found in mathematical handbooks (see, e.g., Abramowitz & Stegun [1]). Therefore,

$$\lambda_{mn} R = \gamma_{mn} \qquad \text{or} \qquad \lambda_{mn} = \frac{\gamma_{mn}}{R}. \tag{6.187}$$

The initial condition for the full problem is $T(r,\mu,0) = 0$. Taking into consideration the decomposition given by Equation (6.176), we come up with

$$\begin{aligned} T_2(r,\mu,0) &= T(r,\mu,0) - T_1(r,\mu) \\ &= 0 - T_1(r,\mu). \end{aligned} \tag{6.188}$$

Making use of the expressions (6.180) and (6.185) in the initial condition (6.188), we obtain

$$-\sum_{n=1,3,5,\ldots}^{\infty} A_n r^n P_n(\mu) = \sum_{n=1,3,5,\ldots}^{\infty} \left[\sum_{m=1}^{\infty} a_{mn} j_n(\lambda_{mn} r)\right] P_n(\mu) \tag{6.189}$$

Equating the coefficients of $P_n(\mu)$ term-by-term,

$$-A_n r^n = \left[\sum_{m=1}^{\infty} a_{mn} j_n(\lambda_{mn} r)\right]. \tag{6.190}$$

The right-hand side is a Fourier-Bessel series on which we may apply orthogonality (with weighting factor r^2; see Example 3.14 on pages 128-129) to yield the coefficients,

$$\begin{aligned} a_{mn} &= \frac{\int_0^R -A_n r^n r^2 j_n(\lambda_{mn} r)\, dr}{\int_0^R r^2 [j_n(\lambda_{mn} r)]^2\, dr} \\ &= -A_n \frac{\int_0^R r^{n+2} j_n(\lambda_{mn} r)\, dr}{\int_0^R r^2 [j_n(\lambda_{mn} r)]^2\, dr}. \end{aligned} \tag{6.191}$$

The integral in the numerator can be found from integral tables (see e.g.,Gradshteyn & Ryzhik [2]) as

$$\int_0^R r^{n+2} j_n(\lambda_{mn} r)\, dr = \frac{R^{n+3} j_{n+1}(\lambda_{mn} R)}{\lambda_{mn} R}. \tag{6.192}$$

Following the earlier development on spherical Fourier-Bessel series in Chapter 3, the denominator in Equation (6.191) can be expressed by Equation (3.151) on page 129, which is repeated it here,

$$\int_0^R r^2\left[j_n(\lambda_{mn}r)\right]^2\,dr = \tfrac{1}{2}R^3\left\{\left[j_n'(\lambda_{mn}R)\right]^2 + \frac{j_n(\lambda_{mn}R)j_n'(\lambda_{mn}R)}{\lambda_{mn}R} + \left(1-\frac{n(n+1)}{\lambda_{mn}^2R^2}\left[j_n(\lambda_{mn}R)\right]^2\right)\right\}. \quad (6.193)$$

With $j_n'(\lambda_{mn}R) = 0$ in the present example, the integral reduces to

$$\int_0^R r^2\left[j_n(\lambda_{mn}r)\right]^2\,dr = \tfrac{1}{2}R^3\left\{\left(1-\frac{n(n+1)}{\lambda_{mn}^2R^2}\right)\left[j_n(\lambda_{mn}R)\right]^2\right\}. \quad (6.194)$$

Using these two integrals in Equation (6.191), along with the expression for A_n given by Equation (6.183) we have for a_{mn},

$$\begin{aligned} a_{mn} &= -A_n\frac{\frac{R^{n+3}j_{n+1}(\lambda_{mn}R)}{\lambda_{mn}R}}{\frac{1}{2}R^3\left\{\left[1-\frac{n(n+1)}{\lambda_{mn}^2R^2}\right]\left[j_n(\lambda_{mn}R)\right]^2\right\}} \\ &= -\left(\frac{2q_0R}{k}\right)\frac{(2n+1)P_n'(0)}{n^2(n+1)}\frac{(\lambda_{mn}R)j_{n+1}(\lambda_{mn}R)}{\left[\lambda_{mn}^2R^2-n(n+1)\right]\left[j_n(\lambda_{mn}R)\right]^2}. \end{aligned} \quad (6.195)$$

With the coefficients a_{mn} completely determined, we can substitute the expression (6.195) into Equation (6.185) and together with the expression (6.184) for $T_1(r,\mu)$, and obtain the explicit expression

$$\begin{aligned} T(r,\theta,t) &= T_1(r,\mu) + T_2(r,\mu,t) \\ &= \left(\frac{q_0R}{k}\right)\sum_{n=1,3,5,\ldots}^{\infty}\frac{(2n+1)P_n'(0)}{n^2(n+1)}\left\{\left(\frac{r}{R}\right)^n\right. \\ &\quad \left. -2\sum_{m=1}^{\infty}\frac{(\lambda_{mn}R)j_{n+1}(\lambda_{mn}R)}{\left[\lambda_{mn}^2R^2-n(n+1)\right]\left[j_n(\lambda_{mn}R)\right]^2}e^{-\alpha\lambda_{mn}^2t}j_n(\lambda_{mn}r)\right\}P_n(\mu). \end{aligned} \quad (6.196)$$

□

EXERCISES 6.4

1. A solid hemisphere is initially at temperature zero. For time $t \geq 0$, a uniform upward heat flux q_0 is applied at the base, while the dome is maintained at $T = 0$. Obtain the temperature distribution $T(r,\mu,t)$ by solving

$$\frac{1}{\alpha}\frac{\partial T}{\partial t} = \frac{\partial^2 T}{\partial r^2} + \frac{2}{r}\frac{\partial T}{\partial r} + \frac{1}{r^2}\frac{\partial}{\partial\mu}\left[\left(1-\mu^2\right)\frac{\partial T}{\partial\mu}\right].$$

Proceed as follows:

(a) First decompose the problem into steady-state and time-dependent parts, i.e.,

$$T(r,\mu,t) = T_1(r,\mu) + T_2(r,\mu,t),$$

and let $T_1(r,\mu)$ take care of the nonhomogeneous boundary condition.

(b) This boundary condition at the base $\mu = 0$ would call for a radial eigenfunction expansion. This is possible but quite complex. Instead, decompose the steady part into

$$T_1(r,\mu) = T_A(z) + T_B(r,\mu),$$

where $T_A(z)$ satisfies the one-dimensional Laplace's equation,

$$\frac{d^2 T_A(z)}{dz^2} = 0,$$

while $T_B(r,\mu)$ satisfies

$$\frac{\partial^2 T}{\partial r^2} + \frac{2}{r}\frac{\partial T}{\partial r} + \frac{1}{r^2}\frac{\partial}{\partial \mu}\left[\left(1-\mu^2\right)\frac{\partial T}{\partial \mu}\right] = 0.$$

(c) Construct a simple solution for $T_A(z)$ so that

$$-k\left.\frac{dT_A(z)}{dz}\right|_{z=0} = q_0.$$

(d) This will not satisfy the zero condition on the dome. Now construct the solution $T_B(r,\mu)$ by eigenfunction expansion in μ with an appropriate boundary condition at $\mu = 0$ so that the uniform flux condition satisfied by $T_A(z)$ is not disrupted.

(e) To satisfy the zero condition on the dome, require that

$$T_1(R,\mu) = \left[T_A(z) + T_B(r,\mu)\right]_{r=R} = 0.$$

In doing so, convert z to spherical coordinates using $z = r\cos\theta = r\mu$.

(f) With the steady state solution obtained, follow the procedure in Example 6.8 from Equation (6.185) onwards to Equation (6.196) on pages 268-269 to obtain the time-dependent part, $T_2(r,\mu,t)$. This procedure actually relies on Example 6.7 on pages 258-265.

Bibliography

[1] M. Abramowitz and I.A. Stegun. *Handbook of Mathematical Functions with Formulas, Graphs and Mathematical Tables*. U.S. Department of Commerce, 1972. National Bureau of Standards Applied Mathematics Series, 55; Tenth printing; LOC Cat. No.: 36-60064.

[2] I.S. Gradshteyn and I.M. Ryzhik. *Table of Integrals, Series and Products*. Academic Press, seventh edition, 2007. Translated from Russian by Scripta Technica, Inc.; ISBN-13: 978-0-12-373637-6.

[3] S.S. Sadhal. Solutions to a class of transport problems with radially dominant convection. *J. Appl. Math. and Phys. (ZAMP)*, 44:314–332, 1993.

[4] M. Murray Spiegel, Seymour Lipschutz, John Schiller, and Dennis Spellman. *Schaum's Outline of Complex Variables*. McGraw-Hill, second edition, 2009. ISBN-13: 9780071615693.

Answers to Selected Problems

Chapter 1

Exercises 1.1

1. $C_1e^{-8x} + C_2e^{-5x}$

3. $C_1e^{7x} + C_2e^{-3x}$

5. $C_1e^{-11x} + C_2e^{-4x}$

7. $e^{-2x}[C_1 \cos 3x + C_2 \sin 3x]$

9. $e^{-x}[C_1 \cos 2x + C_2 \sin 2x]$

11. $e^{3x}[C_1 \cos x + C_2 \sin x]$

13. $C_1e^{x} + C_2e^{2x} + C_3e^{-3x}$

15. $C_1e^{13x} + C_2e^{-4x}$

Exercises 1.2

1. $e^{-2x} + C_1e^{-x} + C_2xe^{-x}$

3. $-\frac{1}{10}e^{x} + C_1e^{6x} + C_2e^{-x}$

5. $-\frac{1}{7}e^{x} + C_1e^{5x} + C_2e^{-2x}$

7. $-\frac{4}{65}\sin 2x - \frac{1}{130}\cos 2x$
$+ C_1e^{-4x} + C_2e^{3x}$

9. $\frac{1}{4}e^{-3x} + e^{-3x}[C_1 \cos 2x + C_2 \cos 2x]$

11. $\frac{1}{2}x^2e^{-x} + C_1e^{-2x} + C_2xe^{-2x}$

13. $-\frac{1}{x}xe^{-5x} + C_1e^{-5x} + C_2e^{-2x}$

15. $-\frac{1}{12}x^3 + \frac{1}{48}x^2 - \frac{109}{288}x + \frac{121}{3456}$
$+ C_1e^{4x} + C_2e^{-3x}$

17. $\frac{1}{14}x^2 + \frac{37}{98}x + \frac{319}{1372}$
$+ C_1e^{2x} + C_2e^{7x}$

19. $-\frac{1}{9}\left(x^3 + x^2 + \frac{2}{3}x\right) + C_1e^{5x} + C_2e^{2x}$

21. $-\left(\frac{1}{6}x^2 + \frac{5}{18}x + \frac{31}{108}\right)$
$+\left(\frac{7}{74}x - \frac{5}{1369}\right)\sin x - \left(\frac{5}{74}x + \frac{199}{2378}\right)\cos x$
$+ C_1e^{-6x} + C_2e^{x}$

Exercises 1.3

1. $e^{x}[\ln x + C_1 + C_2x]$

3. $-\sin x \ln|\cot x + \operatorname{cosec} x|$
$+ C_1 \sin x + C_2 \cos x$

5. $-\frac{1}{9}e^{-x}\sin 3x \ln|\cot 3x + \operatorname{cosec} 3x|$
$+ e^{-x}[C_1 \sin 3x + C_2 \cos 3x]$

7. $e^{-2x}\left[2\cos 2x \ln|\cos x| + 2x \sin 2x - \frac{3}{2}\right]$
$+ e^{-2x}[C_1 \sin 2x + C_2 \cos 2x]$

9. $e^{2x}\left[\frac{1}{16}\sin 2x \ln|\sin x| - \frac{1}{4}x \cos 4x\right]$
$+ e^{2x}[C_1 \sin 2x + C_2 \cos 2x]$

Exercises 1.4

1. $C_1x^{-5} + C_2x^{-8}$

3. $x^2[C_1 \cos(\ln x) + C_2 \sin(\ln x)]$

5. $x^{-2}[C_1 \cos(3\ln x) + C_2 \sin(3\ln x)]$

7. $C_1x^{-4} + C_2x^{3}$

9. $C_1x + C_2x^{6} + C_3x^{-1}$

11. $-\frac{1}{3}x^{-5} + C_1x^{-5} + C_2x^{-2}$

13. $\frac{1}{14}\left[(\ln x)^2 + \frac{37}{7}(\ln x) + \frac{319}{98}\right] + C_1 x^7 + C_2 x^2$

15. $-\frac{1}{16}\left[(\ln x)^2 \cos(4\ln x) - \frac{1}{4}(\ln x)\sin(4\ln x)\right] + C_1\cos(4\ln x) + C_2 \sin(4\ln x)$

Chapter 2

Exercises 2.1

1. $y_1(x) = 1$, $y_2(x) = \frac{1}{2}\ln\left(\frac{1+x}{1-x}\right)$

3. $y_1(x) = \frac{1}{2}\left(3x^2 - 1\right)$, $y_2(x) = \frac{1}{4}\left(3x^2 - 1\right)\ln\left(\frac{1+x}{1-x}\right) - \frac{3}{2}x$

Exercises 2.2

1. $y_1(x) = 1$, $y_2(x) = \ln x + \sum_{n=1}^{\infty} \frac{x^n}{n!n}$

3. $y_1(x) = 2-4x+x^2$, $y_2(x) = (2-4x+x^2)\ln x + \left(10x - \frac{9}{2}x^2\right) + 4\sum_{n=3}^{\infty} \frac{(n-3)!}{n!^2}x^n$

Exercises 2.3

See equations (2.67)-(2.70)

Exercises 2.4

1. $y_1(x) = x^{-2} + x^{-1} + \frac{1}{2}x^2$, $y_2(x) = x^{-2}e^x$

3. $y_1(x) = \sum_{n=0}^{\infty} \frac{(-1)^n x^n}{n!(n+5)!}$, $y_2(x) = y_1(x)\ln x$
$-x^{-3}\left[24 + 6x + x^2 + \frac{1}{6}x^3 + \frac{5}{120}x^4 + \frac{7}{2880}x^6 - \frac{31}{28240}x^7 + \cdots\right]$

5. $y_1(x) = x^{-\frac{1}{4}}e^{-x}$, $y_2(x) = \sum_{n=1}^{\infty} \frac{(-1)^n x^{n+1}}{\Gamma(n+\frac{7}{4})}$

7. $y_1(x) = \sum_{n=0}^{\infty} \frac{(-1)^n x^n}{n!(n+3)!}$,
$y_2(x) = y_1(x)\ln x - x^{-3}\left[-2 - x + \frac{1}{2}x^2 + \frac{5}{96}x^4 - \frac{13}{1600}x^5 \cdots\right]$

Exercises 3.1

1. (a) $f(x) = 2\ell^2 \sum_{n=1}^{\infty}\left[-\frac{3}{n\pi} + \frac{2[(-1)^n - 1]}{n^3\pi^3}\right]\sin\left(\frac{n\pi x}{\ell}\right)$
(b) $f(x) = \frac{4}{3}\ell^2 + 4\ell^2\sum_{n=1}^{\infty}\left[\frac{[2(-1)^n - 1]}{n^2\pi^2}\right]\cos\left(\frac{n\pi x}{\ell}\right)$

4. (a) $f(x) = \sum_{n=1}^{\infty}\left[\frac{1+\cos\left(\frac{1}{2}n\pi\right) - 2(-1)^n}{n\pi}\right]\sin\left(\frac{n\pi x}{2}\right)$
(b) $f(x) = \frac{3}{4} - \sum_{n=1}^{\infty}\left[\frac{\sin\left(\frac{1}{2}n\pi\right)}{n\pi}\right]\cos\left(\frac{n\pi x}{2}\right)$

5. (a) $f(x) = a\sum_{n=1}^{\infty}\left[\frac{(-1)^n-\cos\left(\frac{1}{2}n\pi\right)-2}{n\pi} + \frac{6\sin\left(\frac{1}{2}n\pi\right)}{n^2\pi^2}\right]\sin\left(\frac{n\pi x}{2a}\right)$

(b) $f(x) = \frac{1}{8}a + a\sum_{n=1}^{\infty}\left[\frac{\sin\left(\frac{1}{2}n\pi\right)}{n\pi} + \frac{2\cos\left(\frac{1}{2}n\pi\right)-(-1)^n-1}{n^2\pi^2}\right]\cos\left(\frac{n\pi x}{2a}\right)$

8. (a) $f(x) = 2\sum_{n=1}^{\infty}\frac{\sin(n\pi x)}{n\pi}$ (b) $f(x) = \frac{1}{2} + \sum_{n=1}^{\infty}\frac{1-(-1)^n}{n^2\pi^2}\cos(n\pi x)$

10. (a) $f(x) = 2\sum_{n=1}^{\infty}\left[\frac{2-\cos\left(\frac{1}{2}n\pi\right)-2(-1)^n}{n\pi} - \frac{2\sin\left(\frac{1}{2}n\pi\right)}{n^2\pi^2}\right]\sin\left(\frac{n\pi x}{2}\right)$

(b) $f(x) = \frac{7}{4} + 2\sum_{n=1}^{\infty}\left[\frac{\sin\left(\frac{1}{2}n\pi\right)}{n\pi} + \frac{2\left[(-1)^n-\cos\left(\frac{1}{2}n\pi\right)\right]}{n^2\pi^2}\right]\cos\left(\frac{n\pi x}{2}\right)$

Exercises 3.2

3. $f(x) = \frac{5}{8} + \frac{3}{2}\sum_{n=1}^{\infty}\frac{[1-(-1)^n]}{n^2\pi^2}\cos\left(\frac{n\pi x}{L}\right) - \frac{1}{2}\sum_{n=1}^{\infty}\frac{(-1)^n}{n\pi}\sin\left(\frac{n\pi x}{L}\right)$

4. $f(x) = 8L^2\sum_{n=1}^{\infty}\frac{\sin\left(\frac{1}{2}n\pi\right)}{n^3\pi^3}\cos\left(\frac{n\pi x}{L}\right)$

Exercises 3.3

1. $f(x) = \frac{2}{\pi}\int_0^{\infty}\left[\frac{\pi\sin\lambda\pi}{\lambda} + \frac{(\cos\lambda\pi-1)}{\lambda^2}\right]\cos\lambda x\, d\lambda$

3. $f(x) = \int_0^{\infty}\left[\frac{\sin\lambda\pi\sin\lambda x+(1+\cos\lambda\pi)\cos\lambda x}{\pi(1-\lambda^2)}\right]d\lambda = \int_0^{\infty}\left[\frac{\cos\lambda(\pi-x)+\cos\lambda x}{\pi(1-\lambda^2)}\right]d\lambda$

5. $f(x) = \int_0^{\infty}\left[\frac{(1-\cos\lambda L)\cos\lambda x+(\lambda L-\sin\lambda L)\sin\lambda x}{\pi\lambda^2 L}\right]d\lambda$

7. $f(x,y) = -L^2\sum_{m=1}^{\infty}\left[\frac{(-1)^m}{m\pi}\right]\left\{1 + 4\sum_{n=1}^{\infty}\left[\frac{1-(-1)^n}{(n\pi)^2}\right]\cos\left(\frac{n\pi y}{L}\right)\right\}\sin\left(\frac{m\pi x}{L}\right)$

8. $f(x,y) = \frac{L}{\pi}\int_0^{\infty}\left\{1 - 4\sum_{m=1}^{\infty}\left[\frac{1-(-1)^m}{m^2\pi^2}\right]\cos\left(\frac{m\pi x}{L}\right)\right\}\frac{\cos\lambda y}{(1+\lambda^2)}\,d\lambda.$

Exercises 3.4

1. $f(x) = l\sum_{n=0}^{\infty}\left\{\frac{(-1)^n}{\left(n+\frac{1}{2}\right)\pi} - \frac{1}{\left(n+\frac{1}{2}\right)^2\pi^2}\right\}\cos\left[\frac{\left(n+\frac{1}{2}\right)\pi x}{l}\right]$

Exercises 3.5

1. $f(x) = 2\sum_{m=1}^{\infty}\left[\frac{1}{\lambda_m R} - \frac{4}{\lambda_m^3 R^3}\right]\frac{J_0(\lambda_m r)}{J_1(\lambda_m R)}$

Exercises 3.6

1. $f(x) = 4\sum_{m=1}^{\infty}\left[\sum_{k=0}^{\infty}\frac{(-1)^k\left(\frac{1}{2}\lambda_{mn}a\right)^{2k+n}a^2}{(k+n)!k!(2k+n+2)(2k+n+4)R^2}\right]\frac{J_n(\lambda_{mn}r)}{[J_n'(\lambda_{mn}R)]^2}$

Exercises 3.7

2. $f(r) = \frac{a^3}{R^3} + 2\sum_{m=1}^{\infty} \left(\frac{\sin(\lambda_m a) - \lambda_m a \cos(\lambda_m a)}{\lambda_m R \sin(\lambda_m R)} \right) \frac{\sin(\lambda_m r)}{\lambda_m r}, \qquad \lambda_m R = \tan(\lambda_m R)$

Exercises 3.8

1. $f(r) = \frac{1}{\sqrt{k^2+r^2}} = \int_0^{\infty} e^{-\lambda k} J_0(\lambda r)\, d\lambda$

Exercises 3.9

2. $f(r) = \sum_{n=1,3,5,\ldots}^{\infty} \frac{-2(2n+1)P_n'(0)}{(n-2)n^2(n+1)^2(n+3)} P_n(x)$

Chapter 4

Exercises 4.2

1. $T(x, y) = 4T_0 \sum_{n=1}^{\infty} \frac{1-(-1)^n}{(n\pi)^3} \frac{\sinh(n\pi y/a)}{\sinh(n\pi b/a)} \sin\left(\frac{n\pi x}{a}\right)$

5. $T(x, y) = 2T_0 \sum_{n=0}^{\infty} \frac{\sinh \lambda_n x \sin \lambda_n y}{\lambda_n b[(\lambda_n k/h)\cosh \lambda_n a + \sinh \lambda_n a]}, \qquad \lambda_n = \left(n + \frac{1}{2}\right)\pi/b$

7. $T(x, t) = T_0 \left[1 + 2\sum_{n=0}^{\infty} \left\{ -\frac{1}{\lambda_n l} + \frac{2(-1)^n}{\lambda_n^2 l^2} - \frac{2}{\lambda_n^3 l^3} \right\} e^{-\alpha \lambda_n^2 t} \sin \lambda_n x \right], \qquad \lambda_n = \left(n + \frac{1}{2}\right)\pi/l$

Exercises 4.3

1. $u(x, y, t) = 1024u_0 \sum_{n=0}^{\infty} \sum_{m=0}^{\infty} \frac{(-1)^{m+n} \cos(\beta_{mn} ct)\cos(\lambda_m x)\cos(\mu_n y)}{(\lambda_m a)^3 (\mu_n a)^3}$
   $\lambda_m = (2m+1)\pi/a, \quad \mu_n = (2n+1)\pi/a, \quad \beta_{mn} = \left(\lambda_m^2 + \mu_n^2\right)^{1/2}$

3. $T(x, y, t) = T_1(x, y) + T_2(x, y, t)$
   $T_1(x, y) = \sum_{n=1}^{\infty} A_n \sinh \mu_n x \sin \mu_n y, \quad \mu_n = \frac{n\pi}{b}$
   $A_n = 2T_0 \frac{1-(-1)^n}{n\pi\left[\frac{k\mu_n}{h} \cosh \mu_n b + \sinh \mu_n b\right]}$
   $T_2(x, y, t) = \sum_{n=1}^{\infty} \sum_{m=1}^{\infty} B_{mn} e^{-\alpha(\lambda_m^2 + \mu_n^2)t} \sin \lambda_m x \sin \mu_n y, \quad \tan \lambda_m a = -\frac{k}{h}\lambda_m$
   $B_{mn} = -A_n \frac{\mu_n \cosh \mu_n a \sin \lambda_m a - \lambda_m \sinh \mu_n a \cos \lambda_m a}{(\lambda_m^2 + \mu_n^2)\left[\frac{1}{2}a - \frac{1}{4}\sin(2\mu_n a)\right]}$

Chapter 5

Exercises 5.1

1. $T(r, \theta) = \sum_{n=1}^{\infty} \frac{2[1-(-1)^n]}{n\pi} \left(\frac{\sinh(\lambda_n \theta)}{\sinh(\lambda_n \pi)} \right) \sin\left[\lambda_n \ln\left(\frac{r}{c}\right)\right]$

3. $T(r, \theta) = \sum_{n=1}^{\infty} \frac{(-1)^n}{\left(n+\frac{1}{2}\right)\pi} \left[\frac{\left(\frac{r}{c}\right)^{n+\frac{1}{2}} + \left(\frac{c}{r}\right)^{n+\frac{1}{2}}}{\left(\frac{R}{c}\right)^{n+\frac{1}{2}} + \left(\frac{c}{R}\right)^{n+\frac{1}{2}}} \right] \cos\left(n + \frac{1}{2}\right)\theta$

5. $u(r, \theta) = \frac{8P_0 R^2}{\mu_0} \sum_{n=1}^{\infty} \frac{1-(-1)^n}{n\pi(16-n^2)} \left[\left(\frac{r}{R}\right)^{n+\frac{1}{2}} - \left(\frac{R}{r}\right)^2 \right] \sin\left(\frac{1}{2}n\theta\right)$

Exercises 5.2

1. $T(r,\theta) = 4T_0 \sum_{n=1}^{\infty} \frac{1}{(2n+1)\pi} \left(\frac{(\sinh \lambda_n \theta)}{(\sinh \lambda_n \pi)}\right) \sin\left[\lambda_n \ln\left(\frac{R}{r}\right)\right]$

$= 4T_0 \sum_{n=1}^{\infty} \frac{(-1)^n}{(2n+1)\pi} \left(\frac{(\sinh \lambda_n \theta)}{(\sinh \lambda_n \pi)}\right) \cos\left[\lambda_n \ln\left(\frac{r}{c}\right)\right], \quad \lambda_n = \frac{(n+\frac{1}{2})\pi}{\ln\left(\frac{R}{c}\right)}$

3. $T(r,z) = \frac{2q_0 R}{k} \sum_{m=1}^{\infty} \frac{\sinh[\lambda_m(l-z)]}{(\lambda_m R)^2 \cosh \lambda_m l} \frac{J_0(\lambda_m r)}{J_1(\lambda_m R)}$

Exercises 5.3

1. $T(r,z) = \frac{2q_0 R}{k} \sum_{n=1}^{\infty} \sum_{m=1}^{\infty} \frac{1-(-1)^n}{n\pi} \frac{\sinh[\lambda_{mn}(l-z)]}{\cosh \lambda_{mn} l} \frac{\left[\int_0^R r J_n(\lambda_{mn} r)\, dr\right]}{(\lambda_{mn} R^3) J_n'(\lambda_{mn} R)} J_n(\lambda_{mn} r) \sin n\theta$

Chapter 6

Exercises 6.1

1. $T(r,\mu) = \left(\frac{q_0 R}{k}\right) \sum_{n=1,3,5,\ldots}^{\infty} \frac{(2n+1)\left(1-\mu_0^2\right) P_n'(\mu_0)}{n^2(n+1)} \left(\frac{r}{R}\right)^n P_n(\mu), \quad \mu_0 = \cos\theta_0$

3. $T(r,\mu) = T_0 \sum_{n=1,3,5,\ldots}^{\infty} \frac{(2n+1) P_n'(0)}{n(n+1)} \left[\frac{\left(\frac{r}{c}\right)^n - \left(\frac{c}{r}\right)^{n+1}}{\left(\frac{R}{c}\right)^n - \left(\frac{c}{R}\right)^{n+1}}\right] P_n(\mu)$

Exercises 6.2

1. $T(r,t) = 12T_0 \sum_{n=1}^{\infty} \frac{(-1)^{n+1}}{n^2\pi^2} e^{-\alpha(n\pi/R)^2 t} \frac{\sin(n\pi r/R)}{(n\pi r/R)}$

3. $c(r,t) = c_0 \left[\frac{r_0^3}{R^3} + 2\sum_{n=1}^{\infty} \frac{\sin \lambda_n r_0 - \lambda_n r_0 \cos \lambda_n r_0}{\lambda_n R \sin^2 \lambda_n R} e^{-D\lambda_n^2 t} \frac{\sin \lambda_n r}{\lambda_n r}\right], \quad \tan \lambda_n R = \lambda_n R$

Exercises 6.3

1. $T(r,\theta,t) = T_0 \left\{\frac{1}{2} + \sum_{n=1}^{\infty} \frac{(2n+1)}{n(n+1)} P_n'(0)\right.$

$\times \sum_{m=1}^{\infty} \left(\sum_{k=0}^{\infty} \frac{(-1)^k (k+n)! 2^n (\lambda_{mn} R)^{2k+3+n}}{k!(2k+2n+1)!(2k+n+3)\left[\lambda_{mn}^2 R^2 - n(n+1)\right][j_n(\lambda_{mn} R)]^2}\right)$

$\left.\times e^{-\alpha\lambda_{mn}^2 t} j_n(\lambda_{mn} r) P_n(\mu)\right\}, \qquad j_n'(\lambda_{mn} R) = 0.$

Exercises 6.4

$T(r,\mu,t) = -\left(\frac{q_0 R}{k}\right) \left[\left(\frac{r}{R}\right)\mu + \sum_{n=0,2,4,\ldots}^{\infty} \frac{(2n+1) P_n(0)}{(n+2)(n-1)} \left\{\left(\frac{r}{R}\right)^n\right.\right.$

$\left.\left.-2\sum_{m=1}^{\infty} \frac{(\lambda_{mn} R) j_{n+1}(\lambda_{mn} R)}{[j_n'(\lambda_{mn} R)]^2} e^{-\alpha\lambda_{mn}^2 t} j_n(\lambda_{mn} r)\right\} P_n(\mu)\right], \qquad j_n(\lambda_{mn} R) = 0$

Index

ISBN 978-0-9913683-1-0
US$49.00
9 780991 368310
54900>

Made in the USA
Las Vegas, NV
21 February 2025

18456509R00160

Dr. Sadhal is a Professor of Aerospace & Mechanical Engineering and Ophthalmology at the University of Southern California. As a educator, Professor Sadhal has over 45 years of teachng and research experience and has had the opportunity to share his knowledge and expertise with thousands of graduate and undergraduate students who have gone on to become successful engineers.

This textbook is a succinct introduction to Partial Differential Equations designed for a graduate-level engineering course. It is highly structured with the development of the necessary background material leading chapter-by-chapter to analytically solving linear Partial Differential Equations. The first chapter introduces solution techniques for linear ordinary differential equation and the second chapter goes on to solving more complex linear equations using the power-series method. Here the important classes of special functions (such as Legendre polynomials and Bessel functions) are treated. The third chapter goes on to developing Fourier Series and Integrals, including two-variable expansions, and generalization of the concept of orthogonal series to a wider class of functions (Sturm-Liouville Theory). This background eases in the development of analytical techniques to linear Partial Differential Equations. The fourth fifth and sixth chapters are dedicated to solutions methods in cartesian, cylindrical and spherical coordinates, respectively with examples pertaining to heat/mass transfer and fluid mechanics.

ISBN 9780991368310